# Molt in Neotropical Birds

## Life History and Aging Criteria

# STUDIES IN AVIAN BIOLOGY

*A Publication of The American Ornithological Society*

www.crcpress.com/browse/series/crcstdavibio

Studies in Avian Biology is a series of works founded and published by the Cooper Ornithological Society in 1978, and published by The American Ornithological Society since 2017. Volumes in the series address current topics in ornithology and can be organized as monographs or multi-authored collections of chapters. Authors are invited to contact the Series Editor to discuss project proposals and guidelines for preparation of manuscripts.

Volume 51
**Studies in Avian Biology**
American Ornithological Society

# Molt in Neotropical Birds

## Life History and Aging Criteria

**Erik I. Johnson**
National Audubon Society
Baton Rouge, LA

**Jared D. Wolfe**
Louisiana State University
School of Renewable Natural Resources
Baton Rouge, LA

**CRC Press**
Taylor & Francis Group
Boca Raton London New York

CRC Press is an imprint of the
Taylor & Francis Group, an **informa** business

Cover photo by Erik I. Johnson: Collared Puffbird (*Bucco capensis*). Biological Dynamics of Forest Fragments Project, Amazonas, Brazil, 8 July 2007.

CRC Press
Taylor & Francis Group
6000 Broken Sound Parkway NW, Suite 300
Boca Raton, FL 33487-2742

First issued in paperback 2020

© 2018 by American Ornithological Society
CRC Press is an imprint of Taylor & Francis Group, an Informa business

No claim to original U.S. Government works

ISBN-13: 978-1-4987-1611-6 (hbk)
ISBN-13: 978-0-367-65763-5 (pbk)

**Library of Congress Cataloging-in-Publication Data**

Names: Johnson, Erik I., author. | Wolfe, Jared D.
Title: Molt in Neotropical birds : life history and aging criteria / Erik I. Johnson and Jared D. Wolfe.
Description: Boca Raton : Taylor & Francis, 2017. | Series: Studies in avian biology
Identifiers: LCCN 2016059787 | ISBN 9781498716116 (hardback : alk. paper)
Subjects: LCSH: Feathers--Amazon River Region. | Molting. | Birds--Amazon River Region.
Classification: LCC QL697.4 .J64 2017 | DDC 598.147--dc23
LC record available at https://lccn.loc.gov/2016059787

**Visit the Taylor & Francis Web site at**
**http://www.taylorandfrancis.com**

**and the CRC Press Web site at**
**http://www.crcpress.com**

# CONTENTS

## Part III • Passerines, the Suboscines

## Part IV • Passerines, the Oscines

# AUTHORS

**Erik I. Johnson, PhD,** is Director of Bird Conservation for Audubon Louisiana and an Adjunct Graduate Faculty Member at the School of Renewable Natural Resources, Louisiana State University. Johnson completed his dissertation work studying the effects of forest fragmentation on avian communities at the Biological Dynamics of Forest Fragments Project (BDFFP) in coordination with the Instituto Nacional de Pesquisas da Amazônia (INPA). Through that work, he developed and applied aging criteria to demonstrate how forest fragmentation alters demographic structuring of bird populations and identified novel patterns in molt-breeding overlap across a suite of Neotropical bird species and families. Now primarily working on avian conservation challenges along the Gulf Coast of the United States, Johnson continues to work with students at the BDFFP and helps teach an annual banding workshop offered by INPA. Johnson has more than 15 years of applied ornithological research experience in 5 countries resulting in over 25 scientific publications and much of his work is reflected in this volume. Many of these publications advance or utilize knowledge about aging birds, including the advancement of an age classification system based on plumage and molt patterns that is applicable to birds worldwide.

**Jared D. Wolfe, PhD,** is Research Faculty in the Wildlife Department at Humboldt State University and a Wildlife Ecologist at the U.S. Department of Agriculture (USDA) Forest Service's Pacific Southwest Research Station. In addition to conducting extensive fieldwork in the Amazon, including his PhD at the Biological Dynamics of Forest Fragments Project (BDFFP), Wolfe has broadened his research interests to Central Africa where he regularly conducts fieldwork to better understand and conserve understudied bird communities in Equatorial Guinea. Recognizing the need for an improved method to classify bird age, Wolfe developed a transformative and universal system of age classification for tropical birds based on plumage and molt patterns; the system continually grows in popularity among tropical ornithologists. Wolfe has published numerous articles on bird molt, demography, and behavior, and continues to teach advanced banding courses throughout the Americas and Africa.

# PREFACE

This volume was conceptualized initially as a means for organizing notes from the field and sharing information among researchers and technicians at the Biological Dynamics of Forest Fragments Project (BDFFP), near Manaus, Brazil. We started with little information on the molt sequences of the species involved, so we began by applying a well-developed framework of understanding molt patterns from temperate species. After capturing and collecting data from thousands of birds, and taking thousands of photos, a new understanding of the more common species emerged, which was broadened over the years as we added depth and understanding with more and more species. The resulting volume is a compilation of that work in which we have combined our field experience with specimen reviews and a thorough literature review of the more than 180 focal species detailed here. To help fill a void in our collective understanding of Neotropical bird molt, we are pleased to contribute our research to date on this topic, realizing that much work remains to be done. To document the molt patterns of the 4000-plus species in the Neotropics will take decades and an army of knowledgeable and skilled banders. We hope that this volume will serve as a helpful step toward that goal.

We are extremely thankful to the many colleagues, collaborators, and friends who helped us guide the formative ideas and concepts of this volume. We also would like to thank the many students and banders along the way who helped

shape our understanding of Neotropical bird molt by asking tough questions and providing critical feedback. In particular, Dr. Philip Stouffer, as our graduate adviser, has been tremendously supportive in the development of this work, both practically and intellectually. We would also like to thank the Stouffer Lab, as they allowed themselves to be our guinea pigs for thinking through assumptions and ideas, both for this work and other research endeavors we have been collectively involved in. In particular, Luke Powell, Karl Mokross, Cameron Rutt, Angélica Hernández Palma, Emma DeLeon, Matt Brooks, and Falyn Owens spent many long hours discussing patterns and processes of molt with us. Dr. Gonçalo Ferraz and his lab were also tremendously important in the development of this work, as we tested and taught the patterns we observed to Latin American banding students through a series of five courses and workshops largely organized by Dr. Ferraz.

The inspired ideas of Thomas Lovejoy and Richard O. Bierregaard, Jr., in the early days of the BDFFP, as well as Brazil's Instituto Nacional de Pesquisas da Amazônia and the Smithsonian Institution support, have kept the forest fragments project and particularly the bird monitoring program alive. The depth with which we present an understanding of the annual cycle of many species would not have been possible without their efforts, as well as the many field assistants that spent countless hours collecting data. It is a nearly insurmountable task to list the number of field technicians who have supported this work, but, in particular, Claudier Vargas, Jairo Lopes, Sandra Martins, and Francisco Carvalho "Chico" Diniz worked countless hours in the field helping collect data. In addition, the support and caring friendship of Tatiana Straatman, Leticia Soares, Gabriel McCrate, Aída Rodrigues, Luiza Figueira, Pedro Martins, Joao Vitor "J. B." Compos e Silva, and Gonçalo Ferraz made Manaus that much more special to visit.

Funding for the long-term avian monitoring project has also been provided by the World Wildlife Fund–U.S.; the MacArthur Foundation; the Andrew W. Mellon Foundation; the U.S. Agency for International Development; U.S. National Aeronautics and Space Administration; Brazil's Ministry for Science and Technology; the U.S. National Science Foundation (LTREB 0545491 and LTREB 1257340); the Summit Foundation; Shell Oil; Citibank; Champion International; the Homeland Foundation; the National Geographic Society; and the National Institute of Food and Agriculture, U.S. Department of Agriculture, McIntire Stennis project #94098. This volume was approved for publication by the Director of the Louisiana Agricultural Experiment Station as manuscript number 2016-241-30656. This is publication 707 of the Biological Dynamics of Forest Fragments Project Technical Series and number 41 in the Amazonian Ornithology Technical Series of the INPA Zoological Collections Program.

Other students at the Louisiana State University Museum of Natural Science (Baton Rouge, LA), particularly Ryan Terrill, have been instrumental in advancing and broadening our thinking of the evolutionary, ecological, and life history implications of feathers and molt. The collection managers and staff have also been very generous in allowing us to review specimens to advance the understanding of species we were only rarely able to capture or observe in the wild. We especially are grateful to Peter Pyle for many productive discussions on this topic over the years, and for his critical review of this volume, as well as to Kate Huyvaert and Brett Sandercock for their editorial comments and guidance. We thank Miguel Moreno-Palacios for allowing us to use his figures in the introduction.

Many of the broader concepts of molt processes and the cycle code system have been vetted by volunteers and trainees at the Louisiana Bird Observatory, a program of the Baton Rouge Audubon Society. Volunteer and other support of that program have been critical for solidifying the cycle code system of aging birds. In particular, Daniel A. Mooney has been a leader in that program and we are grateful for his time and energy spent organizing and coordinating banding operations.

We also humbly recognize that this work would not have been possible without the many researchers that have led the way in understanding this once poorly understood aspect of avian life histories. Their collective research and efforts publishing their findings have paved the way toward a more holistic and comprehensive understanding of bird molt. This volume would not be possible if not for the inspirational works created by these talented researchers.

Finally, we thank our families and friends for their support when we were home and during our travels. They have been our rock and we are ever so grateful for their endless patience and love.

# PART ONE

# Overview of the Guide

# CHAPTER ONE

# Introduction to This Guide

Feathers are some of the most unique and fascinating parts of birds. From penguins to high-soaring vultures, feathers help regulate a bird's body temperature and are, in some ways, analogous to the function of hair on mammals. Through natural selection and millions of years of evolution, bird feathers have been modified to create a dizzying array of plumage patterns and ornaments. These modifications have allowed birds to occupy the Earth's most extreme environments—tundra, desert, tropical, alpine, and pelagic habitats. Feathers, and the birds to which they are attached, have broadened the human imagination; when we created machines that fly, we looked to bird feathers as a source of inspiration and design.

Over time an individual bird's feathers fade and wear causing them to lose their structural integrity and attractiveness to potential mates. To mitigate feather fade and wear, birds predictably replace their feathers through a process called molt. The timing, frequency, duration, and extent of molts have presumably evolved in response to environmental and social pressures and have influenced avian life histories across the globe. In North America, Steve Howell (and colleagues 2003, 2010), Peter Pyle (1997, 2008), Sievert Rohwer (and colleagues 1980, 2005, 2009), and other researchers (e.g., Mulvilhill 1993) have revolutionized our understanding of avian molt fashioned on a framework developed by Humphrey and Parkes (1959) where nomenclature reflects perceived evolutionarily homologous (between species) and paralogous molts (within species; Wolfe et al. 2014). Their cumulative work has been developed through scientific papers and field guides designed to help bird banders distinguish the age and sex of captured birds using feather as well as anatomical characteristics. These references have been invaluable resources in avian research and serve as foundations for demographic studies of birds.

Here, we applied known patterns and sequences of molt from northern latitudes to develop age and sex criteria for bird species in the central Amazon, many of which are widespread across the region. Part of our inspiration in developing this guide stems from the general lack of published material for aging tropical species as a whole (although see Guallar et al. 2009; Ryder and Wolfe 2009; Wolfe et al. 2009a,b; Hernández 2012; Wolfe and Pyle 2012 and Pyle et al. 2015, for important reviews and broad regional compilations for neotropical birds). The work presented here is largely based on observations of 12,424 captures of birds between 2007 and 2014, plus another 53,803 historic captures between 1979 and 2001 in the Biological Dynamics of Forest Fragments Project (BDFFP) bird capture database. The volume also includes information from the examination of over 1000 specimens at the Louisiana State University Museum of Natural Science (Baton Rouge, LA). The temporal breadth of the BDFFP long-term capture database allowed us to provide not only age-specific molt sequences and extents, but also estimates of molt and breeding seasonality for the majority of species presented here.

To most effectively use this guide for successfully applying aging criteria and terminology to birds we strongly encourage each reader to thoroughly read this chapter, which briefly covers contemporary molt theory and the cycle code system of bird age categorization, as well as the pertinent literature on the subject (e.g., Humphrey and Parkes 1959, Pyle 1997, Howell et al. 2003, Wolfe et al. 2010, Johnson et al. 2011a). Furthermore, we hope that this volume will inspire careful documentation of relevant molt information during banding efforts and will lead to more extensive examinations of museum specimens to further advance our understanding of molt in other tropical areas in the Western Hemisphere and across the globe.

## MOLT IN THE LIFE CYCLE

Three important avian life cycle events are breeding, migration, and molt. Of course, not all species migrate, and some birds forego breeding in some years for a variety of reasons, but essentially all birds molt at least once per cycle. Despite the importance of feathers, and their regular replacement (molt), ornithological interest has been disproportionately focused on the other phases of the life cycle. As such, much less is known about the environmental stimuli and physiological processes involved in this energetically demanding process across the vast diversity of birds on the planet (Murphy 1996, Bridge 2011). Even so, patterns in the timing and extent of molt have been described in sufficient detail to be useful in a variety of applications, including assisting in the process of roughly determining a bird's age for certain species (e.g., Jenni and Winkler 1994; Pyle 1997, 2008), and also to establish a molt cycle framework in which these patterns can be applied across all species (Humphrey and Parkes 1959, Howell et al. 2003).

In general, a full molt cycle is typically matched to an annual cycle, as molt cycles usually follow annual variations in seasonality, even in the tropics, such as with wet and dry seasons. Each molt cycle is defined as initiating with a prebasic molt, which involves the complete or near-complete replacement of body and flight feathers (*sensu* Howell et al. 2003), but also an endogenous restructuring of protein metabolisms (Thompson and Powers 1924, Murphy and King 1991). Each molt cycle then terminates with the onset of the subsequent prebasic molt, where all or many of the same follicles involved in the previous prebasic molt shed and regrow a new set of feathers. Potential cases where molts may not occur once per annum are likely exceedingly rare (e.g., Summers 1983), and such cases should be closely scrutinized with respect to the molt cycle framework.

The timing, rate, and extent of molts, as well as the plumages they produce, have cascading and interrelated effects on the entire avian life cycle. Because energy available to birds is not limitless, the timely scheduling of breeding, molting, and migration in the annual cycle, relative to the availability of resources, is of critical importance for survival and reproductive success. For example, high reproductive investment can delay molt, decrease feather quality, and reduce parental survival (Siikamäki et al. 1994, Nilsson and Svensson 1996, Dawson et al. 2000). Conversely, molting early can be related to reduced fecundity but higher parental survival (Morales et al. 2007). Such costs associated with molt are believed to prevent its overlap with migration; some migratory species even temporarily suspend molt until migration is completed (Stresemann and Stresemann 1966, Pyle 1997, Leu and Thompson 2002, Pérez and Hobson 2006). Thus, the timing of these life history events is presumably subject to strong evolutionary pressures and is optimized through natural selection to maximize fitness (Dawson et al. 2000, Ricklefs and Wikelski 2002, Moreno 2004).

Despite some documentation that molts in tropical birds can be prolonged, and either aseasonal or seasonal (often occurring after breeding activity), we still know little about the timing of molt relative to other annual cycle events in most tropical species (Pyle 1997, Ricklefs 1997, Dawson 2008, Johnson et al. 2012). It also appears that the temporal overlap of molting and breeding can occur among at least some temperate waterbirds (Thompson and Slack 1983, Howell 2001, Rogers et al. 2014) as well as resident tropical species (Payne 1969, Foster 1975, Echeverry-Galvis 2012). Studies that have examined the timing of molt and breeding found that molt-breeding overlap is highly variable among species, even within a single community, and molt-breeding overlap may occur more at the population level than at the individual level (Johnson et al. 2012, Pyle et al. 2016).

An understanding of when and where birds molt, and the duration and extent of molt, allows researchers to tackle novel ecological and evolutionary questions because differences in molt across species presumably represent adaptive processes that were optimized through selective pressures (Rohwer et al. 2009, Howell 2010). As such, documenting variation in molt strategies among species is a necessary step toward understanding factors that contribute to the evolution of plumage coloration, migration, and social systems as well as other aspects of the avian cycle (Humphrey and Parkes 1959; Rohwer et al. 1980, 2005; Svensson and Hedenström 1999). Knowledge of molt is also useful to population biologists where the predictable and regular extent of molt often varies by age within and across species and allows researchers to place captured individuals into meaningful age classes (e.g., Jenni and Winkler 1994; Pyle 1997, 2008).

## UNDERSTANDING MOLT AND AGING TERMINOLOGY

Humphrey and Parkes (1959) worked to standardize plumage and molt terminology, and their work still serves as a fundamental basis for understanding molt sequences and aging of birds, even in the tropics. Fortunately, the fundamental processes of molt appear largely consistent across the temperate-tropical avian spectrum, and even to a large degree across the 11,000 or so species of birds on this planet. The process of molt appears strongly grounded in an ancient evolutionary history, and strong synapomorphies across distant orders are apparent. For example, soon after hatching from an egg, all birds grow their first set of feathers (except for Kiwis [*Apteryx* spp.], which hatch in their juvenile plumage) and replace these feathers once per cycle, which is often equivalent to a calendar year. Although modifications to the sequence of feather replacement have occurred, these are apparently relatively basal among extant taxa (e.g., Stresemann and Stresemann 1966, Rohwer et al. 2009, Pyle 2013). Even so, modifications to the sequence, timing, and frequency of

feather replacement over the last 150 million years have been molded and crafted, with inserted molts blinking on and off across families, genera, and species, thus highlighting the wonderful plasticity of molt despite still being grounded on an annual ritual of full feather replacement. Add to these scheduled replacements of feathers a diversity of plumage variation across ages, sexes, and species, and one can quickly get consumed by detailing these variations. Variation in molt and plumage sequences among the world's avifauna can appear dizzying and overwhelming, if not also dazzling. Understanding and recognizing these patterns are of fundamental importance in ornithology, as they provide the tools needed to understand demographic and life history processes.

Despite great variation in plumages within and between species, all bird species are believed to exhibit one of only four possible molt strategies, or scenarios, based on the number of molts experienced in each cycle (or year) of life (Figure 1.1; Howell et al. 2003):

1. Simple Basic Strategy—The assumed ancestral system, because of its presence most often in basal avian taxa and potential homology to the shedding of skin in reptiles, describes birds that molt once per year (Howell et al. 2003, Howell and Pyle 2015). This annual molt is called the prebasic molt. The prebasic molt is considered homologous across all bird species and delineates the initiation of each molt cycle. The first prebasic molt is equivalent to what is more familiarly referred to as the prejuvenile molt, which replaces the natal down (when present) with the first

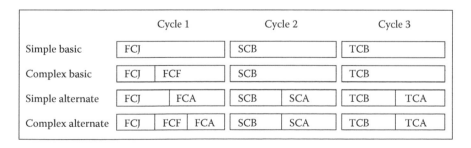

| | Cycle 1 | | | Cycle 2 | | Cycle 3 | |
|---|---|---|---|---|---|---|---|
| Simple basic | FCJ | | | SCB | | TCB | |
| Complex basic | FCJ | FCF | | SCB | | TCB | |
| Simple alternate | FCJ | | FCA | SCB | SCA | TCB | TCA |
| Complex alternate | FCJ | FCF | FCA | SCB | SCA | TCB | TCA |

**Figure 1.1.** The four possible molt strategies exhibited by birds worldwide, redrawn from Howell et al. (2003). In practice, aging as SCB or TCB is only possible in cases of delayed plumage maturation (e.g., some manakins) or incomplete or Staffelmauser prebasic molts (e.g., raptors); in most passerines and also many non-passerines, these would be aged as DCB instead. See Tables 1.2 and 1.3 for a definition of three-letter codes, and Figures 1.3 to 1.10 for more details on practical aging codes according to molt strategy and molt extent.

pennaceous plumage shortly after hatching. A year or so later (i.e., at the end of one cycle), this is followed by the second prebasic molt, followed by the third prebasic molt, and so on, for each (annual) cycle.

2. Complex Basic Strategy—In a slightly modified strategy, a second molt is inserted within the first cycle, called the preformative molt, with subsequent cycles only exhibiting a single prebasic molt; this strategy is referred to as the Complex Basic Strategy. Most birds in this guide exhibit a Complex Basic Strategy, which is generally widespread across passerines and near-passerines, especially in the tropics.

3. Simple Alternate Strategy—Another alternative to the Simple Basic Strategy is for a species to have an inserted molt within all cycles, called a prealternate molt. These birds undergo two molts per cycle (or year), a prebasic and a prealternate, and thus follow a Simple Alternate Strategy. It should be noted that, although we now see an annual inserted molt, which for convenience we refer to as a prealternate molt, it is possible that the evolutionary origin of the first prealternate molt may have been usurped from or merged with an earlier evolved preformative molt, as subsequent annual prealternate molts were inserted, or that a preformative molt has been lost (Pyle 2009). As always, understanding the evolutionary origins of inserted molts may be complex, and provides an opportunity for relatively novel research.

4. Complex Alternate Strategy—Finally, some birds go through three molts in their first cycle (a prebasic, a preformative, and a prealternate) and two molts in all subsequent cycles (a prebasic and a prealternate), and this is called the Complex Alternate Strategy. This last strategy is common among seedeaters (Sporophila) and Nearctic–Neotropic migrants, like many wood-warblers.

To summarize these four strategies and their definitions, it is helpful to realize the distinctions are in the terms "Simple" versus "Complex," and "Basic" versus "Alternate." "Simple" strategies are those without an inserted preformative molt in the first cycle, whereas "Complex" strategies have an inserted preformative molt. "Basic" strategies lack an inserted prealternate molt in each cycle, whereas "Alternate" strategies have an inserted prealternate molt in each cycle. Thus, the combination of these features results in four types of molt strategies (Howell et al. 2003).

Additional complications arise in some species where yet another extra molt is inserted in one or more cycles. Humphrey and Parkes (1959) coined the term "supplemental" for a third plumage in each cycle beyond the alternate plumage, which occurs at least in terns and ptarmigans. The term was subsequently borrowed to also refer to an extra plumage beyond the formative plumage in the first cycle, which has been documented in *Passerina* buntings, Northern Cardinal, and some other North American species (Rohwer 1986, Thompson and Leu 1994, Pyle 1997), as well as *Percnostola rufifrons* (Johnson and Wolfe 2014). This extra first cycle plumage is more appropriately termed an "auxiliary" formative plumage, whereas the use of "supplemental" is still used for third inserted molts in repeating cycles (Howell et al. 2003, Pyle 2008). See the "Cycle Code System for Aging Birds" section for additional details.

Another common term in contemporary literature about molt is "definitive," which is often used interchangeably with "adult" or "adult-like." When one digs into the definition, it becomes clear that this word has taken on a variety of meanings depending on whether it is used in the context of molt or plumage or both (summarized in Wolfe et al. 2014, and Howell and Pyle 2015). Humphrey and Parkes (1959) originally defined "definitive" to describe a plumage aspect that has reached a fully matured or climax state, that is, a plumage that was obtained after a given period of time and is cyclically replaced via molt through the remainder of life. It is worth noting that some individuals may occasionally revert to a preadult plumage aspect (Pyle 2008, Repenning 2012), whereas others may continuously change with age, for example, with some female manakins adding increased male-like plumage to their plumage with age (Graves 1981, Ryder and Durães 2005). Furthermore, in some sexually dichromatic species, again with manakins providing an excellent example, females enter a "definitive plumage" (*sensu* Humphrey and Parkes 1959) usually in the second cycle, whereas males of the same species may variably enter their adult plumage aspect in their third or fourth cycle. Certainly, in this context, the term "definitive" loses any utility of representing homology across, but also sometimes within, species. With regard to molt, "definitive" has been used to refer to the process of undergoing a molt that has homologous counterparts in subsequent cycles (*sensu* Howell et al. 2003). As such, Howell and Pyle (2015) suggest using "definitive" in reference to plumage aspect or molt cycle depending on the context, and to be explicit in that use.

To limit confusion, here we only use the term "definitive" as it pertains to molt cycles (*sensu* Wolfe et al. 2014), although we recognize the utility for a broader use of the word as an adjective to also refer to a mature plumage aspect. For example, because a preformative molt only occurs once in the life of a bird (after the prejuvenile molt in the first cycle) and does not have homologous counterparts in subsequent cycles, it is not part of a definitive molt cycle; however, it can appear adult-like particularly when the preformative molt is complete. In this guide we use the word "adult" instead of "definitive" with regard to plumage maturation, and emphasize that this may not necessarily reflect consistency with breeding maturity, which can occur in the formative, first alternate, second basic, second alternate, or subsequent definitive or alternate plumage, depending on the species or individual within a species. Thus, when we refer to a definitive basic plumage

we are implying nothing about an "adult" plumage or breeding maturity, but are rather unequivocally stating that the plumage resulting from a definitive prebasic molt is a definitive plumage. See Howell and Pyle (2015) for alternative proposals for the broader use of the word "definitive."

## VARIATION IN MOLT EXTENT

Molt extents, or the amount of feathers replaced during molts, should not be confused with molt strategies or sequences. Each molt can range from a limited extent (or even absent in some individuals of a population that otherwise would be scheduled to molt) to complete with a continuous gradation of intermediary extents. For convenience, Pyle (1997) categorized extent as (1) absent, (2) limited, (3) partial, (4) incomplete, and (5) complete (Table 1.1). Note that both prealternate and preformative molts can range from absent to complete, whereas prebasic molts are either complete, or, in some cases (particularly in large species), can be incomplete. It is variation in molt extents, along with age- and sex-related variations in plumage aspect, that creates a vast color palate for birds beyond just the number of species on the planet.

For most passerines and near-passerines, as well as a large variety of non-passerines, a predictable sequence by which feathers are replaced exists. An oversimplified description of the order in

TABLE 1.1

*Range and classification of molt extents, as defined by Pyle (1997, 2008), and the kinds of molts in which these extents appear.*

| Extent | Definition | Prebasic | Preformative | Prealternate |
|---|---|---|---|---|
| Absent | Not present in an individual of a population or species where a molt is otherwise often scheduled. | | X | X |
| Limited | Includes only body feathers, but not wing coverts. | | X | X |
| Partial | Includes most if not all body feathers and at least some wing coverts. In some individuals, it can also include the tertials and one pair or more of rectrices. | | X | X |
| Incomplete | Includes all body feathers, most or all wing coverts, and some primaries, secondaries, and rectrices. | (X) | X | (X) |
| Complete | Includes all body feathers, all wing coverts, and all flight feathers. | X | X | (X) |

NOTE: X = yes, regularly; (X) = yes, occasionally to rarely.

which major feather tracts begin to be replaced is as follows: body feathers, lesser coverts, median coverts, greater coverts, primaries and primary coverts, tertials, and, finally, secondaries (Figure 1.2). Among the flight feathers, both primaries and secondaries begin at the center of the wing (p1 and s1 in North American terminology). Primaries are replaced sequentially toward the tip of the wing, while secondaries are replaced sequentially toward the body. The inner three secondaries, often referred to as the tertials (these are not true tertials in passerines), begin at s8, followed by s9, s7, and sometimes s6; this nomenclature is for species with nine secondaries. Of course, for species with additional secondaries and/or true tertials, the numbering of the innermost flight feathers on the wing is different, depending on the number of feathers, but the

general concept remains with inner secondaries and/or tertials being replaced before the outermost secondary (s1). Molts that do not complete predictably arrest somewhere along this sequence and form molt limits.

Of course there are many well-known exceptions to this generalized pattern. For example, some species have what is called an eccentric molt, in which primaries and secondaries are not replaced in the typical order described earlier. This occurs in some preformative molts, more rarely in prealternate molts, and can result in unusual molt limits, often with the outer 3–6 primaries and/or primary coverts being replaced, but not the inner primaries. Extremes of this can replace all flight feathers in typical sequence, but retain primary coverts, as in many *Myiarchus* and some *Contopus* flycatchers (Pyle 1997). Similarly, falcons and parrots

Figure 1.2. An illustration of the extent of molt at eight stages of a complete molt sequence among wing coverts and flight feathers in *Percnostola rufifrons*. The first (upper left) photo shows two generations of feathers with the lesser, median, and three inner greater coverts being auxiliary formative, and the other wing coverts and flight feathers are juvenile. In the next seven photos, black wing covert feathers with white tips and all-black flight feathers appear—these are feathers of the formative plumage—indicating a generalized sequence of feather replacement exhibited by many passerines and near-passerines, as well as some non-passerines.

generally begin at p4, and have one wave of molt proceeding outward to p10 and another proceeding inward to p1 (Pyle 2013). Another molt pattern, generally known as *Staffelmauser*, refers to blocks of flight feathers being replaced in multiple waves, starting at multiple nodes, which varies across each molt cycle. Often *Staffelmauser* molts arrest before they complete, leading to two or more generations of feathers across the group of flight feathers (Stresemann and Stresemann 1966). *Staffelmauser* often occurs in larger species where it would take too long to replace all feathers in a single molt (Rohwer et al. 2009). Finally, some species replace all flight feathers simultaneously, as in some birds that predominantly swim or walk (Pyle 2008, Howell 2010).

Tail molt begins usually with the central rectrix (r1) and proceeds outward, but occasionally begins with the outer rectrix (r6) and proceeds inward. In some cases, rectrix replacement follows a nonsequential pattern, such as r1 → r2 → r5 → r3 → r4 → r6.

## CYCLE CODE SYSTEM FOR AGING BIRDS

Breeding seasons in tropical and southern latitudes often bridge 1 January, such that the North American calendar-based system of aging birds is not appropriate. The European system based on life-cycle events is also not appropriate across all species where the classic northern temperate life cycle follows a breeding–molt–winter schedule, or where a species' life-cycle schedule is unknown. Indeed, many tropical birds exhibit molt-breeding overlap, although this is often more common at the population level than the individual level (Foster 1975, Echeverry-Galvis 2012, Johnson et al. 2012, Pyle et al. 2016). Furthermore, the complicated and often unknown timing, extent, and predictability of molt and breeding for entire populations or species create a situation in which applying temperate-biased terminologies becomes even more problematic. Surely, this has contributed to a lack of age-specific studies on tropical birds.

Wolfe et al. (2010) introduced a transformative cycle-based aging system (also known as the WRP system), which was further refined by Johnson et al. (2011b) and Pyle et al. (2015), based on the sequence of molts and plumages through which birds progress during their life. With this system, aging codes are not based on calendar years (as in the standard North American aging system) or relative to breeding status (as followed in many European references), but are, instead, derived from Humphrey and Parkes' (1959) molt and plumage terminology modified by Howell et al. (2003). This approach for developing an age code system is driven by a desire to standardize terminology based on perceived molt homologies across each species' annual cycle. Again, we recommend that users of this guide study these landmark papers to familiarize themselves with terminology used here. We will briefly outline some of the key points, but do not wish to recapitulate the extensive, and sometimes complicated, details addressed in these papers.

The cycle-code aging system allows for the maximum accuracy of aging birds based on their molt and plumage state. Like any other avian aging system, it requires knowledge of three important points: (1) the molt strategy (Figure 1.1), (2) the extent of each molt (Table 1.2), and (3) the plumage aspect resulting from each molt. Once these are known, then each bird captured, collected, or observed can be given a cycle code based on its molt and plumage state. Even if one or more of these points is not known, the cycle code system can accommodate uncertainty.

The cycle code system uses a three-letter code, with standardized options available for each letter (Table 1.2). It is perhaps easiest to start with the first and last letter as these refer to the cycle and plumage, respectively, which one wishes to reference. The middle letter provides context of each captured bird relative to the presence of active molt. To apply the cycle code system to an individual bird, imagine a scenario in which one captures a bird in juvenile plumage that is not molting. Because it is in its first cycle, it would be coded as F (first cycle), C (not molting, i.e., *cycle*), and J (*juvenile* plumage), or FCJ (Table 1.2). Because Howell et al. (2003) redefined the juvenile plumage as the first basic, we could actually also call this bird FCB. However, like Howell et al. (2003), we prefer to use the term "juvenile" because of its familiarity and avoid any use of FCB in the cycle-code system. If this same bird were molting into this plumage, we would simply replace the C for P (for *pre*), and age the bird as an FPJ. If we recapture this same bird later, but do not know which plumage it is in (for example, perhaps formative, or perhaps definitive basic, or perhaps first or definitive alternate), we would change the middle

TABLE 1.2
*Available codes for each position in the cycle code aging system (modified from Wolfe et al. 2010, Johnson et al. 2011a).*

| First position | First position definition | Second position | Second position definition | Third position | Third position definition |
|---|---|---|---|---|---|
| F | First cycle | C | Not molting ("C" for cycle) | J | Juvenile |
| S | Second cycle | P | Molting ("P" for pre) | X | Auxiliary formative |
| T | Third cycle | A | After a given plumage | F | Formative |
| 4, 5, 6, etc. | 4th, 5th, 6th cycle, etc. | | | A | Alternate |
| D | Definitive cycle | | | S | Supplemental |
| U | Unknown cycle | | | B | Basic |
| | | | | U | Unknown |

letter to A, meaning it is *after* first juvenile and use FAJ. The letter U (*unknown*) can also be used in the first (cycle) or third (plumage) position, and, in the previous circumstance, we could have also used UCU; because we have more accurate information, we know that it is not in *juvenile plumage*, so we prefer the more specific code FAJ.

Note that UAJ is not an appropriate code. In the cycle code system, the cycle and plumage in the first and third positions, respectively, must be linked such that a bird's age may be "after first cycle juvenile," but not "after an unknown cycle juvenile." This distinction becomes more important in birds with incomplete prebasic molts in which a minimum age (i.e., cycle and plumage) is known, such as when there are multiple generations of flight feathers and none are juvenile. For example, the code TAB would refer to "after third basic plumage." If we were to call this bird "unknown cycle basic plumage," or UAB, we lose the ability to archive whether this bird was a minimum of two, three, four, or more years of age.

One issue not yet resolved is how to code additional plumages inserted into each cycle beyond the formative and alternate, that is, auxiliary formative and supplemental plumages, respectively (see "Understanding Molt and Age Terminology" section). The letter S is a simple enough code to use for *supplemental* for placement in the third position of the cycle code. Auxiliary formative, however, creates a unique challenge. The letter A is already used for *alternate* and F is already used for *formative*. We see four possible ways to treat this for cycle coding:

1. Create a new letter for this molt and plumage, such as X. For example, FPX and FCX.

2. Lump the auxiliary formative with the preformative molt (even if the auxiliary preformative molt arrests briefly before the preformative molt begins), such that FPF is used for auxiliary preformative molts, auxiliary preformative plumage, and the preformative molt, and that FCF remains reserved for the formative plumage.

3. Create a fourth position in the cycle system, for example, FPFa and FCFa, which is similar to use elsewhere prior to the development of the cycle code system (e.g., Howell et al. 2003).

4. Use the nonpreferred term *supplemental* (*sensu* Rohwer 1986, Thompson and Leu 1994, Pyle 1997) instead of auxiliary formative, that is, FPS and FCS.

For now, we use the first option (Table 1.2), although we recognize that, as more practitioners use the cycle code system, a clearer consensus may be reached.

Each of the four molt strategies outlined by Howell et al. (2003) in combination with the extents of each molt (complete versus something less than complete) result in a subset of codes among the entire list of possibilities (see Table 1.3, and Figures 1.3 to 1.10). It may already be clear that some combinations are not possible and nonsensical, such as SCJ (second cycle juvenile) or DPF (definitive preformative). To successfully age birds and to use the cycle code system in practice, understanding why only

## TABLE 1.3
*Cycle codes and descriptions (adapted from Wolfe et al. 2010, Johnson et al. 2011b).*

| Cycle code | Unabbreviated WRP |
|---|---|
| FPJ | Prejuvenile molt |
| FCJ | First cycle juvenile |
| FPX | First cycle auxiliary preformative molt |
| FCX | First cycle auxiliary formative |
| FPF | First cycle preformative molt |
| FCF | First cycle formative |
| FPA | First prealternate molt |
| FCA | First cycle alternate |
| SPB | Second prebasic molt |
| SCB | Second basic plumage |
| SPA | Second prealternate molt |
| SCA | Second cycle alternate plumage |
| DPB | Definitive prebasic molt |
| DCB | Definitive cycle basic |
| DPA | Definitive prealternate molt |
| DCA | Definitive cycle alternate |
| FAJ | After first cycle juvenile |
| SAB | After second cycle basic |
| UPB | Unknown cycle, prebasic molt |
| UCB | Unknown cycle, basic plumage |
| UPA | Unknown cycle, prealternate molt |
| UCA | Unknown cycle, alternate plumage |
| UPU | Unknown cycle, unknown molt |
| UCU | Unknown cycle, unknown plumage |

certain codes are available is important. For several common situations, we provide a list of likely codes that practitioners will encounter. In theory, sequences of codes are slightly different than what we might actually use in practice, which is no different than any other age code system. For example, we may know in theory that the sequence of a Simple Basic Strategy is FPJ → FCJ → SPB → SCB → TPB → TCB → 4PB etc. (where "→" can be thought of as "followed by"), but in practice if the SPB results in an adult-like plumage aspect, our options are thereby limited to FPJ → FCJ → SPB → DCB → DPB → DCB → DPB etc.

Combining the four different molt strategies (Figure 1.1) with molt extents (Table 1.1) creates a practical utility of the cycle code system that essentially results in eight likely scenarios for implementation in most birds, with four of them being most likely in passerines, near-passerines, and even many other non-passerines (see also Pyle et al. 2015). We refer to these eight "groups" throughout the volume in each species account, for easy reference by the reader.

1. Simple Basic Strategy with complete prebasic molts (Figure 1.3): FPJ → FCJ → SPB → DCB → DPB etc. Examples include boobies (*Sula* spp.) and albatrosses (Diomedeidae). Additional codes are available with delayed plumage maturation (beyond the SPB), including SCB, TPB, TCB, 4PB, SAB, and TAB.

2. Simple Basic Strategy with incomplete *Staffelmauser* prebasic molts (Figure 1.4): FPJ → FCJ → SPB → SCB → TPB/DPB → TCB/SAB → 4PB/DPB etc. Examples include large Accipitridae and Cathartidae.

3. Complex Basic Strategy with less-than-complete preformative molts (Figure 1.5): FPJ → FCJ → FPF → FCF → SPB → DCB → DPB → etc. Examples include *Phaeothlypis* spp. (and probably close relatives), many *Myrmotherula* spp., *Thamnomanes* spp., and many Pipridae. Additional codes are available with delayed plumage maturation (beyond the SPB), including SCB, TCB, 4CB, SAB, and TAB. This is "Group 1" in Pyle et al. (2015).

| FPJ | FCJ | SPB | DCB | DPB | Etc. |

**Figure 1.3.** Group 1: Simple Basic Strategy with complete prebasic molts. UPB should be used when SPB and DPB cannot be distinguished in the field, as in some cases especially when these molts are nearly finished.

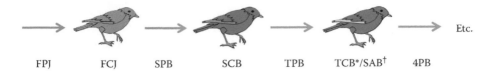

FPJ     FCJ     SPB     SCB     TPB     TCB*/SAB†     4PB     Etc.

**Figure 1.4.** *Group 2: Simple Basic Strategy with incomplete or Staffelmauser prebasic molts. *If the oldest feathers are juvenile, then accurate aging using SCB, TCB, 4CB, etc., is possible. †If the oldest feathers are not juvenile, then "after" aging using SAB, TAB, 4AB, etc., is appropriate. UPB should be used when SPB, TPB, 4PB, etc., cannot be distinguished in the field, as in some cases especially when these molts are nearly finished; or using an "after" code may be preferable to the practitioner.*

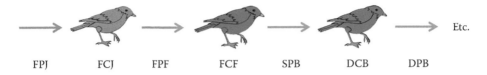

FPJ     FCJ     FPF     FCF     SPB     DCB     DPB     Etc.

**Figure 1.5.** *Group 3: Complex Basic Strategy with less than complete preformative molts and complete prebasic molts. UPB should be used when SPB and DPB cannot be distinguished in the field, as in some cases especially when these molts are nearly finished.*

FPJ     FCJ     FPF     FAJ*     UPB     FAJ     UPB     Etc.

**Figure 1.6.** *Group 4: Complex Basic Strategy with complete preformative molts and complete prebasic molts. *FCF can be used if the skull is unossified (in species where the skull fully ossifies) or where other soft-part or plumage characters may be distinctly immature.*

4. Complex Basic Strategy with complete preformative molts and complete prebasic molts (Figure 1.6): FPJ → FCJ → FPF → FAJ → UPB etc. Examples include *Pithys albifrons*, most Furnariidae, and *Tachphonus coronatus*. UPB is used because it is not possible to distinguish the SPB from DPB. FCF is available in place of FAJ when the skull is unossified in species for which the skull eventually fully ossifies before the SPB. Additional codes are available with delayed plumage maturation, including DCB, SCB, TCB, SPB, TPB, SAB, and TAB. This is "Group 3" in Pyle et al. (2015).

5. Complex Basic Strategy with less-than-complete preformative molts and incomplete prebasic molts (Figure 1.7): FPJ → FCJ → FPF → FCF → SPB → SCB → TPB/DPB → TCB/SAB → 4PB/DPB etc. Examples include Picidae and some Columbidae. This is "Group 4" in Pyle et al. (2015).

6. Simple Alternate Strategy with less-than-complete prealternate molts and complete prebasic molts (Figure 1.8): FPJ → FCJ → FPA → FCA → SPB → DCB → DPA → DCA → DPB etc. Examples include *Larus* spp. In some species, such as Whimbrel (*Numenius phaeopus*), the inserted molt in the first cycle is inferred to be an FPF rather than an FPA based on its phylogenetic position (Pyle 2008), thus it may be more appropriate to use this sequence: FPJ → FCJ → FPF → FCF → SPB → DCB → DPA → DCA → DPB etc. Additional codes are available with delayed plumage maturation (beyond the SPB), including SCB, TCB, SCA, TCA, SAA, TAA, SAB, and TAB.

7. Complex Alternate Strategy with complete preformative molts, less than complete prealternate molts, and complete prebasic molts (Figure 1.9): FPJ →

FPJ   FCJ   FPF   FCF   SPB   SCB   TPB   TCB*/SAB†   4PB   Etc.

**Figure 1.7.** *Group 5:* Complex Basic Strategy with less than complete preformative molts and incomplete or *Staffelmauser prebasic* molts. *If the oldest feathers are juvenile, then accurate aging using SCB, TCB, 4CB, etc., is possible. †If the oldest feathers are not juvenile, then "after" aging using SAB, TAB, 4AB, etc., is appropriate. UPB should be used when SPB, TPB, 4PB, etc., cannot be distinguished in the field, as in some cases especially when these molts are nearly finished; or using an "after" code may be preferable to the practitioner.

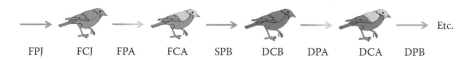

FPJ   FCJ   FPA   FCA   SPB   DCB   DPA   DCA   DPB   Etc.

**Figure 1.8.** *Group 6:* Simple Alternate Strategy with less than complete prealternate molts and complete prebasic molts. UPB should be used when SPB and DPB cannot be distinguished in the field, especially when these molts are nearly finished.

FPJ   FCJ   FPF   FAJ*   UPA   UCA   DPB   FAJ   UPA   Etc.

**Figure 1.9.** *Group 7:* Complex Alternate Strategy with complete preformative molts, less than complete prealternate molts, and complete prebasic molts. *FCF can be used if the skull is unossified (in species where the skull fully ossifies) or where other soft-part or plumage characters may be distinctly immature.

FPJ   FCJ   FPF   FCF   FPA   FCF   SPB   DCB   DPA   DCA   DPB   Etc.

**Figure 1.10.** *Group 8:* Complex Alternate Strategy with less than complete preformative molts, less than complete prealternate molts, and complete prebasic molts. UPB should be used when SPB and DPB cannot be distinguished in the field, as in some cases especially when these molts are nearly finished.

FCJ → FPF → FAJ → UPA → UCA → UPB etc. Examples include *Lanio fulvus* and *Pachyramphus marginatus*. Additional codes are available with delayed plumage maturation (beyond the FPF), including FPA, FCA, DPA, DCA, SAA, SAB, TAA, and TAB, and FCF is available in place of FAJ when the skull is unossified in species for which the skull eventually fully ossifies before the SPB.

8. Complex Alternate Strategy with less-than-complete preformative molts, less-than-complete prealternate molts, and complete prebasic molts (Figure 1.10): FPJ → FCJ → FPF → FCF → FPA → FCA → SPB →

DCB → DPA → DCA → DPB etc. Examples include *Pachyramphus minor*, *Myiarchus* spp., and many migratory Parulidae and Cardinalidae. This is "Group 2" in Pyle et al. (2015).

## TIPS FOR AGING BIRDS IN THE CENTRAL AMAZON AND BEYOND

The diversity of molt patterns and sequences exhibited among Amazonian birds can be intimidating. Luckily for banders working in the Amazon Basin and, really, throughout the Neotropics, most captured land birds exhibit one molt strategy: the Complex Basic Strategy. Most major differences among bird species that exhibit the Complex Basic Strategy is the extent and duration of the preformative molt. In general, and as a starting point, most ovenbirds (Furnariidae) and tanagers (Thraupidae) exhibit complete preformative molts, antbirds (Thamnophilidae) and flycatchers (Tyrannidae) exhibit partial to complete preformative molts, and manakins (Pipridae) exhibit partial preformative molts. The extent and duration of the preformative molt will often dictate how to go about determining ages for a given species (Figures 1.5 and 1.6).

## Species with Complete Preformative Molts

*Pithys albifrons*, for example, a commonly captured obligate antbird, has a distinct juvenile plumage that is replaced with a complete, yet extremely protracted, preformative molt. The preformative molt is complete and the formative plumage is indistinguishable from subsequent plumages, confounding our ability to precisely differentiate First Cycle Formative (FCF) from Definitive Cycle Basic (DCB) birds, thus we would call any nonmolting adult-like birds After First Cycle Juvenile (FAJ; Figure 1.5).

When *P. albifrons* is actively replacing feathers, one can assess age by examining the contrast between old (not yet replaced) and new (molting and replaced) flight feathers. Because the juvenile plumage is readily distinguishable, banders may recognize the distinct retained juvenile feathers being replaced by an adult plumage, which indicates an active preformative molt, results in the age of First Preformative (FPF). Similarly, if the old plumage is unequivocally adult in actively molting birds, then we can conclude that the individual is at least replacing a formative plumage with a second prebasic molt, which cannot be distinguished from subsequent definitive molts and can be aged as Unknown Prebasic (UPB). This same strategy can be applied to all birds that undergo complete preformative molts: look for active molt, identify the old and new feathers as either juvenile or adult, and age the bird appropriately.

If a *P. albifrons* is captured in adult plumage not undergoing molt, we lose the ability to differentiate the formative from subsequent plumages and must age the bird using a less precise code. In this case, all we know is that the bird is not in juvenile plumage, thus it is After First Juvenile (FAJ). It is worth mentioning that in species with rapid, yet complete, preformative molts, unossified skulls can indicate a bird in formative plumage (FCF) where the skull is known to ossify in that species. In contrast, an ossified skull could either be a bird later in formative plumage or in definitive basic plumage, thus we would again not be able to differentiate the plumage and use the FAJ code.

In summary, nonjuvenile molting birds are generally only categorized as FPF or UPB, whereas nonmolting birds are categorized as FCJ or FAJ. In some species with rapid preformative molts, nonjuvenile birds with unossified skulls can be aged as FCF, but beware of cases where small unossified windows can be retained into the definitive cycle.

## Species with Limited, Partial, or Incomplete Preformative Molts

Among birds that exhibit a Complex Basic Strategy, there are many species that undergo limited, partial, or incomplete preformative molts. The formative plumage (FCF) in this group is characterized by the presence of molt limits, or contrast between the more recently replaced formative and older retained juvenile feathers (Figure 1.6). In some species, such as *Thamnomanes ardesiacus*, the male's formative plumage is superficially adult-like, but with obvious retained female-like juvenile feathers. In other species, like *Willisornis poecilinotus*, the formative-plumaged male is female-like, thus determining sex is not possible in the absence of breeding criteria (i.e., presence of a brood patch in females or a cloacal protuberance in males). For these species with delayed plumage maturation, it is critically important to determine the age of captured individuals prior to distinguishing sex to avoid falsely categorizing formative-plumaged males as females.

When individuals undergoing flight feather molt have some retained juvenile flight feathers that can be anticipated to become replaced during this molt, we know the bird must be entering the Second Prebasic molt (SPB). Similarly, if the older plumage actively being replaced is unequivocally definitive, then we can conclude that the individual is at least undergoing the third prebasic molt, which is a definitive molt, and can be aged as Definitive Prebasic (DPB). We follow Pyle et al. (2015) in using UPB in situations where the second prebasic molt cannot be distinguished from subsequent definitive prebasic molts. These are circumstances where it can be difficult to determine whether old feathers are juvenile or definitive.

In summary, nonjuvenile molting birds can be aged as FPF, SPB, or DPB. In particular, if we know that the preformative molt is partial, then any nonjuvenile individual with molting flight feathers must be SPB or DPB. In nonmolting birds, the age code options are typically FCJ, FCF, or DCB. In cases of delayed plumage maturation, like in some manakins, then additional age code options become available, like SAB or FAJ.

## Unusual Molting Strategies

A small subset of species captured in understory mist nets at the BDFFP appears to exhibit a Complex Alternate Strategy, although, realistically, the presence of a limited prealternate molt can be difficult to detect and distinguish from adventitious feather replacement. Regardless, this strategy is often associated with migration as year-round and prolonged exposure to ultraviolet (UV) light (relatively long day lengths both on breeding and wintering grounds) is the primary mechanism thought to drive this, and sexual selection is only secondary if at all influential (see Pyle 2008 and Howell 2010 for additional discussion). As far as we know, however, species exhibiting this strategy at the BDFFP are nonmigratory, yet we posit that high exposure to UV radiation may be driving this pattern, as prealternate molts typically appear in canopy and tropical grassland species.

Relatively few species in tropical forests exhibit Simple Basic or Simple Alternate Strategies, but this should be looked for particularly in non-passerines where preformative molts are thought to be absent in temperate (northern or southern) relatives.

## STUDY SITE: BIOLOGICAL DYNAMICS OF FOREST FRAGMENTS PROJECT

### Background

The 186 species and 37 families included in this guide were selected from a 35-year database of bird captures from the Biological Dynamics of Forest Fragments Project (BDFFP), located about 80 km north of Manaus, Brazil (S 2°30′, W 60°) in the center of the Amazon basin. Many of these species and genera are reasonably widespread

across Amazonia and even Latin America; of the 155 genera found at the BDFFP, 90% are also represented at the Nouragues Field Station, French Guiana, and 71% are found in Cocha Cashu, Peru (Terborgh et al. 1990, Thiollay 1994, Johnson et al. 2011b). Our goal here is to provide an overview and launching point for aging birds in Latin America but, more importantly, provide a comprehensive overview of the process of understanding molt cycles and aging terminology, as well as its application to a variety of species across many families that would be regularly encountered by banders throughout the hemisphere.

The BDFFP was created in 1979 by the World Wildlife Fund–U.S. (WWF) largely because of the ideas, motivation, and effort of Thomas Lovejoy. At the time, Brazilian law required that 50% of forested land remain intact as landowners deforested their land for cattle grazing and agriculture. Taking advantage of this, Lovejoy developed an agreement among the WWF, Brazil's Instituto Nacional de Pesquisas da Amazônia (INPA), the Conselho Nacional de Desenvolvimento Científico e Tecnológico (CNPq), and cooperative ranchers to experimentally isolate fragments as local cattle ranchers cleared their land. This agreement established one of the most important landscape-scale fragmentation experiments on the planet (Bierregaard et al. 2001).

During the early and mid-1980s, changing logistical and economic realities prevented about half of the project from becoming realized. Eleven out of the originally planned 24 fragments were isolated between 1980 and 1990 (10 fragments were isolated by 1984). The surrounding matrix, or dissimilar habitat surrounding the fragment of interest, was used by ranchers for cattle production, but most of this land use ceased within a few years after clearing because governmental incentives for such practices disappeared in the face of an economic downturn during the 1980s (Fearnside 2005). Since then, secondary succession in the matrix has been converting the once open pastures into regenerating secondary forest. The speed of regeneration and composition of the returning vegetation is dependent on how intensively the land was used; burning promoted a *Vismia* (Clusiaceae)-dominated second growth, whereas clear-cutting without burning promoted a *Cecropia* (Cecropiaceae)-dominated second growth (Borges and Stouffer 1999, Lucas et al. 2002), although these second growth communities become more similar as they mature. In most cases, second growth within 100 m of fragments has been occasionally cleared to maintain isolation (Gascon et al. 2001).

The BDFFP remains one of the most important projects evaluating the effects of forest fragmentation; over 700 papers have been published in the BDFFP technical series. Since isolation, monitoring birds and other taxa has documented changes in abundance and diversity in response to fragmentation and second growth regeneration (e.g., Powell and Powell 1987, Bierregaard et al. 2001, Stouffer et al. 2006, Ferraz et al. 2007, Powell et al. 2013). The project has also grown intellectually and today provides one of the most important long-term databases regarding the effects of fragmentation on biodiversity and ecological processes.

One of the results of 35 years of research at BDFFP is an extensive database representing >60,000 bird captures. In the early years, significant progress was made toward understanding the composition of the understory bird community and criteria for identifying species within difficult groups were developed (Cohn-Haft et al. 1997). When we started mist-netting at BDFFP in 2007, we aimed to advance our knowledge of these species by applying aging criteria used in temperate species. This volume is dedicated to this accumulated knowledge, hopefully providing guidance not only to future banders at the BDFFP but also to researchers throughout the Amazon, Latin America, and around the world who are interested in learning about molts and applying aging criteria.

## Climate and Habitat

The BDFFP is located in lowland terra firme wet broadleaf rainforest. The understory is relatively open for a tropical rainforest, which is at least partially due to low nutrient content on the old Guianan shield soils. Even so, the canopy of the forest often exceeds 30 to 35 m high with emergent trees towering to 55 m. The site is bisected by small streams, some of which dry up during the dry season, whereas others flow year-round. There are often short, steep slopes leading to these streambeds, with the elevation of the site fluctuating between 50 and 150 m above sea level. Distinct microhabitats are associated with this topography with moist baixios in low areas having a relatively distinct flora and fauna compared to drier nearby plateaus (Bierregaard et al. 2001).

On average, 2500 mm of rain falls each year with about 100 to 300 mm falling each month, but sometimes over 500 mm rain can fall in a single month and up to 3500 mm of rain has fallen in a year. A fairly distinct wet season lasts from December or January through April or sometimes into early June. The rest of the year is considerably drier at the BDFFP, with the driest months ranging from Jul to Sep.

Seasonality at BDFFP probably has important consequences on the timing of breeding for many species, but their responses to wet and dry seasons can be surprisingly species-specific with some breeding during the dry season, others during the wet season, some overlapping in both seasons for part of the year, and others breeding year-round (Johnson et al. 2012, Stouffer et al. 2013). Most species are resident, particularly those that occupy the understory, but at least one understory species that breeds at the BDFFP, Geotrygon montana, appears to be migratory with a distinct breeding season during the wet season and is mostly absent during the dry season (Stouffer and Bierregaard 1993). The strength of seasonality in the tropics surely influences when breeding and molting occur in tropical species, but the timing of breeding and molting can be highly variable among species within a community, such as the one found at BDFFP.

The majority of understory species, and probably other suites of birds, breed at the BDFFP from the late dry into the early wet seasons (Stouffer et al. 2013). Even so, the exact timing may vary annually according to local conditions, habitat modification (i.e., fragmented versus continuous forest), and other unknown factors. Thus, for any given species, the breeding season may appear to last more than 6 months to year-round, although individuals in their respective population may follow a shorter breeding chronology or may not breed at all in certain years. Certainly, much is yet to be learned about how individuals, populations, and communities in the Neotropics adjust their major life history events to seasonal and annual changes in climate.

## LITERATURE CITED

Bierregaard, R. O., C. Gascon, T. E. Lovejoy, and R. Mesquita. 2001. Lessons from Amazonia: the ecology and conservation of a fragmented forest. Yale University Press, New Haven, CT.

Borges, S. H., and P. C. Stouffer. 1999. Bird communities in two types of anthropogenic successional vegetation in central Amazonia. Condor 101:529–536.

Bridge, E. 2011. Mind the gaps: what's missing in our understanding of feather molt. Condor 113:1–4.

Cohn-Haft, M., A. Whittaker, and P. C. Stouffer. 1997. A new look at the "species-poor" central Amazon: the avifauna north of Manaus, Brazil. Ornithological Monographs 48:205–235.

Dawson, A. 2008. Control of the annual cycle in birds: endocrine constraints and plasticity in response to ecological variability. Philosophical Transactions of the Royal Society B: Biological Sciences 363:1621–1633.

Dawson, A., S. A. Hinsley, P. N. Ferns, R. H. C. Bonser, and L. Eccleston. 2000. Rate of moult affects feather quality: a mechanism linking current reproductive effort to future survival. Proceedings of the Royal Society B 267:2093–2098.

Echeverry-Galvis, M. A. 2012. Molt-breeding overlap in birds: phenology and trade-offs at the individual and the community levels. Ph.D. dissertation. Princeton University, Princeton, NJ.

Fearnside, P. M. 2005. Deforestation in Brazilian Amazonia: history, rates, and consequences. Conservation Biology 19:680–688.

Ferraz, G., J. D. Nichols, J. E. Hines, P. C. Stouffer, R. O. Bierregaard, and T. E. Lovejoy. 2007. A large-scale deforestation experiment: effects of patch area and isolation on Amazon birds. Science 315:238–241.

Foster, M. S. 1975. The overlap of molting and breeding in some tropical birds. Condor 77:304–314.

Gascon, C., R. O. Bierregaard, W. F. Laurance, and J. R.-d. Mérona. 2001. Deforestation and forest fragmentation in the Amazon. Pp. 21–30 in R. O. Bierregaard, C. Gascon, T. E. Lovejoy, and R. C. G. Mesquita (editors), Lessons from Amazonia: the ecology and conservation of a fragmented forest. Yale University Press, New Haven, CT.

Graves, G. R. 1981. Brightly coloured plumage in female manakins (*Pipra*). Bulletin of the British Ornithological Club 101:270–271.

Guallar, S., E. Santana, S. Contreras, H. Verdugo, and A. Gallés. 2009. Paseriformes del Occidente de México: Morfometría, datación y sexado. Monografies del Museu de Ciències Naturals 5 (in Portuguese).

Hernández, A. 2012. Molt patterns and sex and age criteria for selected landbirds of southwest Colombia. Ornitología Neotropical 23:215–223.

Howell, S. N. G. 2001. Molt of the Ivory Gull. Waterbirds 24:438–442.

Howell, S. N. G. 2010. Molt in North American Birds. Houghton Mifflin Harcourt, Boston, MA.

Howell, S. N. G., C. Corben, P. Pyle, and D. I. Rogers. 2003. The first basic problem: a review of molt and plumage homologies. Condor 105:635–653.

Howell, S. N. G., and P. Pyle. 2015. Use of "definitive" and other terms in molt nomenclature: a response to Wolfe et al. (2014). Auk 132:365–369.

Humphrey, P. S., and K. C. Parkes. 1959. An approach to the study of molts and plumages. Auk 76:1–31.

Jenni, L., and R. Winkler. 1994. Moult and ageing of European passerines. Academic Press, London, UK.

Johnson, E. I. and J. D. Wolfe. 2014. Thamnophilidae (antbird) molt strategies in a central Amazonian rainforest. Wilson Journal of Ornithology 126:451–462.

Johnson, E. I., P. C Stouffer, and R. O. Bierregaard. 2012. The phenology of molting, breeding and their overlap in central Amazonian birds. Journal of Avian Biology 43:141–154.

Johnson, E. I., J. D. Wolfe, T. B. Ryder, and P. Pyle. 2011a. Modifications to a molt-based ageing system proposed by Wolfe et al. (2010). Journal of Field Ornithology 82:422–424.

Johnson, E. I., P. C Stouffer, and C. F. Vargas. 2011b. Diversity, biomass, and trophic structure of a central Amazonian rainforest bird community. Revista Brasiliera de Ornitologia 19:1–16.

Leu, M., and C. W. Thompson. 2002. The potential importance of migratory stopover sites as flight feather molt staging areas: a review for Neotropical migrants. Biological Conservation 106:45–56.

Lucas, R. M., M. Honzák, I. Do Amaral, P. J. Curran, and G. M. Foody. 2002. Forest regeneration on abandoned clearances in central Amazonia. International Journal of Remote Sensing 23:965–988.

Morales, J., J. Moreno, S. Merino, J. J. Sanz, G. Tomás, E. Arriero, E. Lobato, and J. Martínez-de la Puente. 2007. Early moult improves local survival and reduces reproductive output in female Pied Flycatchers. Ecoscience 14:31–39.

Moreno, J. 2004. Moult-breeding overlap and fecundity limitation in tropical birds: a link with immunity? Ardeola 51:471–476.

Mulvihill, R. S. 1993. Using wing molt to age passerines. North American Bird Bander 18:1–10.

Murphy, M. E. 1996. Energetics and nutrition of molt. Pp. 31–60 in C. Carey (editor), Avian energetics and nutritional ecology. Chapman & Hall, New York, NY.

Murphy, M. E., and J. R. King. 1991. Ptilochronology—a critical evaluation of assumptions and utility. Auk 108:695–704.

Nilsson, J.-A., and E. Svensson. 1996. The cost of reproduction: a new link between current reproductive effort and future reproductive success. Proceedings of the Royal Society B 263:711–714.

Payne, R. 1969. Overlap of breeding and molting schedules in a collection of African birds. Condor 71:140–145.

Pérez, G. E. and K. A. Hobson. 2006. Isotopic evaluation of interrupted molt in northern breeding populations of the Loggerhead Shrike. Condor 108:877–886.

Powell, A. H., and G. V. N. Powell. 1987. Population dynamics of male euglossine bees in Amazonian forest fragments. Biotropica 19:176–179.

Powell, L. L., P. C Stouffer, and E. I. Johnson. 2013. Recovery of understory bird movement across the interface of primary and secondary Amazon rainforest. Auk 130:459–468.

Pyle, P. 1997. Identification guide to North American birds, Part I. Slate Creek Press, Bolinas, CA.

Pyle, P. 2008. Identification guide to North American birds, Part II. Slate Creek Press, Bolinas, CA.

Pyle, P. 2009. Age determination and molt strategies in North American alcids. Marine Ornithology 37:219–225.

Pyle, P. 2013. Evolutionary implications of synapomorphic wing-molt sequences among falcons (Falconidae) and parrots (Psittaciformes). Condor 115:593–602.

Pyle, P., A. Engilis, and D. A. Kelt. 2015. Manual for ageing and sexing birds of Bosque Fray Jorge National Park and Northcentral Chile, with notes on range and breeding seasonality. Special Publication of the Occasional Papers of the Museum of Natural Science, Louisiana State University, Baton Rouge, LA.

Pyle, P., K. Tranquillo, K. Kayano, and N. Arcilla. 2016. Molt patterns, age criteria, and molt-breeding dynamics in American Samoan landbirds. Wilson Journal of Ornithology 128:56–69.

Repenning, M. 2012. História natural, com ênfase na biologia reprodutiva, de uma população migratória de Sporophila aff. plumbea (Aves, Emberizidae) do sul do Brasil. Pontifícia Universidade Católica do Rio Grande do Sul (in Portuguese).

Ricklefs, R. E. 1997. Comparative demography of New World populations of thrushes (Turdus spp.). Ecological Monographs 67:23–43.

Ricklefs, R. E., and M. Wikelski. 2002. The physiology-life history nexus. Trends in Ecology and Evolution 17:462–468.

Rogers, K. G., D. I. Rogers, and M. A. Weston. 2014. Prolonged and flexible primary moult overlaps extensively with breeding in beach-nesting Hooded Plovers Thinornis rubricollis. Ibis 156:840–849.

Rohwer, S. 1986. A previously unknown plumage of first-year Indigo Buntings and theories of delayed plumage maturation. Auk 103:281–292.

Rohwer, S., L. K. Butler, and D. Froehlich. 2005. Ecology and demography of east-west differences in molt scheduling of Neotropical migrant passerines. Pp. 87–105 in R. Greenberg and P. P. Marra (editors), Adaptations for two worlds. Johns Hopkins University Press, Baltimore, MD.

Rohwer, S., S. D. Fretwell, and D. M. Niles. 1980. Delayed maturation in passerine plumages and the deceptive acquisition of resources. American Naturalist 115:400–437.

Rohwer, S., R. E. Ricklefs, V. G. Rohwer, and M. M. Copple. 2009. Allometry of the duration of flight feather molt in birds. PLoS Biology 7:1–9.

Ryder, T. B., and R. Durães. 2005. It's not easy being green: using molt and morphological criteria to age and sex green-plumage manakins (Aves: Pipridae). Ornitología Neotropical 16:481–491.

Ryder, T. B., and J. D. Wolfe. 2009. The current state of knowledge on molt and plumage sequences in selected Neotropical bird families: a review. Ornitología Neotropical 20:1–18.

Siikamäki, P., M. Hovi, and O. Rätti. 1994. A trade-off between current reproduction and moult in the Pied Flycatcher: an experiment. Functional Ecology 7:476–482.

Stouffer, P. C., and R. O. Bierregaard. 1993. Spatial and temporal abundance patterns of Ruddy Quail-Doves (Geotrygon montana) near Manaus, Brazil. Condor 95:896–903.

Stouffer, P. C., R. O. Bierregaard, C. Strong, and T. E. Lovejoy. 2006. Long-term landscape change and bird abundance in Amazonian rainforest fragments. Conservation Biology 20:1212–1223.

Stouffer, P. C., E. I. Johnson, and R. O. Bierregaard. 2013. Breeding seasonality in Central Amazonian rainforest birds. Auk 130:529–540.

Stresemann, E., and V. Stresemann. 1966. Die Mauser der Vögel. Journal für Ornithologie 107 (Supplement):357–375 (in German).

Summers, R. W. 1983. Moult-skipping by Upland Geese Chloëphaga picta in the Falkland Islands. Ibis 125:262–266.

Svensson, E., and A. Hedenström. 1999. A phylogenetic analysis of moult strategies in Western Palearctic warblers (Aves: Sylviidae). Biological Journal of the Linnean Society 67:263–276.

Terborgh, J., S. K. Robinson, T. A. Parker, C. A. Munn, and N. Pierpont. 1990. Structure and organization of an Amazonian forest bird community. Ecological Monographs 60:213–238.

Thiollay, J. M. 1994. Structure, density and rarity in an Amazonian rain-forest bird community. Journal of Tropical Ecology 10:449–481.

Thompson, B. C., and R. D. Slack. 1983. Molt-breeding overlap and timing of pre-basic molt in Texas Least Terns. Journal of Field Ornithology 54:187–190.

Thompson, C. W., and M. Leu. 1994. Determining homology of molts and plumages to address evolutionary questions: a rejoinder regarding emberizid finches. Condor 96:769–782.

Thompson, T. J., and H. H. Powers. 1924. The variation of certain blood constituents of chickens during the molting season. Poultry Science 4:186–188.

Wolfe, J. D., R. B. Chandler, and D. I. Kin. 2009a. Molt patterns, age, and sex criteria for selected highland Costa Rican resident landbirds. Ornitología Neotropical 20:451–459.

Wolfe, J. D., E. I. Johnson, and R. S. Terrill. 2014. Searching for consensus in molt terminology 11 years after Howell et al.'s "first basic problem." Auk 131:371–377.

Wolfe, J. D., and P. Pyle. 2012. Progress in our understanding of molt patterns in Central American and Caribbean landbirds. Ornitología Neotropical 23:151–158.

Wolfe, J. D., P. Pyle, and C. J. Ralph. 2009b. Breeding seasons, molt patterns, and gender and age criteria for selected northeastern Costa Rican resident landbirds. Wilson Journal of Ornithology 121:556–567.

Wolfe, J. D., T. B. Ryder, and P. Pyle. 2010. Using molt cycles to categorize the age of tropical birds: an integrative new system. Journal of Field Ornithology 81:186–194.

# CHAPTER TWO

# How to Use This Guide

Although species described here are drawn from the 35-year capture database of the Biological Dynamics of Forest Fragments Project (BDFFP), located about 80 km north of Manaus, Amazonas, Brazil, we have attempted to represent a wide variety of tropical families, genera, and species that are widespread across the Neotropics. Clearly, our selection is biased toward understory species that are small or medium sized, but they are representative of birds captured in understory mist-nets over a much broader geographic region.

Our intent is to succinctly provide pertinent information necessary to successfully identify and determine the age and sex of each species at the BDFFP, using photographs as illustrative support. Although the timing of molting and breeding for these species likely varies across the Amazon, in many cases we suspect the progression of molt sequences, molt extents, and plumage aspects is relatively consistent, although we encourage readers to explore and vet this assumption in their work. Surely exceptions exist and we have much to learn about how life history events of Amazonian bird species vary across their range, especially where found on both sides of the equator, as this presents an interesting natural experiment to understand how abiotic cues drive endogenous signals that, in turn, drive physiological processes and life history patterns. Thus, we hope that this volume will serve as a useful starting point from which to assess aging and sexing patterns of Amazonian and Neotropical bird species using standardized terminology (Humphrey and Parkes 1959, Howell et al. 2003).

## FAMILY ACCOUNT

Each chapter begins with a family account that includes a short summary of what is known about the developmental, molting, and breeding strategies of that family, as well as deviations from these strategies. Families for which we provide the most information are usually those that fall into mist-nets (i.e., understory species), so these life history accounts are naturally biased in taxonomic scope. These family accounts are not meant to be exhaustive but, rather, we have tried to emphasize information that is most relevant to banders. We also provide the number of primaries, secondaries, and rectrices found among species in each family, as this information is important for correctly counting the number of feathers in order to assess molt. Among temperate species, less than 1% of individuals across a variety of families can have one or more extra ("supernumerary") primaries or rectrices, and this should be expected to occur occasionally in tropical species as well (e.g., Berger and Mueller 1957, DeRoos 1967, Hammer 1985, Clark et al. 1988).

## SPECIES ACCOUNTS

Each species account begins with the scientific name including subspecies at the BDFFP, if known, using the most up-to-date South American Checklist Committee taxonomy (Remsen et al. 2016) for species names and the *Handbook to the Birds of the World* (del Hoyo et al. 1992–2011) subspecies designations.

## TABLE 2.1
*CEMAVE band codes and associated inside diameters.*

| Letter code | Inside diameter (mm) |
|---|---|
| C | 1.8 |
| D | 2.0 |
| E | 2.4 |
| F | 2.8 |
| G | 3.2 |
| H | 4.0 |
| J | 4.5 |
| L | 5.0 |
| M | 5.5 |
| N | 6.3 |
| P | 7.0 |
| R | 8.0 |
| S | 9.5 |
| T | 11.0 |

## BAND SIZE

The Brazilian CEMAVE (2013) guide to band sizes is often fairly accurate, but may not be entirely representative of Amazonian populations, thus we provide a suggested band size based on measured leg widths in the field. Where we have little of our own data to reference, we provide the CEMAVE (2013) recommendation. Band sizes and associated inside diameters are found in Table 2.1.

## # INDIVIDUALS CAPTURED

This represents simply the number of individuals captured between 1979 and 2014, not including recaptures. This number should only be used to understand the relative frequency of capture for individual species as compared to others, and does not necessarily represent their density or abundance.

## SIMILAR SPECIES

We have tried to point out the most useful characteristics to separate species from potentially similar species at the BDFFP, but not necessarily where similar species exist elsewhere in the Neotropics. Banders should collect photographs and sometimes extensive morphological data for individuals when their identity is not known.

## MEASUREMENTS

In general, we follow techniques for taking measurements as described by Pyle (1997) and urge readers to understand these before interpreting our data. Wing, tail, bill (nares to tip), and tarsus measurements are presented in millimeters (mm), whereas mass is presented in grams (g). For each measurement, we provided the range, arithmetic mean, standard deviation (SD), and sample size. For individuals captured more than once, the mean of each measurement was used for that individual. Measurements beyond 3 SD of the species' mean have been excluded.

We measured the unflattened wing chord to the nearest 0.5 mm. The tail was measured with a ruler inserted between the central rectrices until it firmly rested against the insertion point of the rectrices into the body to the nearest 0.5 mm; the longest rectrices were gently held parallel to, but not pressed against, the ruler. Wing chord or tail length should not be measured if the longest primary or rectrix, respectively, is molting, missing, bent, or severely worn. Mass was historically measured using analog spring scales, often to the nearest 0.25 or 0.5 g, but, since 2007, mass has been measured with an electronic balance to the nearest 0.1 g. Bill length and tarsus were measured with calipers to the nearest 0.01 mm. The bill was measured as the chord (straight line, regardless of bill curvature) between the distal tip of the nares opening to the tip of the bill. The tarsus was measured with the leg bent starting and ending at the outside of each joint. Where distinct sexual dichromatism occurred, we provide measurements for each sex, but note that measurements for males are typically more accurate than for females because young males were sometimes identified as female before aging criteria were fully known (prior to 2007).

## SKULL

Skull ossification can be one of the more challenging pieces of data to collect accurately and its utility can be dependent on bander experience. Viewing the extent of ossification requires good light, the ability of the bander to see close up, and a bird with transparent skin (fortunately a high percentage of species have this). As a bird ages, the skull hardens; in this sense it can be used to identify young birds. In most passerine and near-passerine birds, skull ossification is achieved through the formation of a second layer of bone beneath the natal layer as the bird matures (Miller 1946, Nero 1951, Winkler 1979, Pyle 1997). When ossified, the skull appears to have minute white spots where vertical supports connect the two bone layers, and these are casually known as "stippling." Unossified areas appear pinkish and lack these stipples because it is just one layer of bone. Usually the unossified openings close from the side of the head to the top and in some species the skull never fully ossifies. See Pyle (1997) for a useful review and illustrations. The extent of skull ossification is not useful in many non-passerines (as well as in a few passerines) or at least may only be helpful in cases when it is very early in development because the skull never ossifies, often with large unossified gaps retained for life (McNeil and Burton 1972, Pyle 2008).

## BROOD PATCH

A brood patch can be developed on the belly of birds when they are incubating eggs or brooding young nestlings. This is not to be confused with otherwise unfeathered areas on the belly. An active brood patch can be recognized by the wrinkling of the skin, which is highly vascularized with blood vessels near the skin surface (Jones 1971; Figure 2.1). In nonincubating birds, their skin is usually tight against the belly without wrinkles or blood vessels (although beware of looser skin in some near-passerines like Momotidae and Bucconidae). As the brood patch fades, this skin often begins to flake and lost feathers are replaced. Depending on

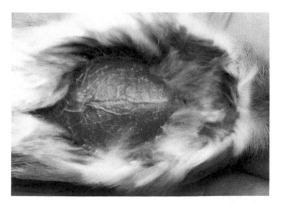

**Figure 2.1.** An example of a brood patch that is not only featherless but also highly vascularized.

the nesting ecology of any given species, males, females, or both can develop a brood patch.

## MOLT

Molt is defined as the regular and predictable replacement of feathers. This can occur between one and four times per year, depending on the species and its age. Understanding molt and plumage sequences is critical to the understanding of aging and sexing birds. We follow the Humphrey-Parkes (H-P) system as modified by Howell et al. (2003), and coded according to Wolfe et al. (2010) and Johnson et al. (2011). In the system, the juvenile plumage is the first basic plumage.

We follow Pyle (1997) in describing the extent of molts (see Chapter 1, Table 1.1). Anything less than a complete molt creates molt limits in which retained feathers contrast in color, wear, size, and/or shape against replaced feathers. This can occur as the result of one or more of three scenarios: (1) incomplete definitive prebasic molts replacing basic feathers; (2) limited, partial, or incomplete preformative molts replacing juvenile feathers; and (3) limited, partial, or incomplete prealternate molts replacing juvenile, formative, or definitive basic feathers. Molt limits can become very complex if both the FPF and FPA molts are less than complete, or if DPB molts are incomplete. In these cases, sometimes three or even four or more generations of feathers can be evident. These complex molts occur in some woodpeckers, owls, hawks, and migratory songbirds, for example, but are unusual among understory birds regularly captured at the BDFFP and elsewhere in lowland Amazonian forests. Rarely, an additional auxiliary formative plumage can be inserted into the first cycle. This is known from only a handful of North American birds (e.g., Rohwer 1986, Thompson and Leu 1994, Pyle 1997) and there is evidence of it in only one species at the BDFFP (*Percnostola rufifrons*; Johnson and Wolfe 2014), although it should be looked for in other understory species, such as certain Cardinalidae like *Cyanocompsa*.

## AGE/SEX

An important rule for determining the age and sex of birds is to determine a bird's age first and then its sex. This is because juvenile and other immature plumages can be female-like in males

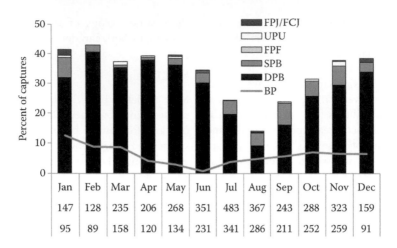

**Figure 2.2.** An example of a standard phenological graph used in this volume showing the frequency at which captured birds have been observed molting or breeding. Sample sizes for each month are provided below the x-axis, with the first line always referring to the number of individuals inspected for molt and the second line referring to those inspected for a brood patch (BP). See Tables 1.2 and 1.3 for cycle code abbreviation descriptions.

of sexually dichromatic species. In our species accounts, we have placed both sexes of a particular age together in the same box. This should help emphasize that identifying the age is to be done before identifying the sex. In reality, and with practice and familiarity with species, age and sex can be determined in rapid succession.

As described earlier in the "Molt" section, the presence of molt limits can help distinguish between birds in their first and definitive cycles. In this section, we describe each successive plumage until the adult-like plumage is achieved. We have intentionally not tried to describe every feature of every plumage, but rather highlight the features that are unique to that age and sex. Obviously, some level of basic bird identification is a prerequisite for aging and sexing.

## NOTES

In this section, we have included notes regarding special handling cautions, taxonomic issues, or other useful pieces of information, where appropriate.

## GRAPHS OF MOLT AND BREEDING FREQUENCY

In some species with enough data, we have graphed the seasonal frequency at which a species has been found molting and/or breeding. The y-axis provides the frequency of birds captured molting body or flight feathers, by month and age, as well as the frequency of birds captured showing evidence of an active brood patch. Sample sizes are provided below the graph, with the first line always referring to the number of individuals inspected for molt, and the second line referring to those inspected for a brood patch (Figure 2.2).

These graphs provide an indication of seasonality in these life history parameters and can also be useful to banders to help narrow aging options. For example, in seasonal species, one would not expect to find a FPJ or FCJ during the nonbreeding season. In contrast, the full realm of aging options would be available in species with highly aseasonal or asynchronous breeding and molting seasons.

## ABBREVIATIONS

| AMNH | American Museum of Natural History |
|---|---|
| BDFFP | Biological Dynamics of Forest Fragments Project |
| LSUMZ | Louisiana State University Museum of Natural Science |
| BP | brood patch |
| juv | juvenile |
| p | primary |
| pp | primaries |
| pp covs | primary coverts |
| rect/rects | rectrix/rectrices |
| s | secondary |

| | |
|---|---|
| **ss** | secondaries |
| **ss covs** | secondary coverts (includes lesser, median, and greater coverts combined) |
| **tert/terts** | tertial/tertials |
| **gr covs** | greater coverts |
| **less covs** | lesser coverts |
| **med covs** | median coverts |
| ♂ | male |
| ♂♂ | males |
| ♀ | female |
| ♀♀ | females |
| **mm** | millimeters |
| **g** | grams |
| **months** | All months are abbreviated to the first three letters (e.g., Jan, Feb, Mar, for January, February, March) |
| > | greater than |
| < | less than |
| (>) | slightly greater than |
| (<) | slightly less than |

## LITERATURE CITED

Berger, D. D., and H. C. Mueller. 1957. Supernumerary rectrices in some raptors. Wilson Bulletin 70:90.

CEMAVE. 2013. Lista das espécies de aves brasileiras com tamanhos de anilha recomendados. Centro Nacional de Pesquisa e Conservação de Aves Silvestres, Cabedelo, Brasil (in Portuguese).

Clark, W., K. Duffy, E. Gorney, M. McGrady, and C. Schultz. 1988. Supernumerary primaries and rectrices in some Eurasian and North American raptors. Journal of Raptor Research 22:53–58.

del Hoyo, J., A. Elliot, J. Sargatal, and D. A. Christie. 1992–2011. Handbook of the birds of the world. Lynx Edicions, Barcelona, Spain.

DeRoos, A. 1967. A swift, *Apus apus*, with twelve rectrices. Bulletin of the British Ornithologists' Club 87:141.

Hammer, D. B. 1985. Abnormal number of tail feathers. Bulletin of the British Ornithologists' Club 105:91–95.

Howell, S. N. G., C. Corben, P. Pyle, and D. I. Rogers. 2003. The first basic problem: a review of molt and plumage homologies. Condor 105:635–653.

Humphrey, P. S., and K. C. Parkes. 1959. An approach to the study of molts and plumages. Auk 76:1–31.

Johnson, E. I., and J. D. Wolfe. 2014. Thamnophilidae (antbird) molt strategies in a central Amazonian rainforest. Wilson Journal of Ornithology 126:451–462.

Johnson, E. I., J. D. Wolfe, T. B. Ryder, and P. Pyle. 2011. Modifications to a molt-based ageing system proposed by Wolfe et al. (2010). Journal of Field Ornithology 82:422–424.

Jones, R. E. 1971. The incubation patch of birds. Biological Reviews 46:315–339.

McNeil, R., and J. Burton. 1972. Cranial pneumatization patterns and bursa of Fabricius in North American shorebirds. Wilson Bulletin 84:329–339.

Miller, A. H. 1946. A method of determining the age of live passerine birds. Bird Banding 17:33–35.

Nero, R. W. 1951. Pattern and rate of cranial "ossification" in the House Sparrow. Wilson Bulletin 63:84–88.

Pyle, P. 1997. Identification guide to North American birds, Part I. Slate Creek Press, Bolinas, CA.

Pyle, P. 2008. Identification guide to North American birds, Part II. Slate Creek Press, Bolinas, CA.

Remsen, J. V., C. D. Cadena, A. Jaramillo, M. Nores, J. F. Pacheco, M. B. Robbins, T. S. Schulenberg, F. G. Stiles, D. F. Stotz, and K. J. Zimmer. [online]. 2016. A classification of the bird species of South America. American Ornithologists' Union. Version November 30, 2016. <http://www.museum.lsu.edu/~Remsen/SACCBaseline.htm>.

Rohwer, S. 1986. A previously unknown plumage of first-year Indigo Buntings and theories of delayed plumage maturation. Auk 103:281–292.

Thompson, C. W., and M. Leu. 1994. Determining homology of molts and plumages to address evolutionary questions: a rejoinder regarding emberizid finches. Condor 96:769–782.

Winkler, R. 1979. Zur pneumatisation des Schädeldachs der Vögel. Ornithologische Beobachter 76:49–118 (in German).

Wolfe, J. D., T. B. Ryder, and P. Pyle. 2010. Using molt cycles to categorize the age of tropical birds: an integrative new system. Journal of Field Ornithology 81:186–194.

# PART TWO

# Non-passerine Land Birds

# CHAPTER THREE

# Tinamidae (Tinamous)

# Species in South America: 45

# Species recorded at BDFFP: 4

# Species captured at BDFFP: 2

Tinamous are an ancient family of birds, thought to be closely related to ratites. They superficially resemble chickens and other gallinaceous birds, but have several structural differences, including a thinner bill and a raised or absent hind toe (hallux). They are mostly terrestrial, although at least *Tinamus major* roosts in trees, and they eat a variety of seeds, fruit, and small invertebrates. Their blood is reptile-like and, among extant birds, their heart is the smallest relative to their body size (Dittmann and Cardiff 2009). Perhaps related to these physiological features, they take considerably longer than most bird species to reach adult size. Tinamou feathers are unique in that they are "joined" together rather than "hooked" by barbs as in most birds, and it is this structure that is believed to be responsible for the noticeable whistling noise tinamous produce during flight (Campbell and Lack 1985, Cabot 1992, Davies 2002). Tinamous have 10 pp, 11–14? ss, and 12? rects (R. Terrill, pers. comm.).

A preformative molt is said to initiate soon after fledging, before the body fully grows into the adult size, progresses rapidly, and is complete (Davies 2002, Dittmann and Cardiff 2009), which we confirmed in an examination of *Tinamus* and *Crypturellus* specimens at Louisiana State University Museum of Natural Science (LSUMZ). Primary replacement begins at p1 and proceeds distally to p10. The progression of the secondaries begins at ~s8 and progresses bidirectionally (Beebe 1925). Reportedly in at least some species, p10 does not emerge during the prejuvenile molt, but emerges for the first time with the apparent preformative molt (Davies 2002). However, in juveniles of *Tinamus* and *Crypturellus* in our specimen examination (LSUMZ) we always found a p10, similarly reduced in size as adults (n = 7 juveniles). Even so, the shape of p10 should be inspected for differences between juvenile, formative, and definitive basic plumages. In at least one *Tinamus guttatus* specimen (LSUMZ), and perhaps also occurring in other large tinamou species, definitive prebasic molts may follow a *Staffelmauser*-like progression, at least in the secondaries, and may not always complete. In other tinamous, sequential molts appear to be complete.

Females average larger than males, but there is extensive overlap, and plumage differences are often subtle and may vary more among subspecies than between sexes (Davies 2002, Tubaro and Bertelli 2003). Similar to waterfowl, male tinamous develop a noticeable penis and females a phallus in their cloaca during the breeding season, which may aid in distinguishing sex (Pyle 2008, Brennan and Prum 2011). In all tinamous, only males incubate and develop a brood patch. Some species are monogamous, but others have females that are polyandrous, laying eggs in multiple males' nests. Males can also be polygamous, enticing additional females to copulate and lay in their nests (Campbell and Lack 1985, Cabot 1992, Davies 2002). Skulls typically ossify in tinamous.

TABLE 3.1
## TABLE 3.1
*Measurements of four species of tinamous found at the BDFFP, Amazonas, Brazil.*

| Species | Wing (mm) Literature | Wing (mm) BDFFP | Mass (g) Literature | Mass (g) BDFFP | Exposed culmen (mm) Literature | Exposed culmen (mm) BDFFP |
|---|---|---|---|---|---|---|
| *Tinamus major* | 216–248[a] | ? | 950–1100[b,c,d] | ? | ? | ? |
| *Crypturellus variegatus* | 170[d,e] | 140–170 | 354–423[f] | 320–440 | 23–28[g] | ? |
| *Crypturellus brevirostris* | 124–140[g] | ? | ? | ? | 18–22[g] | ? |
| *Crypturellus soui* | 117–149[g] | ? | 220–250[b,c] | ? | ? | ? |

[a] Davies 2002; [b] Stiles and Skutch 1989; [c] Hilty 2003; [d] Haverschmidt 1968; [e] Chubb et al. 1916; [f] Cabot et al. 2014; [g] Blake 1977.

## SPECIES ACCOUNTS

### *Tinamus major major*

Great Tinamou • Inhambu-de-cabeça-vermelha

Band Size: T (CEMAVE 2013)

\# Individuals Captured: 1

| | |
|---|---|
| Wing | Unknown at BDFFP |
| | 216–248 mm (mean: 236 mm, n = 36; Davies 2002) |
| Tail | Unknown at BDFFP |
| | 69 mm (Chubb et al. 1916) |
| Mass | Unknown at BDFFP |
| | 1100 g (Stiles and Skutch 1989, Hilty 2003) |
| | ♂ 952–1050 g (Haverschmidt 1968) |
| | ♀ 1013 g (Haverschmidt 1968) |

Similar species: Largest (Table 3.1) and plainest tinamou; brownish overall with faint barring on upperparts and chestnut crown. From *Cypturellus* by large size (Hilty 2003) and rough (not smooth) hind tarsi (Chubb et al. 1916, Skutch 1963).

Skull: Completely ossifies (n = 7, LSUMZ), probably late in the FPF or during the FCF.

Brood patch: Only ♂♂ incubate, but extent and timing of BP unknown.

Sex: ♀♀ are similar to ♂♂, but are slightly larger and whiter below, upperparts less marked and ss browner.

Molt: Group 4 (or sometimes Group 3 or 5?)/Complex Basic Strategy; FPF incomplete(?)-complete, DPBs incomplete(?)-complete. Timing is unknown.

| | |
|---|---|
| FCJ | Darker than FAJ, less distinctly barred ss, and upperparts less barred with sparse buffy spotting (Russell 1964, Cabot 1992). Skull is unossified. |
| FPF | Look for coarsely barred ss and barred body feathers replacing lightly barred ss and spotted body feathers. Relatively small-bodied, approaching or reaching the adult size by the end of the FPF. Skull is unossified or perhaps nearly or completely ossified near the end of the molt. |
| FCF | With one or more retained juvenile ss or pp with much lighter barring than replaced feathers. Skull is variably ossified. |
| FAJ | Without molt limits and distinctly barred ss and ss covs. Note that ss covs are distinctly more olive than the reddish-brown ss, which is not an indication of a molt limit. Skull is ossified. |

UPB    Look for barred ss replacing barred ss and barred body feathers replacing barred body feathers. Skull is ossified.

p1-p2 and s8 new (and also p8?) and with mixed ss covs, presumably in an arrested prebasic molt.

SAB    With two generations of adult-like and coarsely barred ss and/or pp. Skull is ossified.

## *Crypturellus variegatus* (monotypic)

Variegated Tinamou • Inhambu-anhangá

Band Size: N (CEMAVE 2013)

\# Individuals Captured: 12

Wing    140–170 mm (153.5 ± 7.9 mm; n = 11)
        170 mm (Chubb et al. 1916,
           Haverschmidt 1968)

Tail    35–53 mm (43.2 ± 6.4 mm; n = 10)

Mass    320–440 g (376.0 ± 53.4 g; n = 6)
        354–423 g (Cabot et al. 2014)

Similar species: Bill length (exposed culmen?) of
   *C. brevirostris* is 18–22 mm and distinctly shorter than
   *C. variegatus* (23–28 mm), and the wing chord
   averages smaller in *C. brevirostris* (124–140 mm, n = 3;
   Table 3.1; Blake 1977). *C. brevirostris* also differs by
   having a rufous (not blackish) crown and whitish
   (not buffy) underparts (Hilty and Brown 1986). *C.
   brevitrostris* is only recently known from the BDFFP
(Cohn-Haft et al. 1997) and may be more common than previously thought (Johnson et al. 2011), thus may have been misidentified as *C. variegatus* if ever captured. *C. s. soui* has never been captured at the BDFFP, but should be expected in disturbed areas. Small size (wing: 117–149 mm, n = 20; Blake 1977) and mass (220 g, Hilty 2003; 250 g, Stiles and Skutch 1989) and lack of barring on wings and back should separate *C. soui* from *C. variegatus* and *C. brevirostris*, but beware of young birds retaining barring.

Skull: Completely ossifies (n = 5, LSUMZ), probably late in the FPF or during the FCF.

Brood patch: Only ♂♂ incubate, but extent and timing of BP unknown. Fledglings (FPJ) have been captured in May, Jul, and Jan, suggesting at least a dry, but perhaps nearly year-round, breeding season.

Sex: ♀♀ probably average slightly larger than ♂♂, but there is probably overlap in measurements.

Molt: Group 4/Complex Basic Strategy; FPF complete, DPBs complete. Timing is unknown.

FCJ    With bold white spots below and less distinctly barred and lightly marked with pale spots above. Skull is unossified.

| | |
|---|---|
| FPF | Replacing white-spotted juvenile body feathers and flight feathers with adult-like body and flight feathers. Relatively small-bodied, approaching or reaching the adult size by the end of the FPF. Skull is unossified or perhaps nearly or completely ossified near the end of the molt. |

p1–p3 molting, body plumage still mostly juvenile-like, and body size relatively small (182 g). (LSUMZ 131902.)

| | |
|---|---|
| FCF | With one or more retained juvenile ss or pp, with retained ss relatively more faintly barred. Skull is variably ossified. |
| FAJ | Without molt limits and distinctly barred ss and ss covs. Skull is ossified. |
| UPB | Replacing adult-like body and relatively barred flight feathers with adult-like flight feathers. Skull is ossified. |

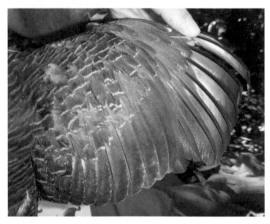

With body feathers and some ss covs relatively new and bright compared to the year old adult-like ss covs; also p1 molting (not visible in the photo).

| | |
|---|---|
| SAB | With two generations of adult-like ss and/or pp, similar in pattern. Skull is ossified. |

---

## LITERATURE CITED

Beebe, W. 1925. The variegated Tinamou *Crypturellus variegatus variegatus* (Gmelin). Zoologia 6:195–227.

Blake, E. R. 1977. Manual of Neotropical birds. University of Chicago Press, Chicago, IL.

Cabot, J. 1992. Family Tinamidae (Tinamous). Pp. 112–138 in J. del Hoyo, A. Elliott, and J. Sargatal (editors), Handbook of the birds of the world: ostrich to ducks. Lynx Edicions, Barcelona, Spain.

Cabot, J., F. Jutglar, and C. J. Sharpe. [online]. 2014. Variegated Tinamou (*Crypturellus variegatus*). In J. del Hoyo, A. Elliot, J. Sargatal, D. A. Christie, and E. de Juana (editors), Handbook of the birds of the world alive. Lynx Edicions, Barcelona, Spain. <http://www.hbw.com>.

Campbell, B., and E. Lack. 1985. A dictionary of birds. Friday Harbor, Washington, DC.

CEMAVE. 2013. Lista das espécies de aves brasileiras com tamanhos de anilha recomendados. Centro Nacional de Pesquisa e Conservação de Aves Silvestres, Cabedelo, Brasil (in Portuguese).

Chubb, C., H. Grönvold, F. V. McConnell, H. F. Milne, P. Slud, and Bale & Danielsson. 1916. The birds of British Guiana: based on the collection of Frederick Vavasour McConnell. Bernard Quaritch, London, UK.

Cohn-Haft, M., A. Whittaker, and P. C. Stouffer. 1997. A new look at the "species-poor" central Amazon: the avifauna north of Manaus, Brazil. Ornithological Monographs 48:205–235.

Davies, S. J. J. F. 2002. Ratites and Tinamous: Tinamidae, Rheidae, Dromaiidae, Casuariidae, Apterygidae, Struthionidae. Oxford University Press, Oxford, UK.

Dittmann, D. L., and S. W. Cardiff. 2009. Tinamous. Pp. 16–17 in T. Harris (editor), National Geographic complete birds of the world. National Geographic Society, Washington, DC.

Haverschmidt, F. 1968. Birds of Surinam. Oliver and Boyd, Edinburgh, UK.

Hilty, S. L. 2003. Birds of Venezuela (2nd ed.). Princeton University Press, Princeton, NJ.

Hilty, S. L., and W. L. Brown. 1986. A guide to the birds of Colombia. Princeton University Press, Princeton, NJ.

Johnson, E. I., P. C. Stouffer, and C. F. Vargas. 2011. Diversity, biomass, and trophic structure of a central Amazonian rainforest bird community. Revista Brasiliera de Ornitologia 19:1–16.

Russell, S. 1964. A distributional study of the birds of British Honduras. Ornithological Monographs 1:1–458.

Skutch, A. F. 1963. Life history of the Little Tinamou. Condor 65:224–231.

Stiles, F. G., and A. F. Skutch. 1989. A guide to the birds of Costa Rica. Comstock Publishing Associates, Ithaca, NY.

Tubaro, P. L., and S. Bertelli. 2003. Female-biased sexual size dimorphism in tinamous: a comparative test fails to support Rensch's rule. Biological Journal of the Linnean Society 80:519–527.

# CHAPTER FOUR

# Odontophoridae (New World Quails)

# Species in South America: 15

# Species recorded at BDFFP: 1

# Species captured at BDFFP: 1

Historically placed within the Phasianidae, New World Quails form a distinct group of mostly ground-dwelling granivores or omnivores that occupy a diverse array of habitats, although they tend to be more closely associated with forests in the tropics. Their short, stout bill has serrations near the tip, which is notably different from other chunky ground-dwelling birds, including pheasants and tinamous. Tropical species are typically less notably sexually dichromatic than temperate taxa (Rosenberg 2009). New World Quails have 10 pp, 12 ss, and 12–14 rects (Pyle 2008).

Most temperate quails follow a Complex Basic Strategy and some follow a Complex Alternate Strategy (Pyle 2008, Howell 2010). The preformative molt in temperate, and probably also tropical species is incomplete; the outer two primaries and all primary coverts are retained, at least in some species (Petrides 1945, Pyle 2008). The preformative molt may be prolonged such that it coincides with a hormone shift, causing feathers replaced earlier in the molt to be more juvenile-like, whereas those replaced later in the molt are more adult-like (Pyle 2008). Replacement of pp is sequential from p1 to p10, and ss are replaced bidirectionally from s3 (Pyle 2008).

A considerable amount of information remains to be gathered about this family, especially in the tropics. We have little information to make statements across the family regarding whether the skull fully ossifies. Apparently in most species, males and females incubate, but the brood patch is probably more extensive in females. In addition, breeding females often develop distended cloacae (Pyle 2008).

## SPECIES ACCOUNT

*Odontophorus gujanensis gujanensis*

Marbled Wood-Quail • Uru-corcovado

Band Size: N (CEMAVE 2013)

# Individuals Captured: 5

| | |
|---|---|
| Wing | 135–144 mm (139.5 ± 6.4 mm; n = 2)<br>130–150 mm (141.1 mm; n = 20; Blake 1977) |
| Tail | 62–68 mm (65.0 ± 4.2 mm; n = 2)<br>61–74 mm (66.6 mm; n = 20; Blake 1977)<br>58 mm (Chubb et al. 1916) |
| Mass | 200–255 g (227.5 ± 38.9 g; n = 2)<br>298–349 g (mean: 322.8 g; n = 4; Haverschmidt and Mees 1994)<br>♂ 313–380 g (Carroll and Kirwan 2013)<br>♀ 298 g (Carroll and Kirwan 2013) |

Similar species: None.

Skull: No data available.

Brood patch: May occur in both sexes, but should be more developed in ♀♀. ♀♀ also often develop distended cloacae (Pyle 2008). Breeds at least during the dry season; fledglings (FPJ) have been caught in Jul and Nov.

Sex: ♂ (>) ♀. ♀♀ may have a paler face, have more coarse mottling brown, buff, and black above, and have more extensive barring below. There is considerable individual variation in plumage in this species, which is probably related to sex and region, so more study of sex-specific differences at the BDFFP is needed.

Molt: Group 3(?)/Complex Basic Strategy; FPF incomplete(?), DPBs complete. PAs unknown, but if present would likely be limited to the head and throat (Pyle 2008, Howell 2010). FPF as in other Odontophoridae (Dickey and Van Rossem 1938, Petrides 1945, as in Pyle 2008, Howell 2010) probably includes most body and flight feathers, but not the pp covs and outer two pp. Molt has been observed about half finished and nearly finished in Nov.

FCJ            Like DCB, but with less vermiculation and orange-red bills (Carroll 1994).

FPF            Perhaps protracted, replacing juvenile body and flight feathers with a mix of juvenile-like and adult-like body and flight feathers. Bills may still have orange-red tones.

FCF            Adult-lilke in aspect, or with a gradient of juvenile-like and adult-like feathers across the body and flight feathers resulting from a protracted that coincides with a hormonal shift during the FPF (Pyle 2008). Probably with molt limits among the outer pp with p9 and p10 retained. All pp covs also retained, contrasting in wear against the replaced gr covs (Dickey and Van Rossem 1938, Petrides 1945, Pyle 2008, Howell 2010).

SPB            Look for three generations of feathers: (1) retained juvenile p9, p10, and pp covs; (2) replaced adult-like inner pp and outer ss; and (3) retained adult-like middle pp and middle ss.

DCB            Without molt limits among the pp and with replaced pp covs. Bill bluish-black or dusky gray (Carroll 1994).

DPB            Look for two generations of feathers: (1) replaced adult-like inner pp, pp covs, and outer ss; and (2) retained adult-like middle pp, pp covs, and middle ss.

---

## LITERATURE CITED

Blake, E. R. 1977. Manual of Neotropical birds. University of Chicago Press, Chicago, IL.

Carroll, J. P. 1994. Family Odontophoridae (New World Quails). Pp. 412–433 in J. del Hoyo, A. Elliott, and J. Sargatal (editors), Handbook of the birds of the world. Lynx Edicions, Barcelona, Spain.

Carroll, J. P., and G. M. Kirwan. [online]. 2013. Marbled Wood-Quail (*Odontophorus gujanensis*). In J. del Hoyo, A. Elliot, J. Sargatal, D. A. Christie, and E. de Juana (editors), Handbook of the birds of the world alive. Lynx Edicions, Barcelona, Spain. <http://www.hbw.com>.

CEMAVE. 2013. Lista das espécies de aves brasileiras com tamanhos de anilha recomendados. Centro Nacional de Pesquisa e Conservação de Aves Silvestres, Cabedelo, Brasil (in Portuguese).

Chubb, C., H. Grönvold, F. V. McConnell, H. F. Milne, P. Slud, and Bale & Danielsson. 1916. The birds of British Guiana: Based on the collection of Frederick Vavasour McConnell. Bernard Quaritch, London, UK.

Dickey, D. R., and A. J. Van Rossem. 1938. The birds of El Salvador. Field Museum of Natural History, Zoological Series, Chicago, IL.

Haverschmidt, F., and G. F. Mees. 1994. Birds of Suriname. VACO, Paramaribo, Suriname.

Howell, S. N. G. 2010. Molt in North American birds. Houghton Mifflin Harcourt, Boston, MA.

Petrides, G. A. 1945. First-winter plumages in the Galliformes. Auk 62:223–227.

Pyle, P. 2008. Identification guide to North American birds, Part II. Slate Creek Press, Bolinas, CA.

Rosenberg, G. H. 2009. New World Quail. Pp. 30–31 in T. Harris (editor), National Geographic complete birds of the world. National Geographic Society, Washington, DC.

# CHAPTER FIVE

# Accipitridae (Hawks, Kites, Eagles, and Allies)

# Species in South America: 59

# Species recorded at BDFFP: 23

# Species captured at BDFFP: 4

The family Accipitridae is a group of raptors that are apex predators in the avian world. Although superficially similar to falcons (Falconidae), these two families are apparently not closely related. In addition to genetic evidence (Hackett et al. 2008), they have different bill structures, contour feather shapes, and flight feather molting sequences. Specifically, Accipitridae lack a "tooth" along the maxilla present in Falconidae, have more rounded contour feathers, and follow a more typical molt sequence (see next paragraph and Chapter 21, this volume). Hawks have 10 pp, 13–19 ss, and 12 rects.

Accipitridae follow a Complex Basic Strategy, usually with an inserted limited preformative molt, but in some individuals this may be absent. Smaller hawks, like *Accipiter* spp., typically have a complete definitive prebasic molt. In larger species, definitive prebasic molts can be incomplete and/or follow *Staffelmauser*, taking up to 4 years to replace all juvenile flight feathers (Pyle 2005a,b). Even so, in many species these incomplete and *Staffelmauser* molts appear to be opportunistic, occurring as a response to time constraints associated with breeding and migration. Consequently, they are more regularly seen in larger species and larger individuals (i.e., females; Pyle 2005a,b, 2006, 2008; Rohwer et al. 2009; Howell 2010),

and by extension may be less common in sedentary tropical species for a given body size. Flight feather molt proceeds from p1 to p10, sometimes in multiple waves in larger species. Ss proceed in two waves proximally from s1 and s5, and bidirectionally from the middle tertial; rectrices often follow a sequence of r1→ r6 → r3 → r4 → r2 → r5, at least in North American hawks.

Brood patches occur in females, but also less extensively in males for some species. Females can also develop distended cloacae. Females are typically much larger than males, often with little to no overlap in body size between sexes. In most, but not all, species plumage aspect is not useful for distinguishing between sexes; it is not useful for any of the species captured thus far at the BDFFP. In some species, there can be two or even three distinct color morphs, which, in addition to individual and age-specific variation, can create substantial identification challenges. Iris color also often changes with age, and these changes can be species-specific and are hard to generalize as the adult iris can be red, orange, yellow, brown, gray, or cream, depending on the species. Knowledge of these changes can be a useful aid in aging hawks and other raptors, but should not be used independently of other characters, like plumage aspect and an examination of molt limits. Perhaps except for the smallest hawks (e.g., small *Accipiter* spp.), they should be banded with lock-on bands rather than butt-end bands.

# SPECIES ACCOUNTS

*Accipiter superciliosus superciliosus*

Band Size: L (♀), Unknown (♂)

Tiny Hawk • Gavião-miudinho

# Individuals Captured: 2

Wing 148–174 mm (n = 2)
♀ = 153–164 mm (n = 8), ♂ = 132–137 mm (n = 12; Hellmayr and Conover 1949)
♀ = 160–167 mm, ♂ = 140–145 mm (Swann 1922)

Tail 98–107 mm (n = 2)
♀ = 106 mm, ♂ = 89 mm (Chubb et al. 1916)

Mass 142 g (n = 1)
♀ > 100 g, ♂ ≈ 75 g (Schulenberg 2010)
♀ 115–134 g, ♂ 61.5–75 g (Bierregaard and Kirwan 2013)

Similar species: Similar to small *Micrastur*, but without bare facial skin and with broad tail bands. This species can be distingued from other small hawks at the BDFFP by relatively long tail and barred underparts in all ages.

Brood patch: May develop in both sexes, but more extensively in ♀♀ (Pyle 2008).

Sex: ♀ > ♂ but degree of overlap unknown at the BDFFP.

Molt: Group 3/Complex Basic Strategy; FPF absent(?)-limited, DPBs complete. The FPF may include a few body feathers or may be absent in some individuals. The DPBs appear to follow a sequential pattern of replacement and are likely complete. Timing of molt unknown.

Notes: Based on skeletal evidence of having a distinct foramen in the procoracoid process (Olson 2006) and molecular evidence (Kocum 2008), this species does not appear closely aligned with *Accipiter* or other small hawks, thus may be better classified in its own genus, with *Hieraspiza* apparently taking priority (Olson 2006).

FCJ Upperparts brownish. Iris variable, possibly pale brown to yellowish. Also more rarely occurs in a rufous morph having rufous-brown upperparts with blackish spotting or scaling (Schulenberg et al. 2007).

FPF Probably involves a few body feathers in at least some individuals, but plumage aspect may not be different than the FCJ.

FCF Look for a few replaced and fresh body feathers, but plumage aspect likely appears similar to FCJ.

SPB Replacing brownish feathers with bluish-gray feathers. Iris may not be completely red.

p1–p4, r1, an inner tert, and a few body feathers replaced; p5 partially grown. The iris color appears relatively advanced, but unless the SCB plumage aspect is similar to the FCJ, this should be an SPB (rather than a TPB).

DCB Upperparts slate and remiges grayish. Iris orangish-red (SCB?) to deep red. Rare rufous morph also likely occurs in DCB (Schulenberg et al. 2007).

DPB Replacing bluish feathers with bluish feathers, but beware of faded definitive basic feathers appearing brownish. Iris deep red.

*Pseudaster albicollis albicollis*

White Hawk • Gavião-branco

Wing   278–350 mm (321.0 ±
           38.0 mm; n = 3)
       ♀ = 365 mm, ♂ = 320–337 mm
           (Swann 1922)

Tail   177–195 mm (186.0 ±
           12.7 mm; n = 2)
       200 mm (Chubb et al. 1916)

Mass   No data from BDFFP
       ♂ 600–670 mm (632 ±
           34.8 mm; n = 4; Haverschmidt
           1948, Brown and Amadon
           1968, Haverschmidt and Mees
           1994)

       No data from BDFFP
       ♀ 710–908 g (mean: 819 g;
           n = 6; Haverschmidt and Mees
           1994, Magnier 2012)

Similar species: Most like a larger *Leucopternis melanops*, but without black mask, with gray (not yellow) cere, and with a white band at the tip of its tail.

Brood patch: Unknown if both sexes incubate *P. a. ghiesbreghti* nests during the dry season in Guatemala and only ♀♀ incubate (Draheim et al. 2012).

Sex: ♀ > ♂; *P. a. albicollis* may be more sexually dimorphic than Central American subspecies (Draheim et al. 2012).

Molt: Group 5/Complex Basic Strategy; FPF absent(?)-limited, DPBs incomplete and *Staffelmauser*. The FPF includes a few body feathers, or may be absent in some individuals. It may be possible to age through first several years of life, as in larger North American raptors (Pyle 2005a,b, 2008). Howell and Webb (1995) suggest that the adult plumage is attained with the SPB at about 2 years of age, which might suggest this species molts every 2 years. More likely, the adult plumage is attained via the TPB, and the SCB is juvenile-like or intermediate between FCJ and TCB. Timing of molt unknown.

FCJ   Similar to DCB, but with streaking on crown and nape (Hilty 2003) and perhaps more mottled black and white above and on flight feathers as in Central American races (Draheim et al. 2012).

FPF   Probably involve only a few body feathers in some individuals.

FCF   Look for a few replaced and fresh body feathers, but plumage aspect would likely appear similar to FCJ.

SPB   Perhaps replaced and retained body and flight feathers are both juvenile-like.

SCB   Perhaps similar to FCJ and/or with less crown streaking, but with two generations of flight feathers, the oldest being juvenile.

TPB   With three generations of remiges: two generations juvenile-like, and a new generation adult-like.

SAB With two generations of adult-like barred black and gray feathers. TCB, TAB, 4AB, and so on, may also be possible aging codes with more study.

Fresher, whiter-tipped, and blacker pp and ss are seen scattered across the wing: at least s2, s6, s9, p2, and p4 appear replaced. With more study, it may turn out that this plumage can be aged as TAB depending on the extent of SPB and TPB molts.

DPB With two or more generations of adult-like feathers.

## *Leucopternis melanops* (monotypic)

Black-faced Hawk • Gavião-de-cara-preta

Band Size: S (CEMAVE 2013); lock-on recommended

\# Individuals Captured: 3

| | |
|---|---|
| Wing | 195 mm (n = 1) |
| | ♀ = 230 mm, ♂ = 210 mm (Swann 1922) |
| | 216 mm (Barlow et al. 2001) |
| Tail | 136 mm (n = 1) |
| | 145 mm (Barlow et al. 2001) |
| | 133 mm (Chubb et al. 1916) |
| Mass | 295 g (n = 1) |
| | 350 g (Barlow et al. 2001) |
| | 297–317 g (Bierregaard and Boesman 2013) |

Similar species: A small version of *Pseudaster albicollis*, but with black mask, thin black streaking on crown, yellow-orange bill, and white tail band.

Brood patch: May develop in both sexes, but more extensively in ♀♀ (Pyle 2008).

Sex: ♀ > ♂ but degree of overlap unknown.

Molt: Group 5/Complex Basic Strategy; FPF absent(?)-limited, DPBs incomplete-complete and *Staffelmauser* based on a single capture and an examination of five specimens (LSUMZ). The FPF is probably limited to a few body feathers, or may be absent in some individuals. Timing unknown.

FCJ Similar to DCB, but with two white bands in tail, and brownish edging to back feathers when fresh that may wear off (Brown and Amadon 1968, Raposo do Amaral et al. 2007, Shrum et al. 2011).

FPF Probably involves only a few body feathers in some individuals.

| | |
|---|---|
| FCF | This plumage would likely be similar to FCJ with a few replaced and fresher-looking body feathers. Beware that brown edging on juvenile back feathers may wear off. |
| SPB | With extensive body, flight, and/or tail feather molt. Mix of juvenile rects with two white bars and adult rects with one white bar. |
| SCB | Look for retained juvenile flight feathers mixed with replaced flight feathers. |
| DCB | Rects with one white bar and without molt limits among the pp or ss. |
| TPB | With three generations of remiges: two generations juvenile-like, and a new generation adult-like. |
| SAB | Upperparts mottled black and white. Head lightly streaked. Underparts white. Tail with one white band. Iris dark brown (Raposo do Amaral et al. 2007). TCB may also be possible aging codes with more study. |

All pp appear replaced and adult-like, but there are at least two generations of adult-like ss (at least s4 and s7–s8 appear replaced).

| | |
|---|---|
| DPB | With two or more generations of adult-like feathers. |

*Spizaetus ornatus ornatus*

Ornate Hawk-Eagle • Apacanim

Band Size: X (CEMAVE 2013); lock-on recommended

# Individuals Captured: 2

| | |
|---|---|
| Wing | 338–370 mm (354.0 ± 22.6 mm; n = 2)<br>♀ = 410 mm, ♂ = 340 – 375 mm (Swann 1922) |
| Tail | 242–350 mm (296.0 ± 76.4 mm; n = 2) |
| Mass | No data from BDFFP<br>♀ = 1452 ± 100.8 g, n = 11; ♂ = 1028 ± 72.4 g, n = 4 (*S. o. vicarius*; Whitacre et al. 2012) |

Similar species: Adult most like immature *Accipiter poliogaster* in plumage, but proportions different being larger overall with relatively shorter tail, has a crest, and has feathered tarsi. Immatures not treated well in field guides, but feathered tarsi will eliminate most species at the BDFFP except *S. tyrannus* and *Spizastur melanoleucus*. Immature *S. ornatus* is distinguished from immature *S. tyrannus* by black flank and thigh barring, spotted wing linings, yellow (not red-orange) cere, no prominent black mask, longer crest, and browner upperparts.

Brood patch: Unknown if BPs develop, but primarily ♀♀ incubate (Whitacre et al. 2012). At least one record of breeding during the dry season at the BDFFP (Klein et al. 1988).

Sex: ♀ > ♂, probably with little or no overlap in measurements.

Molt: Group 5/Complex Basic Strategy; FPF absent(?)-limited, DPBs incomplete and *Staffelmauser*. The FPF is probably limited to a few body feathers or may be absent in some individuals. Apparently a body molt begins several months after fledging, possibly initiating an FPF quickly followed by or overlapping with an SPB. May achieve the adult-like plumage aspect with the TCB (Howell and Webb 1995). Timing unknown.

| | |
|---|---|
| FCJ | Head and upperparts are white, crown is streaked black, and malar strip indistinct if present. Otherwise like adult, but more narrowly and sparsely barred. The tail has six bars (Swann 1922). |

| | |
|---|---|
| FPF | Would probably involve only a few body feathers in some individuals. |
| FCF | Look for a few replaced and fresh body feathers, but plumage aspect would likely appear similar to FCJ. |
| SPB | With extensive body, flight, and/or tail feather molt. Replaced feathers not yet adult-like and retained feathers juvenile. |
| SCB | Apparently like DCB, but paler and less distinctly marked. Potentially look for two generations of remiges and retained juvenile flight feathers in a *Staffelmauser* pattern. |
| TPB | New feathers are probably more heavily barred and adult-like than retained feathers. With three generations of remiges: two generations juvenile-like, and a new generation adult-like. |
| SAB | Brightly patterned head, elongated black crest, and boldly barred flanks and belly. Brown tail with four blackish bands. With two generations of adult-like flight feathers. TCB, TAB, 4AB, and so on, may also be possible aging codes with more study. |
| DPB | With two or more generations of adult-like feathers. |

## LITERATURE CITED

Barlow, J., T. Haugaasen, and C. A. Peres. 2001. Sympatry of the Black-faced Hawk *Leucopternis melanops* and the White-browed Hawk *Leucopternis kuhli* in the Lower Rio Tapajós, Pará, Brazil. Cotinga 18:77–79.

Bierregaard, R. O., and P. Boesman. [online]. 2013. Black-faced Hawk (*Leucopternis melanops*). In J. del Hoyo, A. Elliot, J. Sargatal, D. A. Christie, and E. de Juana (editors), Handbook of the birds of the world alive. Lynx Edicions, Barcelona, Spain. <http://www.hbw.com>.

Bierregaard, R. O., and G. M. Kirwan. [online]. 2013. Tiny Hawk (*Accipiter superciliosus*). In J. del Hoyo, A. Elliot, J. Sargatal, D. A. Christie, and E. de Juana (editors), Handbook of the birds of the world alive. Lynx Edicions, Barcelona, Spain. <http://www.hbw.com>.

Brown, L., and D. Amadon. 1968. Eagles, hawks and falcons of the world. Country Life Books, Feltham, London, UK.

CEMAVE. 2013. Lista das espécies de aves brasileiras com tamanhos de anilha recomendados. Centro Nacional de Pesquisa e Conservação de Aves Silvestres, Cabedelo, Brasil (in Portuguese).

Chubb, C., H. Grönvold, F. V. McConnell, H. F. Milne, P. Slud, and Bale & Danielsson. 1916. The birds of British Guiana: Based on the collection of Frederick Vavasour McConnell. Bernard Quaritch, London, UK.

Draheim, G. S., D. F. Whitacre, A. M. Enamorado, O. A. Aguirre, and A. E. Hernández. 2012. White Hawk. Pp. 120–138 in D. F. Whitacre (editor), Neotropical birds of prey. Cornell University Press, Ithaca, NY.

Hackett, S. J., R. T. Kimball, S. Reddy, R. C. K. Bowie, E. L. Braun, M. J. Braun, J. L. Chojnowski, W. A. Cox, K.-L. Han, J. Harshman, C. J. Huddleston, B. D. Marks, K. J. Miglia, W. S. Moore, F. H. Sheldon, D. W. Steadman, C. C. Witt, and T. Yuri. 2008. A phylogenomic study of birds reveals their evolutionary history. Science 320:1763–1768.

Haverschmidt, F. 1948. Bird weights from Surinam. Wilson Bulletin 60:230–239.

Haverschmidt, F., and G. F. Mees. 1994. Birds of Suriname. VACO, Paramaribo, Suriname.

Hellmayr, C. E., and B. Conover. 1949. Catalogue of birds of the Americas and the adjacent islands. Field Museum of Natural History, Chicago, IL.

Hilty, S. L. 2003. Birds of Venezuela (2nd ed.). Princeton University Press, Princeton, NJ.

Howell, S. N. G. 2010. Molt in North American birds. Houghton Mifflin Harcourt, Boston, MA.

Howell, S. N. G., and S. Webb. 1995. A guide to the birds of Mexico and northern Central America. Oxford University Press, Oxford, U.K.

Klein, B. C., L. J. Harper, R. O. Bierregaard, and G. V. N. Powell. 1988. The nesting and feeding behavior of the Ornate Hawk-Eagle near Manaus, Brazil. Condor 90:239–241.

Kocum, A. 2008. Phylogenie der Accipitriformes (Greifvögel) anhand verschiedener nuklearer und mitochondrialer DNA-Sequenzen. Vogelwarte 46: 141–143 (in German).

Magnier, B. [online]. 2012. White Hawk (*Leucopternis albicollis*). Neotropical birds online. Cornell Lab of Ornithologiy, Ithaca, NY. <http://neotropical.birds.cornell.edu>.

Olson, S. L. 2006. Reflections on the systematics of *Accipiter* and the genus for *Falco superciliosus* Linnaeus. Bulletin of the British Ornithologists' Club 126: 69–70.

Pyle, P. 2005a. First-cycle molts in North American Falconiformes. Raptor Research 39:378–385.

Pyle, P. 2005b. Remigial molt patterns in North American Falconiformes as related to age, sex, breeding status, and life-history strategies. Condor 107:823–834.

Pyle, P. 2006. Staffelmauser and other adaptive strategies for wing molt in larger birds. Western Birds 37:179–185.

Pyle, P. 2008. Identification guide to North American birds, Part II. Slate Creek Press, Bolinas, CA.

Raposo do Amaral, F. S., L. F. Silveira, and B. M. Whitney. 2007. New localities for the Black-faced Hawk (*Leucopternis melanops*) south of the Amazon River and description of the immature plumage of the White-browed Hawk (*Leucopternis kuhli*). Wilson Journal of Ornithology 119:450–454.

Rohwer, S., R. E. Ricklefs, V. G. Rohwer, and M. M. Copple. 2009. Allometry of the duration of flight feather molt in birds. PLoS Biology 7:1–9.

Schulenberg, T. S. [online]. 2010. Tiny Hawk (*Accipiter superciliosus*). Neotropical birds online. Cornell Lab of Ornithology, Ithaca, NY. <http://neotropical.birds .cornell.edu>.

Schulenberg, T. S., D. F. Stotz, D. F. Lane, J. P. O'Neill, and T. A. Parker. 2007. Birds of Peru. Princeton University Press, Princeton, NJ.

Shrum, P. L., W. W. Bowerman, D. G. Olaechea, and R. Amable. 2011. More records of sympatry of Black-faced Hawk (*Leucopternis melanops*) and White-browed Hawk (*L. kuhli*) in Madre de Dios, Peru. Journal of Raptor Research 45:104–105.

Swann, H. K. 1922. A synopsis of the Accipitres (diurnal birds of prey): Comprising species and subspecies described up to 1920, with their characters and distribution (2nd ed.). Wheldon and Wesley, London, UK.

Whitacre, D. F., J. A. Madrid, H. D. Madrid, R. Cruz, C. J. Flatten, and S. H. Funes. 2012. Ornate Hawk-Eagle. Pp. 120–138 in D. F. Whitacre (editor), Neotropical birds of prey. Cornell University Press, Ithaca, NY.

# CHAPTER SIX

# Psophiidae (Trumpeters)

# Species in South America: 3

# Species recorded at BDFFP: 1

# Species captured at BDFFP: 1

The three trumpeter species (or up to five proposed by some authors; e.g., Oppenheimer and Silveira 2009) are confined to Amazonia, and are generally isolated from each other by major river barriers. Although allied with cranes (Gruidae) and rails (Rallidae), trumpeters have a distinctive shape. The head and neck feathers are shortened and velvety, and iridescent feathers on the neck are also unusual among the Gruiformes. Trumpeters have 10 pp, 14? ss, and 12? rects (R. Terrill, pers. comm.).

The molt of trumpeters has been poorly studied, except for a limited description by Sherman (1995) based on tracking a population of 46 banded birds in Cocha Cashu, Peru. In North American Gruiformes (cranes, rails, and limpkin), the predominant molt strategy is the Complex Basic Strategy with limited to partial preformative molts (Pyle 2008, Howell 2010); this seems like a reasonable starting point from which to study molt strategies in trumpeters, although we do not yet know whether preformative molts exist or not. The description that iridescent throat feathers emerging at 9 weeks of age is strongly suggestive (Sherman 1995), although whether this is part of a protracted prejuvenile molt is not certain. Stresemann and Stresemann (1966) indicate the replacement sequence of feathers is typical, starting at the innermost primary and replacing in sequence out to P10, although the process appears to be protracted over the course of a year replacing one or more pp and ss at a time (Sherman and Bonan 2013). Alternatively, they may follow a *Staffelmauser* pattern (P. Pyle, pers. comm.).

Trumpeters have a relatively unique cooperative breeding system (cooperative polyandry) where only a few members in the group breed, but they all share in raising young and multiple members of both sexes share in incubation (Sherman and Bonan 2013), although it is not clear if brood patches or other incubation signals are visible when handling birds. It is unknown if skulls ossify.

# SPECIES ACCOUNT

*Psophia crepitans crepitans*

Gray-winged Trumpeter • Jacamin-cinza

Band Size: T (CEMAVE 2013)

# Individuals Captured: 8

Wing     266 mm (n = 1)
♀ = 267–293 mm (278.6 mm; n = 9),
♂ = 270–286 mm (280.1 mm; n = 9;
Blake 1977)

Tail     118 mm (n = 1)

Mass     No data from BDFFP
Adult mass reached after 12–15 months
of age. ♀ = 1180–1320 g (1256 ± 45.9
g; n = 9), ♂ = 1280–1440 g (1378 ±
59.5 g; n = 7; Sherman 1995).

Tarsus     No data from BDFFP
♀ = 121–125 mm (122.3 mm; n = 9),
♂ = 121–137 mm (128.3 mm; n = 9;
Blake 1977)

Similar species: None.

Skull: Unknown.

Brood patch: Unknown.

Sex: ♂ >(?) ♀

Molt: Group 3 (or Group 1?)/Probably Complex Basic Strategy (or Simple Basic?); FPF absent(?)-limited(?), DPB complete(?). If an FPF is present, it is likely limited based on descriptions by Sherman (1995). Sherman (1996) also reported that feathers in the adult plumage (DPBs) are replaced gradually over the course of a year and may follow a *Staffelmauser* replacement sequence (P. Pyle, pers. comm.).

FCJ     Like adults, but black contour feathers have 1–2 mm of brown at the tips (Sherman 1995).

FPF     Not known whether this occurs during an extended period of feather replacement (such as the emergence of iridescent feathers on the throat at 9 weeks of age) over the first 3 months of age (Sherman 1995).

FCF     Contour feathers with brown tips, but perhaps more iridescence on the throat and head (Sherman 1995).

SPB     Replacing brownish-tipped contour feathers with black contour feathers, but beware of remaining juvenile feathers having tips worn off. Retained feathers should be considerably worn relative to DPB.

DCB     Black contour feathers without brown tips (Sherman 1995).

DPB     Replacing adult-like body and flight feathers with adult-like feathers. Retained feathers should be relatively fresh relative to SPB.

# LITERATURE CITED

Blake, E. R. 1977. Manual of Neotropical birds. University of Chicago Press, Chicago, IL.

CEMAVE. 2013. Lista das espécies de aves brasileiras com tamanhos de anilha recomendados. Centro Nacional de Pesquisa e Conservação de Aves Silvestres, Cabedelo, Brasil (in Portuguese).

Howell, S. N. G. 2010. Molt in North American birds. Houghton Mifflin Harcourt, Boston, MA.

Oppenheimer, M., and L. F. Silveira. 2009. A taxonomic review of the Dark-winged Trumpeter *Psophia viridis* (Aves: Gruiformes: Psophiidae). Papéis Avulsos Zoologia 49:547–555.

Pyle, P. 2008. Identification guide to North American birds, Part II. Slate Creek Press, Bolinas, CA.

Sherman, P. T. 1995. Breeding biology of White-winged Trumpeters (*Psophia leucoptera*) in Peru. Auk 112:285–295.

Sherman, P. T. 1996. Family Psophiidae (Trumpeters). Pp. 96–107 in J. del Hoyo, A. Elliott, and J. Sargantal (editors), Handbook of the birds of the world. Lynx Edicions, Barcelona, Spain.

Sherman, P. T., and A. Bonan. [online]. 2013. Psophiidae (Trumpeters). In J. del Hoyo, A. Elliott, J. Sargatal, D. A. Christie, and E. de Juana (editors), Handbook of the birds of the world alive. Lynx Edicions, Barcelona, Spain. <http://www.hbw.com>.

Stresemann, E., and V. Stresemann. 1966. Die Mauser der Vögel. Journal für Ornithologie 107 (Supplement):357–375 (in German).

# CHAPTER SEVEN

# Columbidae (Pigeons and Doves)

# Species in South America: 51

# Species recorded at BDFFP: 7

# Species captured at BDFFP: 3

Pigeons and doves are easily distinguished by their relatively small heads, bulky bodies, short legs, and fleshy nasal membranes. In tropical forests, they occupy all strata from the floor to the canopy. Sexual dichromatism occurs in many, but not all, species and, although males generally average larger, there is often too much overlap to distinguish males from females by their measurements. Pigeons and doves have 10 pp, 11–12 ss, and 12 or 14 rects.

Most, if not all, species apparently follow the Complex Basic Strategy (Howell 2010). If prealternate molts are present in some species, they may be particularly challenging to identify given how readily the contour plumage can be shed and then replaced adventitiously. The preformative molt is often complete or nearly so (often with just a few retained pp and ss), being variable especially within species. In some cases, definitive prebasic molts are also incomplete, but presumably less often and involving fewer retained feathers than in preformative molts (Pyle 1995). In general, larger doves more likely have incomplete DPBs, and can sometimes follow a *Staffelmauser* replacement pattern (Pyle et al. 2016).

Generally, the skull does not completely ossify in larger species and may only be useful in cases of individuals exhibiting a small amount of ossification, but in some smaller species it may complete. This can make it problematic to identify individual birds in formative plumage that underwent a complete preformative molt. Until we learn more about tropical species, including how pervasive incomplete preformative molts are and how extensively skulls ossify, it is best to age birds without molt limits as after first juvenile (FAJ). Brood patches and cloacal protuberances are poorly developed and generally should not be used to sex birds or identify breeding. With experience, banders can examine birds for cloacal conical papillae in males or oviductal openings in females (see Pyle 2008 for more details).

# SPECIES ACCOUNTS

## *Columbina passerina griseola*

Common Ground-Dove • Rolinha-cinzenta

Band Size: G, H (CEMAVE 2013)

\# Individuals Captured: 2

Wing     76–79 mm (77.5 ± 2.1 mm; n = 2)
         77–83 mm (Chubb et al. 1916)

Tail     56–57 mm (56.5 ± 0.7 mm; n = 2)
         55 mm (Chubb et al. 1916)

Mass     32.5–42.0 g (37.3 ± 6.7 g; n = 2)
         24–50 g (Baptista et al. 2014a)

Similar species: Only likely to be confused with ♀ *C. talpacoti*, never before captured at the BDFFP, which is ruddy above and in ss and ss covs, lacks scaly head and underparts, has black (not pinkish) bill, and spotted scapulars. *C. minuta*, not yet recorded at the BDFFP, lacks scaly neck and breast and has a dark (not pale) bill (Baptista et al. 2013b).

Skull: Probably does not complete, but may be useful in cases of low ossification during FPF (Pyle 1997).

Sex: ♂ has more blue in the nape and more pink or rose in the chest than ♀.

Molt: Groups 3, 4, and 5/Complex Basic Strategy; FPF incomplete-complete, DPBs incomplete-complete. At least in North American subspecies of *C. passerina*, FPF can be protracted and sometimes does not complete with the 1–4 outer pp and pp covs, and 1–6 ss retained. Occasionally DPB does not complete, with 1–3 ss retained (Pyle 1997). One bird was caught finishing wing molt in Sep, and another in Feb was not molting.

FCJ     Like adult ♀, but scaled whitish-buffy on breast, and fainter dusky markings on breast (Stiles and Skutch 1989). Pp covs with more extensive and paler rufous bases (Pyle 1995). Skull relatively unossified.

FPF     Juvenile feathers being replaced with adult-like feathers. Skull relatively unossified.

FCF     Skull variably ossified with retained juvenile outer pp, pp covs, or ss, but beware of heavily worn DCB with similar molt limits.

FAJ     Without molt limits in flight feathers or ss covs. Skull relatively ossified.

SPB     Actively molting with a mix of not yet replaced juvenile and formative pp, pp covs, or ss. Skull relatively ossified.

FAJ     Without molt limits in flight feathers or ss covs. Skull relatively ossified.

DCB     With retained non-juvenile (formative or definitive basic) outer pp, pp covs, or ss; beware of heavily worn DCB with retained feathers that falsely appear juvenile. Skull relatively ossified. SCB and SAB may rarely be possible, which would have three generations of flight feathers: one generation of retained juvenile in SCB and no retained juvenile feathers in SAB.

s4–s6 retained and adult-like (formative or definitive basic). (Courtesy of Lindsay Wieland.)

UPB     Adult-like feathers being replaced with adult-like feathers. Skull relatively ossified.

*Leptotila verreauxi brasiliensis*

White-tipped Dove • Juriti-pupu

| | |
|---|---|
| Wing | 120–129 mm (123.7 ± 4.7 mm; n = 3) |
| | 127–130 mm (Chubb et al. 1916) |
| Tail | 92–106 mm (96.7 ± 8.1 mm; n = 3) |
| Mass | 130.0–43.0 g (136.5 ± 9.2 g; n = 2) |
| | 165 g (Hilty and Brown 1986) |
| | 96–157 g (Baptista et al. 2013a) |

Similar species: *Leptotila* distinguished from other large columbids by bright cinnamon or rufous underwing covs. *L. rufaxilla* has not been recorded at the BDFFP, but is common in Manaus and might eventually be encountered in areas of second growth. It has less extensive white on the tail tips, blue-gray (not grayish-brown) crown, rufescent buff (not pinkish gray) sides of head and neck (Stiles and Skutch 1989), and averages slightly larger (115–183 g; Baptista et al. 2014b).

(Courtesy of Lindsay Wieland.)

Skull: Probably not complete in adults, but may be useful in cases of low ossification (Pyle 1997).

Brood patch: Undeveloped and not useful for sexing (Pyle 1997). Timing of breeding not known.

Sex: ♂♂ are more blue in the nape, more pink or rose in the chest, and more metallic on the back than ♀♀.

Molt: Groups 3, 4, and 5/Complex Basic Strategy; FPF incomplete-complete, DPBs incomplete-complete (see also Pyle 1997, Wolfe et al. 2009). The FPF may be protracted (Dickey and Van Rossem 1938) and in at least in North American subspecies of *L. verreauxi*, FPF and DPB can sometimes not complete, retaining 1–2 outer pp and pp covs, and 1–5 ss (Pyle 1997). One bird was starting molt in Jun, but one Jul and one Jan capture were not molting.

| | |
|---|---|
| FCJ | Slightly duller than adults and have buff- or rufous-tipped ss, scapulars, and ss covs (Pyle 1997). Pp covs washed rufous (Pyle 1995). Skull is unossified. |
| FPF | Juvenile feathers being replaced with adult-like feathers. Skull is unossified but may nearly ossify toward completion of FPF. |
| FCF | With retained juvenile outer pp, pp covs, or ss, but beware of heavily worn DCB with similar molt limits. The flanks may average buffier than in the DCB as in *L. v. bangsi* (Dickey and Van Rossem 1938). Skull is relatively ossified. |
| FAJ | Without molt limits in flight feathers or covs. Pp covs washed brown (Pyle 1995). Skull is relatively ossified. |
| SPB | Actively molting, with a mix of not-yet-replaced juvenile and formative pp, pp covs, or ss. Skull is relatively ossified. |
| DCB | With retained nonjuvenile (formative or definitive basic) outer pp, pp covs, or ss; beware of heavily worn DCB with retained feathers that falsely appear juvenile. The flanks may average less buffy than in the FCF as in *L. v. bangsi* (Dickey and Van Rossem 1938). Skull is relatively ossified. SCB and SAB may rarely be possible, which would have three generations of flight feathers: one generation of retained juvenile in SCB and no retained juvenile feathers in SAB. |
| UPB | Adult-like feathers being replaced with adult-like feathers. Skull is relatively ossified. |

*Geotrygon montana montana*

Ruddy Quail-Dove • Pariri

Band Size: H, L

# Individuals Captured: 513

Wing    ♂    120.0–145.0 mm (132.0 ± 4.0 mm; n = 165)

       ♀    119.0–144.0 mm (129.8 ± 4.5 mm; n = 129)

Tail    ♂    56.0–83.0 mm (73.3 ± 4.5 mm; n = 147)

       ♀    57.0–90.0 mm (71.4 ± 5.7 mm; n = 129)

Mass    ♂    78.0–150.0 g (116.4 ± 13.6 g; n = 187)

       ♀    74.0–160.0 g (110.1 ± 15.2 g; n = 174)

Bill    ♂    6.5–8.6 mm (7.5 ± 0.7 mm; n = 7)

       ♀    6.8–8.6 mm (7.5 ± 0.6 mm; n = 7)

Tarsus    ♂    30.9–33.5 mm (32.4 ± 1.0 mm; n = 7)

       ♀    29.8–32.4 mm (31.1 ± 0.9 mm; n = 7)

Similar species: None, but see *Leptotila verreauxi*.

Skull: Probably nearly completely ossifies in adults (n = 23) and may be useful to identify FPF or FCF when relatively unossified (Pyle 1997).

Brood patch: As in other columbids, probably poorly developed and may not be reliable for sexing. Breeding season occurs mainly during the wet season (Jan–May) and then most migrate away during the dry season (Stouffer and Bierregaard 1993), although a few may breed during the dry season (especially Aug–Sep?).

Sex: ♂ is reddish overall with a more distinct mustachial streak than ♀. ♀ are also olive-brownish overall.

Molt: Groups 3, 4, and 5/Complex Basic Strategy; FPF incomplete-complete, DPBs incomplete-complete. During the FPF, all feathers are replaced, except for 1–5 ss, 0–several ss covs, and sometimes an alula; however, the FPF may occasionally complete, and the DPB may occasionally be incomplete as in the FPF, retaining 1–3 ss, and/or several ss covs. The peak of molting appears to be between Apr and Jun, presumably after breeding, which occurs during the wet season. Does molt complete before migration?

(Courtesy of Lindsay Wieland.)

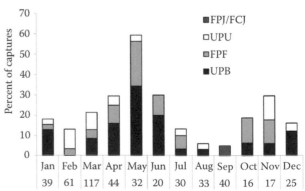

FCJ    Like adult ♀, but with dark bars and pale spots on back, ss covs, and flight feathers (Hilty and Brown 1986). Iris is dull brown.

FPF    Actively molting with juvenile feathers being replaced by adult-like formative feathers. Iris dull brown.

p10 growing, s1–s3, s9–s11 replaced, and s4, s7, and s8 growing.

FCF    Like DCB, but with molt limits among especially ss (but sometimes also outer pp and/or ss covs) with retained juvenile feathers that have buffy tips. Retained juvenile ss are also notably shorter than replaced feathers. Iris is dull brown but should become more golden-brown with age. Some FCF without molt limits and with brown iris may be safely aged as FCF.

s6 and s11 retained (top); s4–s8 retained (bottom).

FAJ — Without molt limits in flight feathers or ss covs. Iris is golden-brown.

SPB — Actively molting with a mix of not yet replaced juvenile and formative pp, pp covs, or ss. Iris is golden-brown.

DCB — Up to 3(?) retained formative or definitive ss, or 1 or more inner or outer pp. In these birds, the retained ss should lack buffy tips and be roughly the same length as replaced feathers. Beware of pseudolimits among the ss and do not confuse retained juvenile feathers that have reduced or worn buffy tips with retained adult-like feathers. Iris is golden-brown. SCB and SAB may rarely be possible, which would have three generations of flight feathers: one generation of retained juvenile in SCB and no retained juvenile feathers in SAB.

s3–s5 retained.

UPB — Adult-like feathers being replaced with adult-like feathers. Iris is golden-brown.

p1–p3 new, p4 growing; no apparent molt limits in the unreplaced ss or pp.

## LITERATURE CITED

Baptista, L. F., P. W. Trail, H. M. Horblit, and P. Boesman. [online]. 2013a. White-tipped Dove (*Leptotila verreauxi*). In J. del Hoyo, A. Elliott, J. Sargatal, D. A. Christie, and E. de Juana (editors), Handbook of the birds of the world alive. Lynx Edicions, Barcelona, Spain. <http://www.hbw.com>.

Baptista, L. F., P. W. Trail, H. M. Horblit, and P. Boesman. [online]. 2014a. Common Ground-Dove (*Columbina passerina*). In J. del Hoyo, A. Elliott, J. Sargatal, D. A. Christie, and E. de Juana (editors), Handbook of the birds of the world alive. Lynx Edicions, Barcelona, Spain. <http://www.hbw.com>.

Baptista, L. F., P. W. Trail, H. M. Horblit, A. Bonan, and P. Boesman. [online]. 2014b. White-fronted Dove (*Leptotila rufaxilla*). In J. del Hoyo, A. Elliott, J. Sargatal, D. A. Christie, and E. de Juana (editors), Handbook of the birds of the world alive. Lynx Edicions, Barcelona, Spain. <http://www.hbw.com>.

Baptista, L. F., P. W. Trail, H. M. Horblit, G. M. Kirwan, and P. Boesman. [online]. 2013b. Plain-breasted Ground-Dove (*Columbina minuta*). In J. del Hoyo, A. Elliott, J. Sargatal, D. A. Christie, and E. de Juana (editors), Handbook of the birds of the world alive. Lynx Edicions, Barcelona, Spain. <http://www.hbw.com>.

CEMAVE. 2013. Lista das espécies de aves brasileiras com tamanhos de anilha recomendados. Centro Nacional de Pesquisa e Conservação de Aves Silvestres, Cabedelo, Brasil (in Portuguese).

Chubb, C., H. Grönvold, F. V. McConnell, H. F. Milne, P. Slud, and Bale & Danielsson. 1916. The birds of British Guiana: based on the collection of Frederick Vavasour McConnell. Bernard Quaritch, London, UK.

Dickey, D. R., and A. J. Van Rossem. 1938. The birds of El Salvador. Field Museum of Natural History, Zoological Series, Chicago, IL.

Hilty, S. L., and W. L. Brown. 1986. A guide to the birds of Columbia. Princeton University Press, Princeton, NJ.

Howell, S. N. G. 2010. Molt in North American birds. Houghton Mifflin Harcourt, Boston, MA.

Pyle, P. 1995. Incomplete flight feather molt and age in certain North American non-passerines. North American Bird Bander 15:15–26.

Pyle, P. 1997. Identification guide to North American birds, Part I. Slate Creek Press, Bolinas, CA.

Pyle, P. 2008. Identification Guide to North American Birds, Part II. Slate Creek Press, Bolinas, CA.

Pyle, P., K. Tranquillo, K. Kayano, and N. Arcilla. 2016. Molt patterns, age criteria, and molt-breeding dynamics in American Samoan Landbirds. Wilson Journal of Ornithology 128:56–69.

Stiles, F. G., and A. F. Skutch. 1989. A guide to the birds of Costa Rica. Comstock Publishing Associates, Ithaca, NY.

Stouffer, P. C., and R. O. Bierregaard. 1993. Spatial and temporal abundance patterns of Ruddy Quail-Doves (*Geotrygon montana*) near Manaus, Brazil. Condor 95:896–903.

Wolfe, J. D., P. Pyle, and C. J. Ralph. 2009. Breeding seasons, molt patterns, and gender and age criteria for selected northeastern Costa Rican resident landbirds. Wilson Journal of Ornithology 121:556–567.

# CHAPTER EIGHT

# Cuculidae (Cuckoos, Roadrunners, and Anis)

# Species in South America: 23

# Species recorded at BDFFP: 8

# Species captured at BDFFP: 1

Cuculidae contains three major groups in the Americas: tree cuckoos, ground cuckoos, and anis; another two groups are found in the Old World. Medium to large birds, unifying features include a short, stout, slightly decurved bill, and zygodactylous feet. Most are forest-dwelling, although a few live in drier habitats, such as the famous roadrunner. Few plumage features distinguish cuckoos, although many have long tails, bold eyerings, or brightly colored irises. Some species are metallic bronzy or green, others are boldly colored, whereas still others are dull and cryptically plumaged (Clement 2009). Cuckoos have 10 pp (the 10th is reduced), 9–13 ss, and 8 (in anis) to 10 rects.

New World cuckoos apparently follow the Complex Basic Strategy, even those that are migratory or live in open environments. Even so, Dickey and Van Rossem (1938) suggested that *Piaya cayana* follows a Complex Alternate Strategy, certainly unusual in a resident tropical species and not consistent with temperate cuckoos (Pyle 1995, 1997), thus we are suspect of this claim. In cuckoos, the sequence of wing molt can be highly irregular, appearing to start in multiple points and proceeding forward or backward, skipping some feathers, but simultaneously growing others. Rohwer and Broms (2013) showed that Yellow-billed Cuckoo (*Coccyzus americanus*) and Common Cuckoo (*Cuculus canorus*) molt primaries in three or four series and rectrices in two series, conceptually not all that different from a *Staffelmauser* replacement pattern but resulting in complete replacement. The juvenile plumage can be like adults, or dramatically different, depending on the species. Preformative molts are often complete and protracted, but may be incomplete with a few ss or rects retained especially in larger species. In *Piaya cayana*, however, preformative molts are thought to be partial (Dickey and Van Rossem 1938). Prebasic molts, too, can be incomplete but probably less regularly than during preformative molts (Pyle 1997, Howell 2010).

Old World cuckoos are often brood parasites, laying their eggs in nests of other species; however, New World cuckoos build and lay eggs in their own nests. Incubation is shared between the sexes and both develop brood patches and, in some species, males play more of an incubation role than females (Payne 1997, Pyle 1997). Body size in females typically averages larger than in males, but there is often considerable overlap. Skulls probably do not completely ossify in most species, but the percentage of ossification may help with aging, if visible (the skin in cuckoos is often dark; Pyle 1997).

# SPECIES ACCOUNT

*Piaya melanogaster* (monotypic)

Black-bellied Cuckoo • Chincoã-de-bico-vermelho

Band Size: Unknown

\# Individuals Captured: 1

| | |
|---|---|
| Tail | 215 mm (n = 1) |
| | 224.8 mm (Stone 1908) |
| Mass | 90 g (n = 1) |
| Bill | 22.7 mm (n = 1) |
| Tarsus | 39.2 mm (n = 1) |

Similar species: *P. cayana*, which is common in second growth areas at the BDFFP but has not yet been captured, is easily distinguished by having a pale (not blackish) belly, yellow (not red) bill, lack of loral spot, and lack of gray cap.

Skull: Probably not complete in adults but may be useful in cases of relatively reduced ossification (Pyle 1997). The skin may be dark, however, making this difficult to determine.

Brood patch: Probably develops in both sexes as in other Cuculidae (Pyle 1997). Timing unknown.

Sex: ♀ = ♂.

Molt: Group 3 (and sometimes Group 5?)/Complex Basic Strategy; FPF partial(?), DPBs incomplete(?)-complete(?). The extent of FPF replacement may be similar to *P. cayana* (Dickey and Van Rossem 1938). DPBs may be complete or perhaps are occasionally to regularly incomplete, and likely follow an irregular sequence as in other Cuculidae (Pyle 1997, Rohwer and Broms 2013). Timing unknown.

| | |
|---|---|
| FCJ | Similar to adult, but look for duller bill, iris, and orbital skin. |
| FPF | Actively molting with juvenile body feathers and ss covs being replaced by adult-like feathers. |
| FCF | Like DCB, but with molt limits among the ss covs. |
| SPB | Uncertain how to distinguish from DPB (use UPB), other than perhaps replacing narrower juvenile rects with broader adult rects. |
| SCB | With a mix of relatively shorter and worn juvenile, and relatively longer and fresh definitive basic flight feathers. |
| DPB | Uncertain how to distinguish from SPB (use UPB). Adult-like feathers (note especially rects) being replaced with adult-like feathers. |
| DCB | Adult-like and without molt limits among the flight feathers. |
| SAB | With a mix of retained and replaced definitive basic flight feathers of the same length and less contrast in wear than SCB. |

## LITERATURE CITED

Clement, P. 2009. Cuckoos. Pp. 135–138 in T. Harris (editor), National Geographic complete birds of the world. National Geographic Society, Washington, DC.

Dickey, D. R., and A. J. Van Rossem. 1938. The birds of El Salvador. Field Museum of Natural History, Zoological Series, Chicago, IL.

Howell, S. N. G. 2010. Molt in North American birds. Houghton Mifflin Harcourt, Boston, MA.

Payne, R. 1997. Family Cuculidae (Cuckoos). Pp. 508–607 in J. del Hoyo, A. Elliott, and D. A. Christie (editors), Handbook of the birds of the world. Lynx Editions, Barcelona, Spain.

Pyle, P. 1995. Incomplete flight feather molt and age in certain North American non-passerines. North American Bird Bander 15:15–26.

Pyle, P. 1997. Identification guide to North American birds, Part I. Slate Creek Press, Bolinas, CA.

Rohwer, S., and K. Broms. 2013. Replacement rules for the flight feathers of Yellow-billed Cuckoos (*Coccyzus americanus*) and Common Cuckoos (*Cuculus canorus*). Auk 130:599–608.

Stone, W. 1908. A review of the genus *Piaya* (Lesson). Proceedings of the Academy of Natural Sciences of Philadelphia 60:492–501.

# CHAPTER NINE

# Strigidae (Typical Owls)

# Species in South America: 43

# Species recorded at BDFFP: 8

# Species captured at BDFFP: 3

Strigidae, or typical owls, are familiar predatory birds that occur on every continent except Antarctica. Most, but not all, species are active at night. Typical owls are socially monogamous and are generally characterized by fluffy, cryptic plumage and strong, zygodactylous feet. Molt in these birds is quite variable across taxa and size, and is not well studied outside of temperate zones. Typical owls have 10 pp, 11–17 ss, and 12 rects (Pyle 1997a).

Typical owls in North America and Europe have a partial FPF and a variable DPB; most large owls undergo an incomplete DPB, whereas smaller owls typically exhibit a complete DPB (Cramp 1985, Pyle 1997a,b). These same patterns have been found in the Amazonian Pygmy-Owl (*Glaucidium hardyi*) at our study site and Great Horned Owl (*Bubo virginianus*) throughout South America (Traylor 1958), and probably occur in other Neotropical species. A word of caution: molt limits can be quite cryptic as complicated plumage patterns can create a seemingly endless combination of pseudolimits. Some researchers have successfully used ultraviolet (UV) light to show differences in feather generations when UV light is applied to the ventral surfaces of primaries and secondaries, where newer generations of feathers reflect greater intensities of fluorescence compared to older generations (Weidensaul et al. 2011). Limited or partial prealternate molts are thought possibly to occur in two North American species, although this remains uncertain and maybe even unlikely (see Pyle 2008, Telfry and Holroyd 2012), and should probably be absent in most resident Neotropical owls. In complete feather replacement patterns, molt typically proceeds distally from s1 and s5, and proximally from the second tert, typical in diastataxic species (lacking the ss corresponding to the fifth gr cov; Pyle 2008). Pp replacement starting points appear variable across genera (and perhaps also within?), starting from p7 in *Bubo*, p5 in *Strix*, p3 in *Athene*, and either p2 or p1 in smaller genera (Pyle 1997a,b).

Skull ossification in North American owls typically does not complete in larger species (but possibly does in smaller species) and the percentage of completion may be useful for aging (Pyle 1997a). Cloacal protuberances are typically not well developed; however, brood patches are developed by females in North America and males are not known to share in incubation (Pyle 1997a, Marks et al. 1999). Like other families of raptors, owls exhibit reverse sexual size dimorphism with females being larger than males, although there are exceptions (e.g., *Athene cunicularia*; Marks et al. 1999).

# SPECIES ACCOUNTS

## *Megascops watsonii watsonii*

Tawny-bellied Screech-Owl • Corujinha-orelhuda

Band Size: N (CEMAVE 2013)

# Individuals Captured: 9

| | |
|---|---|
| Wing | 165.0–180.0 mm (172.0 ± 5.3 mm; n = 7) |
| Tail | 81.0–100.0 mm (90.3 ± 6.2 mm; n = 8) |
| Mass | 118.0–167.0 g (135.9 ± 15.9 g; n = 8) |
| | 115–155 g (Holt et al. 2013a) |

Similar species: *M. choliba* has never been captured at the BDFFP as it is locally rare, and probably only occurs in or near second growth. *M. watsonii* is darker overall, has darker eyebrows, and has tawny-colored (not white) underparts.

Skull: May complete (n = 1), but at least the percentage may be useful in determining age (Pyle 1997a).

Brood patch: BPs exhibited by ♀♀. Timing unknown.

Sex: ♀ > ♂.

Molt: Like temperate *Megascops*? If so, Group 3/Complex Basic Strategy; FPF partial, DPBs complete. FPF may include only body feathers and no ss covs (Pyle 1997a). Molts at least Jan–May.

| | |
|---|---|
| FCJ | Undescribed (Holt et al. 2013a). |
| FPF | Unknown, but would only likely be replacing body feathers. |
| FCF | Look for molt limits among the less or med covs and a single generation of flight feathers. |
| SPB | Actively molting, replacing juvenile flight feathers with adult flight feathers. |
| DCB | Nonjuvenile without molt limits. |
| DPB | Actively molting, replacing adult-like feathers with adult feathers. |

## *Lophostrix cristata cristata*

Crested Owl • Coruja-de-crista

Band Size: Unknown

# Individuals Captured: 1

| | |
|---|---|
| Wing | Unknown at BDFFP |
| | 303 mm (Chubb et al. 1916) |
| Tail | Unknown at BDFFP |
| | 184 mm (Chubb et al. 1916) |
| Mass | Unknown at BDFFP |
| | 400–600 g (Holt et al. 1999) |

Similar species: No other medium-sized owl at the BDFFP has such distinctive white ear tufts contrasting against a dark face and body.

Skull: May complete, but if not at least the percentage may be useful in determining age (Pyle 1997a,b).

Brood patch: BPs only developed by ♀♀. Apparently breeds dry to early wet season (Holt et al. 1999).

Sex: ♀ > ♂.

Molt: Group 5(?)/Complex Basic Strategy; FPF limited(?)-partial(?) (Storer 1972), DPBs perhaps regularly incomplete. Timing unknown.

| | |
|---|---|
| FCJ | All white body, ear tufts, and head with rufous and black face patch. Ss covs are white distally and barred basally. Flight feathers are darker, similar to DCB (Storer 1972). |
| FPF | White juvenile body feathers replaced with darker adult-like feathers (Storer 1972). |
| FCF | May have a few retained white body feathers (Storer 1972) as well as molt limits among the ss covs. |
| SPB | Actively molting, replacing juvenile flight feathers with adult flight feathers. |
| SCB | With one or more retained juvenile ss and/or pp. |
| DPB | Actively molting, replacing adult feathers with adult feathers. TPB may be possible with a third generation of feathers replacing a mix of definitive basic and juvenile flight feathers. |
| SAB | With two generations of adult-like flight feathers mixed among ss and/or pp. It may also be possible to age to TAB and 4AB with three and four generations of adult-like flight feathers, respectively. |

## Glaucidium hardyi (monotypic)

Amazonian Pygmy-Owl • Caburé-da-amazônia

Wing    85.0–95.0 mm (90.8 ± 3.0 mm; n = 14)

Tail    44.0–51.5 mm (46.9 ± 2.0 mm; n = 12)

Mass    44.5–63.0 g (55.0 ± 5.1 g; n = 13)

Similar species: Only possibly confused with *G. brasilianum*, which has never been captured at the BDFFP, and is apparently locally rare and probably only occurs in areas with second growth. *G. brasilianum* has streaking (not spotting) on crown and nape, usually lacks white barring in tail, and has a relatively longer tail, which may be useful for separating juvenile birds.

Skull: May complete (n = 1), but at least the percentage may be useful in determining age (Pyle 1997a).

Brood patch: Cloacal protuberances not developed, but BPs exhibited by ♀♀. Based on the timing of molting and as suggested by Holt et al. (2013b), perhaps mainly during the dry season.

Sex: ♀ > ♂.

Molt: Group 3/Complex Basic Strategy; FPF limited-partial, DPBs complete. Based on one capture, the FPF appears to include at least some body feathers, as well as a few less covs, but no other ss covs, consistent with other *Glaucidium* (see also Pyle 1997a, Pyle et al. 2004). Rects may often molt simultaneously in DPB. Molts at least Dec–May.

(Courtesy of Philip C. Stouffer.)

FCJ     As in other *Glaucidium*, crown spots are less distinct or lacking in juveniles (Pyle 1997a). Generally more brown and less gray on crown, back, and wings than adult. The skull should be relatively unossified.

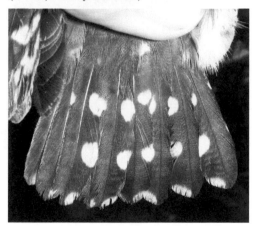

Juvenile rects appear slightly more pointed than in DCB.

FPF     Gray head and body feathers replacing brown juvenile feathers. This molt may be difficult to distinguish from early stages of SPB, but probably does not include gr covs, and should never include rects or flight feathers. The skull should be relatively unossified.

| FCF | Mix of brown and gray crown feathers and molt limits on the back, among the less covs, or between the less covs and med covs. |
|---|---|

A mix of adult-like (grayer with bolder white spots) and juvenile (browner with less distinct buffy spots) on the crown (left). Mixed adult-like grayer body and less covs contrast against retained juvenile ss covs and flight feathers (right).

| SPB | Gray flight feathers replacing juvenile brown flight feathers, but beware of DPB birds with heavily worn flight feathers that may also appear brown. Skull relatively ossified. |
|---|---|
| DCB | Gray crown with distinct white spots and gray on the back and wings. Skull relatively ossified. |

Uniformly grayish-brown flight feathers show no molt limits (left; Courtesy of Philip C. Stouffer) and definitive basic rects appear more rounded than in FCJ (right).

| DPB | Gray flight feathers replacing adult grayish-brown(?) flight feathers, perhaps very similar to SPB. Skull relatively ossified. |
|---|---|

---

## LITERATURE CITED

CEMAVE. 2013. Lista das espécies de aves brasileiras com tamanhos de anilha recomendados. Centro Nacional de Pesquisa e Conservação de Aves Silvestres, Cabedelo, Brasil (in Portuguese).

Chubb, C., H. Grönvold, F. V. McConnell, H. F. Milne, P. Slud, and Bale & Danielsson. 1916. The birds of British Guiana: based on the collection of Frederick Vavasour McConnell. Bernard Quaritch, London, UK.

Cramp, S. 1985. Birds of the Western Palearctic (Vol. 4). Oxford University Press, Oxford, UK.

Holt, W., R. Berkley, C. Deppe, P. R. Enriquez, J. L. Petersen, J. L. R. Salazar, K. P. Segars, and K. L. Wood. [online]. 1999. Crested Owl (*Lophostrix cristata*). In J. del Hoyo, A. Elliott, J. Sargatal, D. A. Christie, and E. de Juana (editors), Handbook of the birds of the world alive. Lynx Edicions, Barcelona, Spain. <http://www.hbw.com>.

Holt, W., R. Berkley, C. Deppe, P. R. Enriquez, J. L. Petersen, J. L. R. Salazar, K. P. Segars, K. L. Wood, and E. de Juana. [online]. 2013a. Tawny-bellied Screech-Owl (*Megascops watsonii*). In J. del Hoyo, A. Elliott, J. Sargatal, D. A. Christie, and E. de Juana (editors), Handbook of the birds of the world alive. Lynx Edicions, Barcelona, Spain. <http://www.hbw.com>.

Holt, W., R. Berkley, C. Deppe, P. R. Enriquez, J. L. Peterson, J. L. R. Salazar, K. P. Segars, K. L. Wood, A. Bonan, and E. de Juana. [online]. 2013b. Amazonian Pygmy-Owl (*Glaucidium hardyi*). In J. del Hoyo, A. Elliott, D. A. Christie, and E. de Juana (editors), Handbook of the birds of the world alive. Lynx Edicions, Barcelona, Spain. <http://www.hbw.com>.

Marks, J., R. J. Cannings, and H. Mikkola. 1999. Family Strigidae (Typical Owls). Pp. 76–242 in J. del Hoyo, A. Elliott, and J. Sargatal (editors), Handbook of the birds of the world, Barn-owls to Hummingbirds. Lynx Edicions, Barcelona, Spain.

Pyle, P. 1997a. Identification guide to North American birds, Part I. Slate Creek Press, Bolinas, CA.

Pyle, P. 1997b. Flight-feather molt patterns and age in North American owls. Monographs in Field Ornithology 2:1–32.

Pyle, P., A. McAndrews, P. Veléz, R. L. Wilkerson, R. B. Sigel, and D. F. DeSante. 2004. Molt patterns and age and sex determination of selected southeastern Cuban landbirds. Journal of Field Ornithology 75:136–145.

Pyle, P. 2008. Identification Guide to North American Birds, Part II. Slate Creek Press, Bolinas, CA.

Storer, R. W. 1972. The juvenal plumage and relationships of *Lophostrix cristata*. Auk 89:452–455.

Telfry, H. E., and G. L. Holroyd. 2012. Molt in Burrowing Owls (*Athene cunicularia*). North American Bird Bander 37:4–10.

Traylor, M. A. 1958. Variation in South American great horned owls. Auk 75:143–149.

Weidensaul, C. S., B. A. Colvin, D. F. Brinker, and J. S. Huy. 2011. Use of ultraviolet light as an aid in age classification of owls. Wilson Journal of Ornithology 123:373–377.

# CHAPTER TEN

# Nyctibiidae (Potoos)

# Species in South America: 6

# Species recorded at BDFFP: 5

# Species captured at BDFFP: 2

Nyctibiidae is a family composed of a monotypic genus with seven species restricted to the Neotropics. Potoos forage on insects at dusk and night; during daylight hours potoos adopt postures that resemble stumps or broken branches. All seven potoos are garbed in cryptic, bark-patterned plumages allowing individuals to easily blend into the substrate upon which they roost, even out in the open. Although potoos may superficially resemble nightjars, potoos possess a unique "tooth" on their upper mandible. All potoos lack rictal bristles, except Rufous Potoo (N. bracteatus), and instead have fine filamentous feathers above the nares (Cleere 1998, Cohn-Haft 1999). Potoos also have unique slits in their eyelids that allow them to detect approaching dangers while their eyes remain closed (Borrero 1974). The White-winged Potoo (Nyctibius leucopterus), relatively recently rediscovered at BDFFP (Cohn-Haft 1993), produces a song of a single-mournful descending whistle, which many local people believe is the nocturnal call of Curupira, a rambunctious and troublesome mythical creature. Potoos have 10 pp, 10–11 ss, and 12 rects.

Although potoos are frequently heard singing on moonlit nights in forested and deforested areas, they are rarely captured by regular daily passive mist-netting operations. Molt patterns of potoos are essentially unknown, but may be like nightjars (Caprimulgidae) and follow a Complex Basic Strategy; the extent of the preformative molt is unknown, but may be partial again like many Caprimulginae (nightjars) and Strigidae. It may also be valuable to use ultraviolet light to assess molt limits as in Strigidae (Weidensaul et al. 2011). Juvenile plumages are often distinct, either being paler as in Common Potoo (Nyctibius griseus) and Great Potoos (N. grandis), or notably darker, as in Rufous Potoos (N. bracteatus; Cohn-Haft 1999). Flight feather replacement occurs in multiple waves as in a Staffelmauser pattern among the pp and ss (Cleere 1998), and based on a review of specimens at LSUMZ (among N. bracteatus, N. leucopterus, and N. griseus) at least often results in incomplete replacement of flight feathers as in many Caprimulgidae and Strigidae (Figure 10.1).

Sexes are similar in appearance and pairs appear to be socially monogamous. Both sexes incubate a single egg, which is placed on a branch or stump, with males often incubating during the day and the female at night (Cohn-Haft 1999). Based on a review of specimens among four species at LSUMZ, potoos appear to completely ossify their skulls.

Figure 10.1 Nyctibius griseus SAB with block of retained definitive basic ss. (s4–s9; LSUMZ tissue B-80881.)

## SPECIES ACCOUNTS

### Nyctibius aethereus longicaudus

Long-tailed Potoo • Mãe-da-lua-parda

Band Size: Unknown

\# Individuals Captured: 2

| | |
|---|---|
| Wing | 310.0–320.0 mm (315.0 ± 7.1 mm; n = 2)<br>281 mm (Chubb et al. 1916)<br>284–311 mm (Cohn-Haft and Kirwan 2012a) |
| Tail | 265.0–270.0 mm (267.5 ± 3.5 mm; n = 2)<br>241 mm (Chubb et al. 1916)<br>241–270 mm (Cohn-Haft and Kirwan 2012) |
| Mass | 272.0–280.0 g (276.0 ± 5.7 g; n = 2)<br>280–447 g (Cohn-Haft and Kirwan 2012) |

Similar species: Relatively long and rounded tail compared to other Nyctibius spp. and also with a distinctive buffy shoulder patch (Cohn-Haft and Kirwan 2012).

Skull: Completely ossifies, but timing unknown.

Brood patch: Brood patch may develop in both sexes. Timing unknown, but reported to be middle of wet season to dry season in French Guiana (Pelletier et al. 2006).

Sex: ♀ = ♂.

Molt: Group 5 (or Group 3?)/Complex Basic Strategy; FPF partial, DPBs incomplete(-complete?). Unknown timing.

Notes: N. a. aethereus of the Atlantic forest may be a distinct species (Cohn-Haft and Kirwan 2012).

| | |
|---|---|
| FCJ | Apparently similar to adult, but paler and more cinnamon (Holyoak 2001). Skull is unossified. |
| FPF | Replacing paler juvenile body feathers with darker adult-like body feathers. Skull is unossified. |
| FCF | Presumably with molt limits among the ss covs, although details are not known. Skull is unossified or ossified. |
| SPB | Unknown how to distinguish from DPB (use UPB), but retained juvenile feathers may appear relatively cinnamon compared to incoming feathers. Skull is ossified. |
| SCB | With molt limits among the ss, pp, and/or rects with retained feathers appearing juvenile. In particular, ss would appear shorter than surrounding feathers. |
| DCB | Without molt limits among the ss covs or flight feathers. Skull is ossified. |
| DPB | Unknown how to distinguish from SPB (use UPB), but retained feathers may be similar in ground color (i.e., lacking a more cinnamon-toned effect) to the incoming feathers. Skull is ossified. TPB may be a possible code where three generations of feathers are evident, with the oldest being juvenile. |
| SAB | With one or more retained definitive basic ss and/or pp. Skull is ossified. It may be possible to age a small percentage of birds as TAB if three generations of flight feathers are evident. |

### Nyctibius bracteatus (monotypic)

Rufous Potoo • Urutau-ferrugem

Band Size: Unknown

\# Individuals Captured: 1

| | |
|---|---|
| Wing | 161 mm (n = 1)<br>162 mm (Chubb et al. 1916) |
| Tail | 123 mm (n = 1)<br>128 mm (Chubb et al. 1916) |
| Mass | 50 g (n = 1)<br>46–58 g (Cohn-Haft and Kirwan 2012) |

Similar species: Distinct among local Nyctibius spp. by small size and overall rich rufous coloration.

Skull: Completely ossifies, perhaps during FCF.

Brood patch: Brood patch may develop in both sexes. Breeding may occur mainly during the middle to late dry season (Cohn-Haft and Kirwan 2012).

Sex: ♀ = ♂.

Molt: Group 5 or Group 3/Complex Basic Strategy; FPF partial, DPBs incomplete-complete. As the smallest potoo, it may more regularly have complete DPBs than other species. Unknown timing.

FCJ         Iris paler yellow. Upperparts more heavily barred and vermiculated, and head, nape, and ss covs more boldly spotted black (Ingels et al. 2008). Skull is ossified.

FPF         Replacing more barred and spotted juvenile body feathers with adult-like body feathers. Skull is unossified.

FCF         Iris may be pale yellow to bright yellow. Otherwise unknown, but look for molt limits among the body feathers and/or ss covs. Skull may be ossified or unossified.

Note retained paler outer gr and med covs. (LSUMZ 52972.)

SPB         Unknown how to distinguish from DPB, but look for flight feather replacement of juvenile-like feathers with adult-like feathers. Skull is ossified.

SCB         With molt limits among the ss, pp, and/or rects with retained feathers appearing juvenile. In particular, juvenile ss appear shorter than surrounding feathers and distinctly less rufous, especially along the leading edge, with more internal pale markings near the tip.

A partially folded wing showing replaced inner 3 ss longer and and darker than next 2 paler ss. (LSUMZ 114641.)

DCB         Without molt limits among the ss covs or flight feathers. Skull is ossified.

DPB         Unknown how to distinguish from SPB, but look for flight feather replacement of adult-like feathers with adult-like feathers. Skull is ossified. TPB should be a possible code where three generations of feathers are evident, with the oldest being juvenile.

SAB         With molt limits among the ss, pp, and/or rects with retained feathers appearing adult. Size, color, and pattern should be similar between old and replaced feathers. Skull is ossified.

## LITERATURE CITED

Borrero, J. 1974. Notes of the structure of the upper eyelid of potoos (Nyctibius). Condor 76:210–211.

Chubb, C., H. Grönvold, F. V. McConnell, H. F. Milne, P. Slud, and Bale & Danielsson. 1916. The birds of British Guiana: based on the collection of Frederick Vavasour McConnell. Bernard Quaritch, London, UK.

Cleere, N. 1998. Nightjars: a guide to the Nightjars, Nighthawks, and their relatives. Yale University Press, New Haven, CT.

Cohn-Haft, M. 1993. Rediscovery of the White-winged Potoo (Nyctibius leucopterus). Auk 110:391–394.

Cohn-Haft, M. 1999. Family Nyctibiidae (Potoos). Pp. 288–301 in J. Del Hoyo, A. Elliott, and J. Sargantal (editors), Handbook of the birds of the world. Lynx Edicions, Barcelona, Spain.

Cohn-Haft, M., and G. M. Kirwan. [online]. 2012a. Long-tailed Potoo (Nyctibius aethereus). In J. del Hoyo, A. Elliott, J. Sargatal, D. A. Christie, and E. de Juana (editors), Handbook of the birds of the world alive. Lynx Edicions, Barcelona, Spain. <http://www.hbw.com>.

Cohn-Haft, M., and G. M. Kirwan. [online]. 2012b. Rufous Potoo (Nyctibius bracteatus). In J. del Hoyo, A. Elliott, J. Sargatal, D. A. Christie, and E. de Juana (editors), Handbook of the birds of the world alive. Lynx Edicions, Barcelona, Spain. <http://www.hbw.com>.

Holyoak, D. T. 2001. Nightjars and their allies: The Caprimulgiformes. Oxford University Press, Oxford, UK.

Ingels, J., N. Cleere, V. Pelletier, and V. Héquet. 2008. Recent records and breeding of Rufous Potoo Nyctibius bracteatus in French Guiana. Cotinga 29:144–148.

Pelletier, V., A. Renaudier, O. Claessens, and J. Ingels. 2006. First records and breeding of Long-tailed Potoo Nyctibius aethereus for French Guiana. Cotinga 26:69–73.

Weidensaul, C. S., B. A. Colvin, D. F. Brinker, and J. S. Huy. 2011. Use of ultraviolet light as an aid in age classification of owls. Wilson Journal of Ornithology 123:373–377.

# CHAPTER ELEVEN

# Caprimulgidae (Nighthawks and Nightjars)

# Species in South America: 33

# Species recorded at BDFFP: 5

# Species captured at BDFFP: 2

There are two groups of nightjars (or "goatsuckers"): Chordeilinae (nighthawks) and Caprimulginae (nightjars), although these two groups may not be reciprocally monophyletic (AOU 1998, Barrowclough et al. 2006, Han et al. 2010). Ecologically, these two groupings are convenient: "nighthawks" are typically fast-flying aerial insectivores that have very short rictal bristles, whereas "nightjars" sally large insects from the ground or low branches and generally have long rictal bristles. Both groups have large eyes to aid nocturnal foraging and very large mouths with a lower jaw that expands both vertically and horizontally. They have short legs and the middle toe has a pectinate (comb-like) claw, apparently used to preen feathers. Nightjars and nighthawks have cryptic plumage, acting as camouflage for resting during the day (Dittmann and Cardiff 2009). Caprimulgids have 10 pp, 12 (sometimes 13) ss, 10 (occasionally 12 in *Nyctidromus*) rects (Pyle 1997).

To our knowledge, most, if not all, caprimulgids follow a Complex Basic Strategy. Preformative molts are often partial, but may be highly variable within species with some individuals replacing only body feathers and others replacing a considerable number of secondary coverts, but rarely flight feathers. In some species (such as *Chordeiles minor*), the tail and sometimes tertials are molted during the preformative molt, but this appears to be uncommon (Pyle 1997, Howell 2010). Primary and secondary (including tertial) molt often follows the typical replacement sequence (see Chapter 1), although in some species a *Staffelmauser* replacement pattern occurs, perhaps being more likely in larger and/or tropical species (Rohwer 1971; Pyle 1995, 1997).

Most species are sexually dichromatic, with males having more extensive white or other coloration in the throat, tail, and/or wings that is used for courtship, but these patterns are often concealed while the bird is resting. Females primarily incubate the eggs and brood the young, developing brood patches, and the timing of egg laying is often thought to be synchronized with lunar cycles (Murton and Westwood 1977, Holyoak 2001). Cloacal protuberances, however, are poorly developed. Skulls typically do not completely ossify, but the extent of ossification may be useful for aging. Males average larger than females, and adults typically have substantially longer primaries than first-cycle birds (Pyle 1997, Howell 2010).

# SPECIES ACCOUNTS

## *Nyctidromus albicollis albicollis*

Common Pauraque • Bacurau

Band Size: G–H

\# Individuals Captured: 7

| | | |
|---|---|---|
| Wing | ♂ | 140.0–144.5 mm<br>(142.3 ± 3.2 mm; n = 2) |
| | ♀ | 139.0–152.5 mm<br>(144.1 ± 6.2 mm; n = 4) |
| Tail | ♂ | 127.0–139.0 mm<br>(133.0 ± 8.5 mm; n = 2) |
| | ♀ | 127.0–134.0 mm<br>(131.0 ± 3.6 mm; n = 3) |
| Mass | ♂ | 53.0–57.3 g (55.2 ± 3.0 g; n = 2) |
| | ♀ | 47.0–55.2 g (49.8 ± 3.7 g; n = 4) |

Similar species: From other caprimulgids by extremely long tail projecting well beyond wing tips.

Skull: Ossification does not complete, but percentage may be useful in aging (Pyle 1997).

Brood patch: Probably only develops in ♀♀ (Pyle 1997). Timing unknown.

Sex: ♂ has white r4, white inner web of r5, and partially white inner web on r3. ♀ has only white tips on outer 2 or 3 rects. White crescent on pp is broader in ♂♂ than ♀♀.

Molt: Group 5 and Group 3/Complex Basic Strategy (based on N. American birds); FPF partial, DPBs incomplete (~35% of birds) to complete (Pyle 1997). FPF includes a variable number of ss covs, from none to all. DPB molts may more regularly be incomplete in tropical populations, resulting from a *Staffelmauser* replacement pattern. One bird was finishing molt in Jan and birds captured between Jun and Oct were not molting, suggesting a late dry to early wet season molt.

Notes: Definitive basic pp covs in *D. a. albicollis* do not appear to have the buff markings described in U.S. subspecies.

| | |
|---|---|
| FCJ | Upperparts are spotted (not streaked) and brown scapulars lack bold black and buff pattern (Pyle 1997). Pp may be 5–10 mm shorter than in adults as in *N. a. intercedens* (Dickey and Van Rossem 1938). Skull is unossified. |
| FPF | Black and buff scapulars replacing brown scapulars. This may be difficult to distinguish from early stages of SPB but does not include flight feathers, and the skull should be relatively open. Skull is unossified. |
| FCF | Molt limits among the body feathers, scapulars, and/or ss covs (Pyle 1997). Skull is relatively ossified. |
| SPB | Actively molting, replacing juvenile feathers with adult feathers. Skull is relatively ossified. |
| SCB | With retained and notably shorter juv ss, juv pp, and/or juv pp covs (Pyle 1997). Skull is relatively ossified. |
| DPB | Actively molting, replacing adult feathers with adult feathers. Skull is relatively ossified. |
| DCB | Black and buff scapulars without molt limits among the ss, pp, or pp covs (Pyle 1997). Skull is relatively ossified. |

SAB    With molt limits of retained adult ss, adult pp, and/or pp covs that appear similar in length to replaced
       flight feathers. Skull is relatively ossified.

s3 and s4 retained.                                    p9 and greater alula retained.

## *Nyctipolus nigrescens* (monotypic)

Blackish Nightjar • Bacurau-de-lajeado

Band Size: F

# Individuals Captured: 2

| Wing | ♂ | Unknown at the BDFFP |
|------|---|----------------------|
|      |   | 150 mm (Chubb et al. 1916) |
|      | ♀ | 146.0 mm (n = 1) |
|      |   | 136 mm (Chubb et al. 1916) |
| Tail | ♂ | Unknown at the BDFFP |
|      | ♀ | 102.0 mm (n = 1) |
|      |   | 88 mm (Chubb et al. 1916) |
| Mass | ♂ | Unknown at the BDFFP |
|      |   | 32–42 g (Cleere and de Juana 2013) |
|      | ♀ | 36.6 g (n = 1) |
|      |   | 32–50 g (Cleere and de Juana 2013) |

Similar species: No other nightjar at the BDFFP has such reduced white markings.

Skull: Ossification may not complete, but percentage may be useful in aging (Pyle 1997).

Brood patch: Apparently both sexes incubate and probably develop BPs (Cleere and de Juana 2013). Timing unknown, but probably breeds at least during the mid to late dry season at BDFFP based on nests, consistent with suggested timing of breeding in Guyana and Surinam (Cleere and de Juana 2013).

Sex: ♂ has a small, narrow white band in inner pp and white tips to outer rects, which the ♀ lacks.

Molt: Group 5 and Group 3(?)/Complex Basic Strategy; FPF partial-incomplete, DPBs incomplete-complete(?). The FPF may include a variable amount of ss covs and rects. The DPB may complete, but may more regularly be incomplete as in other Caprimulgidae. Timing unknown, but probably follows breeding, during the late dry to wet season.

FCJ    Very downy body plumage. Back and ss covs each with thick tawny edging and spotting. Skull is relatively unossified.

This example is finishing completing the FPJ molt.

FPF    Feathers with intricate buffy markings replacing brown juvenile feathers. This may be difficult to distinguish from early stages of SPB, but does not include flight feathers. Skull is relatively unossified.

FCF    Molt limits among the less, med, gr covs, and/or rects with retained outer juvenile feathers with tawny edging and replaced inner formative feathers brown with intricate buffy spots or vermiculations. Skull may be relatively unossified to almost or occasionally completely ossified.

SPB    Actively molting, replacing juvenile feathers with adult feathers. Skull is relatively unossified.

SCB    With retained juv ss, sometimes with retained juv pp and/or pp covs. Skull is relatively unossified.

DCB    Without molt limits among the ss, pp, or pp covs. Skull is relatively unossified.

DPB    Actively molting, replacing adult feathers with adult feathers. Skull is relatively unossified.

SAB    With retained adult ss, sometimes with retained adult pp and/or pp covs. Skull is relatively unossified.

s3–s4 retained and adult-like (left) and outer three rects retained and adult-like (right).

---

## LITERATURE CITED

AOU. 1998. Check-list of North American birds (7th ed.). American Ornithologists' Union, Washington, DC.

Barrowclough, G. F., J. G. Groth, and L. A. Mertz. 2006. The RAG-1 exon in the avian order Caprimulgiformes: phylogeny, heterozygosity, and base composition. Molecular Phylogenetics and Evolution 41: 238–248.

Chubb, C., H. Grönvold, F. V. McConnell, H. F. Milne, P. Slud, and Bale & Danielsson. 1916. The birds of British Guiana: based on the collection of Frederick Vavasour McConnell. Bernard Quaritch, London, UK.

Cleere, N., and E. de Juana. 2013. Blackish Nightjar (*Nyctipolus nigrescens*). In J. del Hoyo, A. Elliott, J. Sargatal, D. A. Christie, and E. de Juana (editors), Handbook of the birds of the world alive. Lynx Edicions, Barcelona, Spain. <http://www.hbw.com>.

Dickey, D. R., and A. J. Van Rossem. 1938. The birds of El Salvador. Field Museum of Natural History, Zoological Series, Chicago, IL.

Dittmann, D. L., and S. W. Cardiff. 2009. Nighthawks and Nightjars. Pp. 147–149 in T. Harris (editor), National Geographic complete birds of the world. National Geographic Society, Washington, DC.

Han, K. L., M. B. Robbins, and M. J. Braun. 2010. A multi-gene estimate of phylogeny in the nightjars and nighthawks (Caprimulgidae). Molecular Phylogenetics and Evolution 55:443–453.

Holyoak, D. T. 2001. Nightjars and their allies: the Caprimulgiformes. Oxford University Press, Oxford, UK.

Howell, S. N. G. 2010. Molt in North American birds. Houghton Mifflin Harcourt, Boston, MA.

Murton, R. K., and Westwood, N. J. 1977. Avian breeding cycles. Clarendon Press, Oxford, UK.

Pyle, P. 1995. Incomplete flight feather molt and age in certain North American non-passerines. North American Bird Bander 20:15–26.

Pyle, P. 1997. Identification guide to North American birds, Part I. Slate Creek Press, Bolinas, CA.

Rohwer, S. A. 1971. Molt and the annual cycle of the Chuck-will's-widow (*Caprimulgus carolinensis*). Auk 88:485–519.

# CHAPTER TWELVE

# Trochilidae (Hummingbirds)

# Species in South America: 258

# Species recorded at BDFFP: 15

# Species captured at BDFFP: 10

Hummingbirds are a familiar New World group of small, high energy, fast-flying nectivores (that also regularly eat insects). Many species are not particularly sensitive to forest fragmentation at the BDFFP, probably instead positively responding to increased flowering along forest edges (Stouffer and Bierregaard 1995). Color and shape are quite varied across the family, but a long, thin bill used for foraging on flower nectar is typical. Many have iridescent plumage, especially in the throat (gorget) feathers; even hermits have some iridescence on their rectrices. Most species are obviously sexually dichromatic, and a recent detailed examination of rectrix patterns in hermits has suggested evidence of subtle differences between sexes with pale outer rectrix margins reduced in males and immature females (Piacentini 2011), but we have found this hard to see in our specimen examinations. Other species, including some longer-billed hermits but also more famously in the Purple-throated Carib Hummingbird (*Eulampis jugularis*), are sexually dimorphic in bill curvature with females having more curved bills (Temeles et al. 2000, Temeles et al. 2010). Forked tails are widespread across the family, estimated to have evolved independently at least 26 times within the family (Clark 2010). Hummingbirds have 10 pp, 6 ss, and 10 rects.

Despite great interest in hummingbirds, both in scientific research and by bird-watchers, their molt strategies are poorly understood. Howell et al. (2003) reported that hummingbirds only follow the Complex Basic Strategy, but *Phaethornis longirostris* has been suggested to lack a preformative molt (Stiles and Wolf 1974, Schuchmann 1999). Conflicting with that, Wolfe et al. (2009) reported complete preformative molts in *P. longirostris* and *P. striigularis* from Costa Rica. We examined 59 specimens of *P. bourceiri* from Peru, Ecuador, Guyana, and Suriname, and 14 specimens of *P. superciliosus* from Suriname and Bolivia, and conclude that the preformative molt is partial in these species (see species account for more details). We suspect partial molts may be more widespread or even universal across the genus, especially given that preformative molts are partial in several other basal Trochilidae (Zimmer 1950, Hu et al. 2000). Even so, the preformative molt is complete in other taxa (Pyle 1997, Pyle et al. 2015). Ruby-throated Hummingbirds (*Archilochus colubris*) of North America undergo a prealternate molt (Dittmann and Cardiff 2009), and this should be looked for in other migratory hummingbird species, although it may not be likely in resident tropical species. We refer readers to Dittmann and Cardiff (2009) and Pyle (2013) describing scenarios for interpreting limited/partial fall molts and complete winter molts. Molt can last up to 4 to 5 months in some species, including a suspension (e.g., North American migrants), but may progress more rapidly in others (Schuchmann 1999). In many temperate and Central American species, and probably regularly across the family, p10 is replaced before p9. Ss molt can be quite variable, proceeding distally, proximally, both, or irregularly. The outermost rectrix (r5) is also typically replaced before the last rectrix is replaced and tail molt is often bilaterally asymmetrical (Stiles 1995, Pyle 1997). In some species the juvenile plumage is distinct between sexes, but this appears to be an exception among sexually dichromatic hummingbird species (Schuchmann 1999).

Understory hummingbirds regularly fall into ground-level mist-nets and occasionally canopy species do as well, although there are several more efficient methods available for targeting hummingbirds at flowers and feeders (Russell and Russell 2001). As far as is known, all species of hummingbirds can be aged by examining the bill for corrugations (Ortiz-Crespo 1972, Baltosser 1987, Yanega et al. 1997). In birds less than roughly 5–9 months old, corrugations are found along the ramphotheca and appear as many minute furrows. Good light is required to see these furrows without magnification, and using a small hand lens will greatly improve precision in estimating the proportion of the bill covered in corrugations. These corrugations run the length of the bill in the youngest birds and are lost roughly proximally as the bird ages. Many adult birds can retain 10%–20% of corrugations for perhaps their entire life. In general, birds with corrugations covering at least 50% of the bill are certainly less than 9 months old (Ortiz-Crespo 1972). This is a conservative cutoff and Pyle (1997) suggests that 10% is a reasonable standard for North American species. Because of the uncertainty in the primary literature regarding an appropriate cutoff, we follow a conservative approach based on our observations and select 30% as an appropriate cutoff until more information from the vast majority of hummingbirds, as of yet undocumented, becomes available. Skull ossification occurs differently in hummingbirds than in other near-passerines and passerines. Instead of the skull forming a double-layer, it becomes thicker and less transparent with time (Pyle 1997). With experience, the condition of the skull can be useful to age some birds (Stiles 1972, Pyle 1997), although this has been questioned (Ortiz-Crespo 1972). In any case, bill corrugations provide a much easier and more reliable method of age determination such that we do not include a skull section for hummingbird accounts. Only females incubate and rear young, but we have had little success identifying brood patches because the belly is normally unfeathered and glossy, thus we also leave this section out of species accounts. Molt-breeding overlap has been documented in Costa Rica (Stiles and Wolf 1974), but is unknown at the BDFFP because of difficulties in identifying active brood patches.

## SPECIES ACCOUNTS

*Topaza pella pella*

Crimson Topaz • Beija-flor-brilho-de-fogo

Band Size: A (CEMAVE 2013)

# Individuals Captured: 14

| | | |
|---|---|---|
| Wing | ♂ | 69.5–81.5 mm (75.5 ± 5.4 mm; n = 5) |
| | ♀ | 69.0–75.0 mm (71.7 ± 3.1 mm; n = 3) |
| Tail | ♂ | 43.0–48.0 mm (45.7 ± 2.5 mm; n = 3 w/out streamer) 92.0–93.0 mm (92.5 ± 0.7; n = 2 w/ streamer) |
| | ♀ | 40.0–42.0 mm (41.0 ± 1.4 mm; n = 2) |
| Mass | ♂ | 10.0–13.0 g (11.9 ± 1.2 g; n = 5) |
| | ♀ | 9.0–11.0 g (10.3 ± 0.7 g; n = 6) |
| Bill | ♂ | 20.5–21.0 mm (20.8 ± 0.3 mm; n = 2) |
| | ♀ | 21.2–22.3 mm (21.8 ± 0.8; n = 2) |

Similar species: None.

Sex: ♂ is iridescent red overall with a green throat and ♀ is iridescent green overall with a red-orange throat.

Molt: Group 3/Complex Basic Strategy; FPF limited-partial (Hu et al. 2000), DPBs complete (or occasionally incomplete?). FPF includes many or all body feathers and ss covs, but not flight feathers. Unknown timing, but two birds observed completing pp molt in Feb.

(Courtesy of Cameron Rutt.)

FCJ     ♂ is ♀-like, but with bronzy hue throughout and without distinct gorget. ♀ is like DCB ♀, but with green throat and grayish feathers throughout body (Nicholson 1931, Ruschi 1986, Restall et al. 2006). Bill corrugations >30%.

FPF     ♂ replacing orange/green feathers with formative iridescent red feathers. ♀ replacing grayish feathers with green feathers. Probably with >30% bill corrugations.

FCF     ♂ with mixed orange/green retained juvenile body feathers and replaced iridescent red and orange adult-like feathers. Black on head may be duller, undertail covs may be greener with less yellow, and uppertail covs may be less coppery. R2 streamers may not be developed. ♀ gorget with mix of grayish feathers and yellow-green iridescent feathers and underparts with a mix of gray and green feathers (Hu et al. 2000). With or without bill corrugations.

SPB     Replacing juvenile flight feathers with adult flight feathers. ♂ perhaps with some green body feathers remaining. ♀ perhaps with some gray body feathers remaining. Bill corrugations <30%.

DCB     ♂ without green body feathers; compared to FCF, the head may be darker black, undertail covs may be more yellow, and uppertail covs may be more coppery. R2 streamers long, but beware of them being broken off. ♀ (not illustrated) without gray feathers in gorget and underparts. Bill corrugations <30%.

DPB     Replacing adult flight feathers with adult flight feathers. Bill corrugations <30%.

DCB plumage aspect in males (left) and females. (Right; Courtesy of Cameron Rutt.)

## *Florisuga mellivora mellivora*

White-necked Jacobin • Beija-flor-azul-de-rabo-branco

Band Size: A (CEMAVE 2013)

# Individuals Captured: 41

Wing    62.0–71.0 mm (66.5 ± 2.5 mm; n = 31)

Tail    31.0–40.0 mm (36.1 ± 2.3 mm; n = 25)

Mass    5.0–8.0 g (6.3 ± 0.7 g; n = 37)

Bill    18.3 mm (n = 1)

Similar species: Only the "speckle-throated" ♀ may be confused with other hummingbirds, but note the tail pattern and heavy iridescent chevrons on the throat.

Sex: ♀♀ are dichromatic with "speckle-throated" and "blue-headed" morphs; adult ♂♂ probably only have a blue-headed form (see Zimmer 1950 for a detailed description). The only way to safely separate ♂♂ and ♀♀ in this blue-headed plumage is by the tail pattern, although ♀♀ also average less iridescent throughout. In ♂♂ all five rects are white with a narrow brown margin that is often absent in r5 (note that the upper tail covs are nearly as long as the tail and bright metallic green). Blue-headed ♀♀ typically have at least the central rects with broad dark blue tips, sometimes with green; the brown border of r5 is often incomplete or irregular in width, but not typically absent. Speckle-throated birds without bill corrugations are apparently always ♀♀; this plumage can have a few to many blue feathers as well. FCJ ♂♂ and ♀♀ are like blue-headed form, but with buffy-orange feathers in the malar, chin, back and rump, and although ♂♂ may have more white in tail feathers on average, there may be considerable overlap between sexes. ♀♀ apparently have shorter wings, longer bills, and weigh less on average, but there may be considerable overlap (Stiles et al. 2014). It is not clear what the juvenile speckle-throated ♀ looks like; none of the 21 speckle-throated ♀ specimens (LSUMZ) had >20% bill corrugations. It is possible that ♀♀ progress through a blue-headed form as FCJ and FCF first (Zimmer 1950), but certainly more study on this complex series of plumages is needed.

Molt: Group 3/Complex Basic Strategy; FPF partial, DPBs complete (Zimmer 1950). FPF appears to include all body and ss covs. Timing unknown; the only bird seen in molt was during Jun, but birds have not been examined during the wet season.

FCJ    Only known as the blue-headed form, but with a distinct orange malar streak, chin, back, and rump (Zimmer 1950, Stiles et al. 2014, LSUMZ). With >30% bill corrugations.

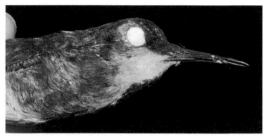

(LSUMZ 116582.)

FPF    Probably with >30% bill corrugations, and replacing orange in malar, chin, back, and rump feathers with adult-like blue and green feathers.

FCF    Probably typically with <30% bill corrugations. No known plumage or feather shape criteria known, as the retained juvenile rects are similar in shape to definitive basic feathers. Best to use FAJ.

FAJ    Without bill corrugations and without orange malar, chin, back, and rump.

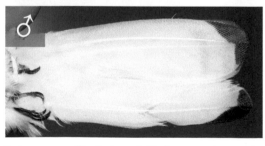

"Blue-headed" ♀ with dark margins of each rect (upper left), "speckle-throated" ♀ with mostly dark rects (upper right), and ♂ with mostly white-margined rects. (Lower left; LSUMZ 109389.)

SPB    Perhaps not readily distinguishable from DPB (use UPB). Bill corrugations <30%.

DCB    Unknown how to distinguish from FCF; use FAJ instead for birds with bill corrugations <30%. With more study, differences between FCF and DCB pp covs, rects, and/or flight feathers may become apparent.

DPB    Unknown if how to distinguish from SPB (use UPB). Bill corrugations <30%.

*Phaethornis bourcieri bourcieri*

Straight-billed Hermit • Rabo-branco-de-bico-reto

Band Size: A (CEMAVE 2013)

# Individuals Captured: 635

Wing    50.0–60.0 mm
         (55.4 ± 2.2 mm; n = 372)

Tail     52.0–65.0 mm
         (58.9 ± 2.5 mm; n = 310)

Mass    3.0–5.5 g
         (4.1 ± 0.4 g; n = 423)

Bill     24.6–32.2 mm
         (28.7 ± 1.9 mm; n = 54)

Similar species: From *P. superciliosus* by straight (not curved) bill. All measurements average shorter than *P. superciliosus*, although there is some overlap.

Sex: ♂ (>) ♀. Bill length shows a bimodal distribution based on 55 nonjuveniles, but more study is needed to determine if this is related to sex. Bill curvature is 43% greater in ♀♀ than ♂♂ (Temeles et al. 2010), but given that bills are relatively straight in this species, differences are subtle and have been hard to confirm with a visual (qualitative) examination of specimens. Bill and tail lengths average slightly larger in ♂♂, but these measures show substantial overlap between sexes according to an examination of specimens across the range (Piacentini 2011). This overlap between sexes, however, may be reduced if considering individuals from a narrower region or at a single site.

Molt: Group 3/Complex Basic Strategy; FPF partial, DPBs complete. The FPF may be delayed as corrugations are minimal once it completes; it includes most or all body covs and ss covs, but usually not the upper tail coverts. Molt is most frequent (>10%) between Oct and Feb and is least frequent (<5%) between Mar and Jul.

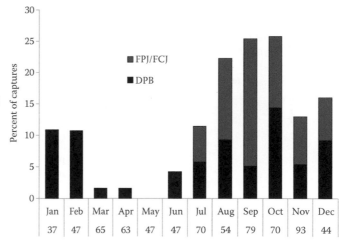

FCJ    Crown and back feathers with broad tawny edging. Pp covs may average duller than DCB, inner ss (especially s5 and s6) have narrow buffy edging, and upper tail covs are relatively weakly structured. The outer four rects are relatively pointed. With >30% bill corrugations.

FPF     With active body molt, and bill corrugations usually >30%. Retains juvenile-type buffy-tipped inner ss
        and relatively pointed rects.

FCF     With or without minimal tawny edging
        to crown and back feathers. Retains
        juvenile-type buffy-tipped inner ss
        and relatively pointed rects. Bill
        corrugations <30%.

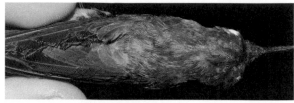

(LSUMZ 172794.)

SPB     With active body and flight feather molt. Look for retained juvenile-type buffy-tipped inner ss and
        relatively pointed outer rects. Bill corrugations <30%.

DCB     Pp covs may average brighter metallic green,
        rump feathers are relatively robustly
        structured and brightly colored, and inner
        ss without buffy edging when fresh (or
        occasionally with faint buff tip to s6). Outer
        rects relatively rounded. Bill corrugations
        <30%.

DPB     With active body and flight feather molt.
        Retained inner ss without buffy tip. Bill
        corrugations <30%.

p5–p10 retained, p1–p4 missing, s1–s3 missing, s4–s6
retained and lacking buffy tips.

*Phaethornis superciliosus muelleri*                          Band Size: A (CEMAVE 2013)

Long-tailed Hermit • Rabo-branco-de-bigodes                   # Individuals Captured: 958

Wing    51.0–63.0 mm
        (57.6 ± 2.3 mm; n = 532)

Tail    55.0–72.0 mm
        (64.7 ± 2.9 mm; n = 465)

Mass    3.7–7.0 g (5.5 ± 0.6 g; n = 603)

Bill    30.6–36.8 mm
        (33.3 ± 1.4 mm; n = 89)

Similar species: From *P. bourcieri* by curved
  (not straight) bill.

Sex: ♂ (>) ♀. DCB ♂♂ may have relatively narrow buffy margins on the rects (Piacentini 2011). Hilty (2003) reports the cord of the exposed culmen to be 7 mm larger in ♂♂ than in ♀♀, and Hinkelmann et al. (2013) repeat this, but Piacentini (2011) found overlap in an examination of specimens from across the range. Based on 70 adults at the BDFFP, bill length spans about 6 mm and appears to have a unimodal distribution, suggesting a lack of sex-related variation similar to the closely related *P. longirostris* (Stiles and Wolf 1974, Piacentini 2011). Piacentini (2011) also found minimal differences between sexes in tail length, but wing length separated slightly more across the range.

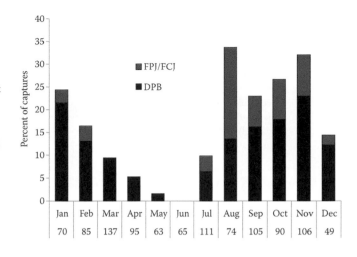

Molt: Group 3/Complex Basic Strategy; FPF partial, DPBs complete. The FPF includes most or all body covs and ss covs, but usually not the upper tail coverts. Molt is most frequent (>10% of individuals) between Aug and Feb, and <5% of individuals in molt between Apr and Jun.

FCJ    Crown and back feathers with broad buffy edging. Pp covs may average duller than DCB, inner ss (especially s5 and s6) have narrow buffy edging, and upper tail covs are relatively weakly structured. The outer four rects are relatively pointed. With >30% bill corrugations.

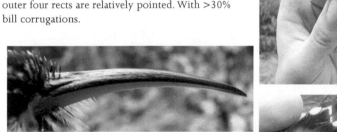

FPF    With active body molt, and bill corrugations usually >30%. Retains juvenile-type buffy-tipped inner ss and relatively pointed rects.

FCF    Without or with minimal tawny edging to crown and back feathers. Retains juvenile-type buffy-tipped inner ss and relatively pointed rects. Bill corrugations <30%.

SPB    With active body and flight feather molt, and bill corrugations <30%. Look for retained juvenile-type buffy-tipped inner ss and relatively pointed outer rects.

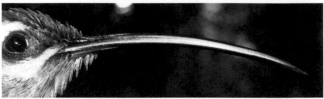

DCB | Pp covs may average brighter metallic green, rump feathers are relatively robustly structured and brightly colored, and inner ss without buffy edging when fresh (or occasionally with faint buff tip to s6). Outer rects relatively rounded. Bill corrugations <30%.

DPB      With active body and flight feather molt; bill corrugations <30%. Retained inner ss without buffy tip.

## *Heliothryx auritus auritus*

Black-eared Fairy • Beija-flor-de-bochecha-azul

Band Size: A (CEMAVE 2013)

# Individuals Captured: 35

| | |
|---|---|
| Wing | 50.0–67.0 mm (63.4 ± 1.7 mm; n = 27) |
| Tail | 40.0–67.0 mm (55.1 ± 7.0 mm; n = 26) |
| Mass | 3.5–6.0 g (4.9 ± 0.6 g; n = 31) |
| Bill | 14.9–19.2 mm (17.1 ± 1.8 mm; n = 4) |

Similar species: No other green hummingbird has the combination of short bill, black auriculars, and all white underparts and undertail.

Sex: ♀ has dusky spotting in the throat, a black band at the base of outer three rects and lacks purple iridescence in the auriculars. FAJ ♂ has no spotting in the all-white throat, lacks a black band at the base of outer rects, and has purple iridescence in the auriculars. Unusual among hummingbirds, ♀♀ have longer tails than ♂♂ (Clark 2010).

Molt: Group 4/Complex Basic Strategy; FPF probably complete(?), DPBs complete. Two birds with ≥30% bill corrugations have been observed in molt and neither were molting flight feathers, suggesting a partial FPF; however, the unique pattern to the rects not appearing in adult-like birds suggests a complete replacement of feathers during the FPF. Only three individuals have been observed in wing molt and all were early to mid stages of pp molt in Nov.

FCJ     Has brownish-cinnamon edges to green head and back feathers, but this probably wears quickly with age. May not be possible to determine sex because both sexes probably like FAJ ♀♀ with dusky spotting in throat (Hilty 2003), although spotting may be reduced in ♂♂. Study of rect pattern and length may prove useful. Bill with >30% corrugations.

(Courtesy of Angelica Herñandez Palma.)        (Courtesy of Angelica Herñandez Palma.)

FPF     Body and possibly wing/tail molt with bill corrugations >30%.

FAJ     Without molt limits among the ss covs. If the FPF is complete, then this would become FAJ. If the FPF is partial, then this would become DCB. Bill with <30% corrugations.

UPB     Body and wing molt with bill corrugations <30%.

## *Campylopterus largipennis largipennis*

Gray-breasted Sabrewing • Asa-de-sabre-cinza

Band Size: A (CEMAVE 2013)

\# Individuals Captured: 198

| | |
|---|---|
| Wing | 65.0–78.0 mm (72.0 ± 2.6 mm; n = 153) |
| Tail | 46.0–54.0 mm (51.0 ± 1.7 mm; n = 126) |
| Mass | 6.0–11.0 g (8.4 ± 1.0 g; n = 167) |
| Bill | 21.7–28.0 mm (24.7 ± 1.2 mm; n = 35) |

Similar species: Large size, gray underparts, and large white outer tail spots distinct.

Sex: ♂♂ and ♀♀ are similar except that ♂♂ have a thicker and more flattened p10 rachis compared to ♀♀ in all age classes based on an examination of 70 specimens at LSUMZ (n = 4 *C. l. largipennis*, n = 66 *C. l. aequatorialis*; Table 12.1). Plumage, bill length, and general morphology does not appear to be sexually dimorphic, but ♂♂ average slightly longer-winged (Grantsau 1988).

### TABLE 12.1
*Variation in p10 rachis width in* Campylopterus largipennis *by age and sex.*

| p10 rachis width | FCJ/FCF | DCB |
|---|---|---|
| Male | 1.37–1.76 mm (1.58 ± 0.10 mm; n = 22) | 2.64–3.36 mm (3.05 ± 0.20 mm; n = 18) |
| Female | 0.93–1.14 mm (1.04 ± 0.07 mm; n = 11) | 1.24–1.68 mm (1.42 ± 0.12 mm; n = 19) |

Molt: Group 3/Complex Basic Strategy; FPF limited-partial(?), DPBs complete. Birds with >50% corrugations have not been observed molting (n = 7) at our study sites, but fresh-plumaged birds (some with bursas) lacking copper-edged head feathers (and some recorded with having body molt), retained copper-edged juvenile feathers on the rump, and relatively pointed rects are found at the LSUMZ collection (n = 4 *C. l. longipennis*, n = 66 *C. l. aequatorialis*); collectively this suggests a limited FPF molt. Molt occurs between Oct and Dec (–Mar?). Of 97 birds observed between Apr and Sep, only one had wing molt. Given the fairly short and distinct molting period and the low percentage of birds in molt at any given time, molt probably proceeds rapidly.

FCJ     Contour feathers have copper edging. ♂♂ with slightly wider p10 rachi than ♀♀ (Table 12.1). Rects relatively pointed, particularly obvious in r1 and r2 (and note the outer white-tipped feathers may wear faster than inner dark-tipped feathers). With >30% bill corrugations.

FPF     Body molt replacing copper-edged feathers with metallic green feathers. Bill corrugations variable.

FCF     Without broad tawny edging to the head and back feathers. Look for subtle molt limits among the ss covs or between the body feathers and ss covs. Some juvenile rump feathers may also be retained with worn copper edges (although beware that some DCB have slightly buffy-tipped rump feathers). Rects pointed and bluish-green relative to greener and more rounded DCB rects. Sex determination as in FCJ (Table 12.1). Bill corrugations variable.

It appears that all ss covs are retained (left); note pointed and relatively bluish rects. (Right; LSUMZ 178329.)

SPB     Replacing relatively pointed juvenile-like rects with more rounded adult-like rects. Probably otherwise not readily distinguishable from DPB, except in ♂♂ with incoming p6–p9 with relatively wide rachi compared to retained juvenile p10. Note p9 might be the last flight feather replaced, after p10. Bill with <30% corrugations.

DCB     Without molt limits between the body feathers and ss covs or within the ss covs. ♂♂ with broad p10 rachis are DCB (Table 12.1). ♀♀ may be more difficult to distinguish from both sexes of FCF, but rects are relatively green and rounded relative to blue-green and more pointed FCF rects. Bill with <30% corrugations.

(Right; LSUMZ 175350.)

DPB     Replacing rounded adult-like rects with rounded adult-like rects. Otherwise not readily distinguishable from SPB, except perhaps in some males (see SPB). Bill with <30% corrugations.

*Thalurania furcata furcata*

Fork-tailed Woodnymph • Beija-flor-tesoura-verde

Band Size: A (CEMAVE 2013)

# Individuals Captured: 480

| Wing | ♂ | 46.0–57.0 mm (52.5 ± 1.6 mm; n = 113) |
| | ♀ | 44.0–55.0 mm (49.1 ± 1.6 mm; n = 128) |
| Tail | ♂ | 26.0–41.0 mm (35.8 ± 2.9 mm; n = 107) |
| | ♀ | 25.0–37.0 mm (28.4 ± 1.9 mm; n = 105) |
| Mass | ♂ | 3.0–5.0 g (4.1 ± 0.4 g; n = 131) |
| | ♀ | 2.8–5.2 g (3.8 ± 0.4 g; n = 166) |
| Bill | ♂ | 16.8–19.1 mm (17.7 ± 0.8 mm; n = 8) |
| | ♀ | 17.3–19.5 mm (18.3 ± 0.7 mm; n = 12) |

Similar species: Only ♀ likely confused; differs from *Amazilia versicolor* by bronzy-green (not iridescent blue) on head and nape, tail mostly metallic green and dark blue (not mostly gray), and all black bill (without red on mandible). Smaller than *Campylopterus*.

Sex: ♂ is violet iridescent overall with metallic green gorget. ♀ has metallic green upperparts and white underparts and are "generic-looking" hummingbirds.

Molt: Group 3/Complex Basic Strategy, FPF partial, DPBs complete. The FPF may overlap with the FPJ, and appears to include most or all body feathers and ss covs. Molt peaks Sep–Feb.

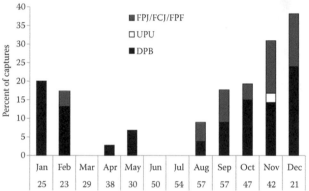

Percent of captures

Legend: ■ FPJ/FCJ/FPF  □ UPU  ■ DPB

| | Jan | Feb | Mar | Apr | May | Jun | Jul | Aug | Sep | Oct | Nov | Dec |
|---|---|---|---|---|---|---|---|---|---|---|---|---|
| | 25 | 23 | 29 | 38 | 30 | 50 | 54 | 57 | 57 | 47 | 42 | 21 |

FCJ    Both sexes like adult of respective sexes, but with copper edging to crown feathers. ♂ also differs by being dull gray underneath. This plumage probably quickly transitions to FPF. Rects relatively pointed. Bill corrugations >30%.

| FPF | With body molt. ♀ otherwise like DCB ♀ except perhaps with some copper edging to crown feathers (or worn off, or being replaced with all-green crown feathers). ♂ with incoming metallic green gorget and purple underpart feathers. Probably toward the end of molt, most birds should be more safely aged UPU. Bill with corrugations >30%. |

| FCF | Both sexes adult-like, but with relatively pointed rects. ♂♂ can show varying degrees of gorget completeness, with some showing a couple of grayish, suggesting perhaps a difference between FCF and DCB, but more study is needed. Bill with <30% corrugations (or rarely with >30%?). |
| SPB | Replacing relatively pointed rects with more rounded rects. May often be difficult to distinguish from DPB (use UPB). Bill with <30% corrugations. |
| DCB | Like FCF, but with relatively rounded rects. Bill with <30% corrugations. |
| DPB | Replacing relatively rounded rects with rounded rects. May often be difficult to distinguish from SPB (use UPB). Bill with <30% corrugations. |

## *Amazilia versicolor nitidifrons/milleri?*

Band Size: A (CEMAVE 2013)

Versicolored Emerald • Beija-flor-de-banda-branca

\# Individuals Captured: 5

| Wing | 47.0–51.0 mm (48.6 ± 1.8 mm; n = 5) |
| Tail | 27.0–30.0 mm (28.4 ± 1.1 mm; n = 5) |
| Mass | 3.0–4.0 g (3.4 ± 0.4 g; n = 5) |
| Bill | 13.9–14.2 mm (14.0 ± 0.2 mm; n = 2) |

Similar species: From *Thalurania furcata* by blue (not bronzy-green) iridescence on head and nape, mostly grayish-green (not dark blue) tail with a subterminal bluish band, and red base of mandible.

Sex: ♂ averages larger than ♀, but probably with overlap. ♀ may also have more extensive gray tips to outer rectrices (Weller et al. 2014).

Molt: Group 4/Complex Basic Strategy; FPF complete(?), DPBs complete. Timing unknown.

| FCJ | With rufous edging to crown, back, and rump (Weller et al. 2014), but probably wears with age. With >30% bill corrugations. |
| FPF | Replacing juvenile like body (and perhaps also flight feathers and rects) with adult-like feathers. Bill corrugations likely >30%. |
| FAJ | Without rufous edging to crown, back, or rump. Bill corrugations <30%. |
| UPB | Bill with <30% corrugations with wing and/or tail molt. |

## Amazilia fimbriata fimbriata

Glittering-throated Emerald • Beija-flor-de-garganta-verde

Band Size: A (CEMAVE 2013)

# Individuals Captured: 4

Wing    44.0–57.0 mm (50.3 ± 6.5 mm; n = 3)

Tail     26.0–32.0 mm (29.3 ± 3.1 mm; n = 3)

Mass    4.0–4.5 g (4.2 ± 0.3 g; n = 3)

Similar species: Combination of green throat and white belly separate it from other green hummingbirds.

Sex: ♀ reported to have white subterminal bars on gorget feathers unlike ♂ (Weller et al. 2013), although we found this difficult if not impossible to discern in a sample of five specimens (three ♀ and two

 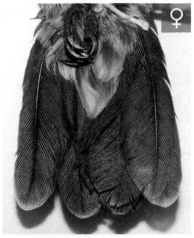

(LSUMZ 78047.)        (LSUMZ 109397.)

♂). ♀ also averages more gray and less green than ♂ in the tips of the outer rects, which is most visible in the underside of the folded tail (Weller et al. 2013).

Molt: Group 4/Complex Basic Strategy; FPF complete(?), DPBs complete. Timing unknown.

FCJ      More grayish-brown in the belly than the adult (Weller et al. 2013). Bill corrugations >30%.

FPF     Replacing juvenile like body (and perhaps also flight feathers and rects) with adult-like feathers. Bill corrugations likely >30%.

FAJ      Relatively white belly. Bill corrugations <30%.

(LSUMZ 109397.)

UPB     Bill with <30% corrugations with wing and/or tail molt.

## Hylocharis sapphirina (monotypic)

Rufous-throated Sapphire • Beija-flor-safira

Band Size: A (CEMAVE 2013)

# Individuals Captured: 7

Wing    43.0–53.0 mm (49.6 ± 4 mm; n = 5)
51 mm (Chubb 1916)

Tail     26.0–29.0 mm (27.6 ± 1.2 mm; n = 5)
25 mm (Chubb 1916)

Mass    3.5–4.5 g (4.1 ± 0.5 g; n = 6)

Bill      15.8 mm (n = 1)

Similar species: Combination of rufous tail and at least some red on the bill and chin (including first-cycle birds?) separates all plumages from other species.

(Courtesy of Cameron Rutt.)

Sex: ♂ has violet-blue throat and extensive red on bill. ♀ has little red on lower mandible and white throat with light reddish-brown chin.

Molt: Group 4/Complex Basic Strategy; FPF complete(?), DPBs complete. Timing unknown.

FCJ    Like adult ♀, but with rufous edging to head feathers. ♂♂ average brighter rufous chins (Schuchmann et al. 2013). Back less intensely blue-green and rects with broad dark dusky-purplish edging and tips. Bill corrugations >30%.

View of underside of folded tail. (LSUMZ 136879.)

FPF    Replacing juvenile body and flight feathers with adult-like feathers. Bill corrugations likely >30%.

(LSUMZ 136879.)

FAJ    Bill corrugations <30%.

(LSUMZ 172800.)                        (LSUMZ 172800.)

UPB    Bill with <30% corrugations with wing and/or tail molt.

## LITERATURE CITED

Baltosser, W. H. 1987. Age, species, and sex determination of four North American hummingbirds. North American Bird Bander 12:151–166.

CEMAVE. 2013. Lista das espécies de aves brasileiras com tamanhos de anilha recomendados. Centro Nacional de Pesquisa e Conservação de Aves Silvestres, Cabedelo, Brazil (in Portuguese).

Chubb, C. 1916. The Birds of British Guiana, vol. 1. Bernard Quaritch, London, UK.

Clark, C. J. 2010. The evolution of tail shape in hummingbirds. Auk 127:44–56.

Dittmann, D. L., and S. W. Cardiff. 2009. The alternate plumage of the Ruby-throated Hummingbird. Birding 41:32–35.

Grantsau, R. 1988. Os beija-flores do Brasil. Expressão e Cultura, Rio de Janeiro, Brasil (in Portuguese).

Hilty, S. L. 2003. Birds of Venezuela (2nd ed.). Princeton University Press, Princeton, NJ.

Hinkelmann, C., G. Kirwan, and P. Boesman. [online]. 2013. Long-tailed Hermit (*Phaethornis superciliosus*). *In* J. del Hoyo, A. Elliott, J. Sargatal, D. A. Christie, and E. de Juana (editors), Handbook of the birds of the world alive. Lynx Edicions, Barcelona, Spain. <http://www.hbw.com>.

Howell, S. N. G., C. Corben, P. Pyle, and D. I. Rogers. 2003. The first basic problem: a review of molt and plumage homologies. Condor 105:635–653.

Hu, D.-S., L. Joseph, and D. Agro. 2000. Distribution, variation, and taxonomy of *Topaza* hummingbirds (Aves: Trochilidae). Ornitología Neotropical 11:123–142.

Nicholson, E. M. 1931. Field-notes on the Guiana king hummingbird. Ibis 13:534–553.

Ortiz-Crespo, F. I. 1972. A new method to separate immature and adult hummingbirds. Auk 89:851–857.

Piacentini, V. de Q. 2011. Taxonomia e distribuição geográfica dos representantes do gênero Phaethornis Swainson, 1827 (Aves: Trochilidae). Ph.D. dissertation, Universidade de São Paulo, São Paulo, Brasil (in Portuguese).

Pyle, P. 1997. Identification guide to North American birds, Part I. Slate Creek Press, Bolinas, CA.

Pyle, P. [online]. 2013. Appearance. Molts. Plumages. In S. T. Weidensaul, R. Robinson, R. R. Sargent, and M. B. Sargent, Ruby-throated Hummingbird (*Archilochus colubris*). In A. Poole (editor), The birds of North America online. Cornell Lab of Ornithology, Ithaca, NY. <https://birdsna.org>.

Pyle, P., A. Engilis, and D. A. Kelt. 2015. Manual for ageing and sexing birds of Bosque Fray Jorge National Park and Northcentral Chile, with notes on range and breeding seasonality. Special Publication of the Occasional Papers of the Museum of Natural Science, Louisiana State University, Baton Rouge, LA.

Restall, R. L., C. Rodner, and R. M. Lentino. 2006. Birds of northern South America: an identification guide. Yale University Press, New Haven, CT.

Ruschi, A. 1986. Aves do Brasil. Expressão e Cultura, Rio de Janeiro, Brasil (in Portuguese).

Russell, S. M., and R. O. Russell. 2001. The North American banders' manual for banding hummingbirds. North American Banding Council, Point Reyes, CA.

Schuchmann, K. L. 1999. Family Trochilidae (Hummingbirds). Pp. 468–680 in J. del Hoyo, A. Elliott, and D. A. Christie (editors), Handbook of the birds of the world: Barn-owls to Hummingbirds. Lynx Editions, Barcelona, Spain.

Schuchmann, K. L., G. Kirwan, and P. Boesman. [online]. 2013. Rufous-throated Hummingbird (*Amazilia sapphirina*). In J. del Hoyo, A. Elliott, J. Sargatal,

D. A. Christie, and E. de Juana (editors), Handbook of the birds of the world alive. Lynx Edicions, Barcelona, Spain. <http://www.hbw.com>.

Stiles, F. G. 1972. Age and sex determination of Rufous and Allen's Hummingbirds. Condor 74:25–32.

Stiles, F. G. 1995. Intraspecific and interspecific variation in molt patterns of some tropical hummingbirds. Auk 112:118–132.

Stiles, F. G., G. Kirwan, and P. Boesman. [online]. 2014. White-necked Jacobin (*Florisuga mellivora*). In J. del Hoyo, A. Elliott, J. Sargatal, D. A. Christie, and E. de Juana (editors), Handbook of the birds of the world alive. Lynx Edicions, Barcelona, Spain. <http://www.hbw.com>.

Stiles, F. G., and L. L. Wolf. 1974. A possible circannual molt rhythm in a tropical hummingbird. American Naturalist 108:341–354.

Stouffer, P. C., and R. O. Bierregaard. 1995. Effects of forest fragmentation on understory hummingbirds in Amazonian Brazil. Conservation Biology 9:1085–1094.

Telemes, E. J., J. S. Miller, J. L. Rifkin. 2010. Evolution of sexual dimorphism in bill size and shape of hermit hummingbirds (Phaethornithinae): a role for ecological causation. Philosophical Transactions of the Royal Society B: Biological Sciences 365:1053–1063.

Temeles, E. J., I. L. Pan, J. L. Brennan, and J. N. Horwitt. 2000. Evidence for ecological causation of sexual dimorphism in a hummingbird. Science 289:441–443.

Weller, A. A., G. Kirwan, and P. Boesman. [online]. 2013. Glittering-throated Emerald (*Amazilia fimbriata*). In J. del Hoyo, A. Elliott, J. Sargatal, D. A. Christie, and E. de Juana (editors). Handbook of the birds of the world alive. Lynx Edicions, Barcelona, Spain. <http://www.hbw.com>.

Weller, A. A., G. Kirwan, and P. Boesman. [online]. 2014. Versicolored Emerald (*Amazilia versicolor*). In J. del Hoyo, A. Elliott, J. Sargatal, D. A. Christie, and E. de Juana (editors), Handbook of the birds of the world alive. Lynx Edicions, Barcelona, Spain. <http://www.hbw.com>.

Wolfe, J. D., P. Pyle, and C. J. Ralph. 2009. Breeding seasons, molt patterns, and gender and age criteria for selected northeastern Costa Rican resident landbirds. Wilson Journal of Ornithology 121:556–567.

Yanega, G. M., P. Pyle, and G. R. Geupel. 1997. The timing and reliability of bill corrugations for ageing hummingbirds. Western Birds 28:13–18.

Zimmer, J. T. 1950. Studies of Peruvian Birds: the genera *Eutoxeres, Campylopterus, Eupetomena,* and *Florisuga*. American Museum Novitates 56:1–14.

# CHAPTER THIRTEEN

# Trogonidae (Trogons)

---

# Species in South America: 18

# Species Recorded at BDFFP: 5

# Species Captured at BDFFP: 3

Trogons are sexually dichromatic subcanopy omnivores that infrequently fall into mistnets. Males are brightly patterned metallic green or bluish-purple with a bright yellow or red belly. Females have duller upperparts with grays or browns and can be somewhat difficult to identify, but nonoverlapping measurements can help separate most syntopic species. All trogons have two toes oriented forward and two back (heterodactyl) and have very short, weak legs for their body size; their leg muscle to body mass ratio is 3%, the lowest of any known bird (Collar 2001). Holding birds in the "photo pose" should be avoided or only done with the utmost care. Like many passerines, trogons have 10 pp, 9 ss, and 12 rects.

The most common molt strategy is apparently the Complex Basic Strategy (Pyle 1997). Although a prealternate molt is suggested by Dickey and Van Rossem (1938) from two of three species they describe in El Salvador, this does not appear to be evident in the species at the BDFFP. Furthermore, a prealternate molt seems unlikely in this family; perhaps as a consequence of relatively low sample sizes, Dickey and Van Rossem (1938) may have confused protracted or adventitious body feather replacement with a prealternate molt. The juvenile plumage, including tail pattern, is female-like at least in Neotropical trogons, and secondary coverts have diagnostic buff-tipped feathers. Trogons typically have partial to incomplete FPF molts and often retain female-like juvenile rectrices, thus rectrix shape and pattern may be useful for aging with juvenile rects being more pointed (Collar 2001). Definitive prebasic molts are typically complete, but may sometimes be incomplete (Pyle 1997). Replacement patterns of flight feathers proceed distally from p1 to p10, and ss replacement also appears similar to passerines.

The skull probably does not completely ossify in most cases (Pyle 1997), but may be useful for aging; plumage characteristics are probably easier in most situations. In subtropical North America, only the female incubates eggs, thus males should not develop brood patches (Pyle 1997), but it appears that nests of most Neotropical species are tended by both parents (Collar 2001), thus brood patches may be evident in males and females, although they might be expected to be better developed in females. The breeding season in the Manaus region probably peaks during the dry season, as juveniles have only been captured at the BDFFP from Oct to Dec, but more study is needed.

## TABLE 13.1

Range of measurements of Trogon spp. at the BDFFP; T. melanurus is unknown.

| Species | Wing (mm) | Tail (mm) | Mass (g) |
|---|---|---|---|
| T. viridis | 137.0–147.0 | 160.0 | 76.9–84.3 |
| T. violaceus | 105.0–114.0 | 134.0–142.0 | 45.0–56.5 |
| T. rufus | 102.0–120.0 | 119.0–153.0 | 42.0–57.0 |

NOTE: These measurements are based on very small sample sizes (see Species Accounts) but reveal important differences in size.

## TABLE 13.2

Plumage characteristics useful in identification of "yellow-bellied" trogons at the BDFFP; ♂ and ♀ T. melanurus have red bellies.

| Species | ♂ Body | ♂ Undertail | ♀/Juv body | ♀/Juv undertail |
|---|---|---|---|---|
| T. viridis | Bluish-purple | Solid white | Dark gray | Barred white |
| T. violaceus | Bluish-purple | Barred white | Dark gray | Barred white |
| T. rufus | Green | Barred white | Brown | Barred white |

## SPECIES ACCOUNTS

### Trogon viridis (monotypic)

Green-backed Trogon • Surucuá-grande-de-barriga-amarela

Band Size: H

# Individuals Captured: 3

Wing     137.0–147.0 mm (141.0 ± 5.3 mm; n = 3)

Tail      160.0 mm (n = 2)

Mass    76.9–84.3 g (80.6 ± 5.2 g; n = 2)

          69.0–99.0 g (Collar 2001)

Similar species: ♀ from ♀ T. violaceus by larger size (Table 13.1) and by complete (not broken) eyering. ♂ from ♂ T. violaceus by larger size, blue (not yellow) complete eyering, and underside of tail solid (not barred) white (Table 13.2).

Skull: Probably does not always complete, but percentage may be useful in determining age (Pyle 1997).

Brood patch: Probably only ♀♀ develop BPs, but possible some ♂ develop less extensive BPs (Pyle 1997). Timing uncertain, but observed breeding at least during the dry season.

Sex: ♂ from ♀ by metallic blue (not dark gray) on head and back and underside of tail solid (not barred) white.

Molt: Group 3 (or rarely also Group 5?)/Complex Basic Strategy; FPF partial, DPBs incomplete(?)-complete. FPF may include most or all less, med, and gr covs, and possibly one or more terts. DPBs may rarely not complete as in other Trogons (Pyle 1997); look for one or a few retained ss. Timing unknown, but one SPB began molting in Aug.

| | |
|---|---|
| FCJ | Like adult ♀, but less gray and more brown overall and belly feathers are dull tawny-orange. Ss and med covs with broad, buffy bars, at least in birds of SE Brazil (LSUMZ). Pp covs and alula unpatterned and dark brown. Rects, especially r6, are more pointed and less squared-off than adults and the shape of the ss may be useful (Pyle 1997). ♂ may have less extensive white barring on the outer webbing of the outer rect than ♀. Skull relatively unossified. |
| FPF | With body and/or ss cov molt. Look for scattered adult-like feathers on the upperparts to determine sex as in adults. Skull relatively unossified. |
| FCF | Molt limits among the ss covs and/or ss. Look for buff-tipped covs and/or juvenile-like ss. ♂ dull, with ♀-like retained feathers. Skull variable. |

Rects retained from the juvenile plumage are worn, relatively narrow, and ♀-like in both sexes.

| | |
|---|---|
| SPB | Replacing juv flight feathers with adult flight feathers. May be difficult in ♀♀ to distinguish from DPB. Skull relatively ossified. |

One central rect replaced (adventitiously?; left) and with a mix of duller formative and more metallic blue body feathers (right).

| | |
|---|---|
| SCB | Like DCB, but with one or more retained juv ss. Skull relatively ossified. |

**DCB** Without molt limits among ss covs, ss, pp, and rects with unpatterned black pp covs and alula, bright yellow belly, and narrow white barring on gr, less, and med covs. Rects are relatively broad and truncate. Skull relatively ossified.

**DPB** Replacing adult flight feathers with adult flight feathers. May be difficult in ♀♀ to distinguish from SPB. Skull relatively ossified. TPB should be possible if adult feathers are replacing a mix of adult and juvenile flight feathers.

p1–p3 (and corresponding pp covs) and s9 replaced and other pp, pp covs, and ss retained, perhaps in suspension.

**SAB** Like DCB, but with one or more retained adult ss. Skull relatively ossified.

s6 retained; note many gr covs are missing, but it was not molting (adventitious replacement). (Courtesy of Angelica Hernández Palma.)

*Trogon violaceus* (monotypic)                                            Band Size: G, H?

Guianan Trogon • Surucuá-violáceo                                         # Individuals Captured: 6

Wing       105.0–116.0 mm (112.0 ± 4.2 mm; n = 5)
           115–125 mm (n = 6; Trinidad and Tobago, Junge and Mees 1961)

Tail       134.0–142.0 mm (137.8 ± 3.3 mm; n = 4)
           112–126 mm (n = 6; Trinidad and Tobago, Junge and Mees 1961)

Mass       45.0–56.5 g (49.1 ± 4.0 g; n = 6)
           38.0–57.0 g (Collar 1999)
           48.5–53.0 g (n = 6; Trinidad and Tobago; Junge and Mees 1961)

Bill       11.0 mm (n = 1)

Tarsus     13.9 mm (n = 1)

Similar species: From *T. viridis* by small size (Table 13.1). ♂ from *T. viridis* also by yellowish (not whitish-blue eyering) and ♀ from *T. viridis* by incomplete eyering (note that eyering color is the same as in ♀ *T. viridis*). ♂ from ♂ *T. rufus* by bluish-purple (not green) metallic plumage on head and back, and ♀ from ♀ *T. rufus* by dark gray (not rufous) plumage on head and back (Table 13.2).

Skull: Probably does not always complete, but percentage may be useful in determining age (Pyle 1997).

Brood patch: Probably only ♀♀ develop BPs, but possible some ♂♂ develop less extensive BPs (Pyle 1997). Timing unknown.

Sex: ♂ from ♀ by metallic bluish-purple (not dark gray) on head and back.

Molt: Group 3 (or rarely also Group 5?)/Complex Basic Strategy; FPF partial-incomplete, DPBs incomplete(?)-complete. FPF may include most or all less, med, and gr covs, and possibly one or more terts. DPBs may rarely not complete as in other Trogons (Pyle 1997); look for one or a few retained ss. PAs unknown, but perhaps limited as in closely related *T. violaceus sallaei* (Dickey and Van Rossem 1938), although there is no evidence of this at the BDFFP. Timing unknown, but one DPB captured in middle stages of molt in Oct.

Notes: Birds at the BDFFP seem to have relatively long tails for their body compared to other published accounts (e.g., Trinidad and Tobago; Junge and Mees 1961).

FCJ        Unclear given recent splitting of subspecies, but probably like closely related *T. ramonianus* and/or *T. caligatus*. Therefore, probably like adult ♀, but less gray and more brown overall. Ss and med covs probably with broad, buffy bars are more pointed and less squared-off than adults and the shape of the ss may be useful (Pyle 1997). Skull relatively unossified.

FPF        With body and/or ss cov molt. Look for scattered adult-like feathers on the upperparts to determine sex as in adults. Skull relatively unossified.

FCF        Molt limits among the ss covs and/or ss. ♂ with female-like retained feathers. Skull variable.

SPB        Replacing juv flight feathers with adult flight feathers. May be difficult in ♀♀ to distinguish from DPB. Skull relatively ossified.

SCB        Like DCB, but with one or more retained juv ss. Skull relatively ossified.

DCB        Without molt limits among ss covs and/or ss. Pp covs and alulas in both sexes are unmarked black. Skull relatively ossified. Or with molt limits in which the age of retained feathers cannot be determined.

DPB        Replacing adult flight feathers with adult flight feathers. May be difficult in ♀♀ to distinguish from SPB. Skull relatively ossified. TPB should be possible if adult feathers are replacing a mix of adult and juvenile flight feathers.

SAB        Like DCB, but with one or more retained adult ss. Skull relatively ossified.

*Trogon rufus rufus*

Black-throated Trogon • Surucuá-de-barriga-amarela

Band Size: G

# Individuals Captured: 54

| | | |
|---|---|---|
| Wing | ♂ | 108.0–117.0 mm (110.1 ± 3.3 mm; n = 7) |
| | ♀ | 102.0–120.0 mm (110.6 ± 4.6 mm; n = 24) |
| Tail | ♂ | 119.0–143.0 mm (134.1 ± 9.0 mm; n = 9) |
| | ♀ | 121.0–152.0 mm (136.3 ± 8.4 mm; n = 24) |
| Mass | ♂ | 46.0–53.0 g (48.3 ± 2.5 g; n = 11) |
| | ♀ | 42.0–57.0 g (50.5 ± 3.3 g; n = 32) |
| Bill | ♂ | No data at the BDFFP |
| | ♀ | 11.6–12.2 mm (11.9 ± 0.4 mm; n = 2) |
| Tarsus | ♂ | No data at the BDFFP |
| | ♀ | 15.3–15.7 mm (15.5 ± 0.3 mm; n = 2) |

Similar species: Brown-backed ♀ not likely confused at the BDFFP. ♂ from *T. viridis* and *T. violaceus* by green (not blue) metallic plumage on head and from *T. viridis* by small size (Tables 13.1 and 13.2).

Skull: Probably does not always complete (n = 7), but percentage may be useful in determining age (Pyle 1997).

Brood patch: Probably only ♀♀ develop BPs, but possibly some ♂ develop less extensive BPs (Pyle 1997). One BP was observed in a ♀ in Aug, and, with the timing of molt, suggests an early to mid dry season breeding schedule.

Sex: ♂ from ♀ by metallic green (not rufous) on head and back.

Molt: Group 3 (or rarely also Group 5?)/Complex Basic Strategy; FPF probably partial-incomplete, DPBs incomplete(?)-complete. FPF may include most or all less, med, and gr covs, and possibly one or more terts. DPBs may rarely not complete as in other Trogons (Pyle 1997); look for one or a few retained ss. Molt appears to begin during the mid to late dry season (Oct–Dec) and finish in the late wet season (May or Jun), but there are few data available.

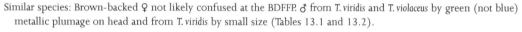

FCJ    Like adult ♀, but without intricate vermiculations to ss covs and instead tipped with buff. A ♂ had bronzy-green coloration toward the base of its juvenile central rects, a feature of adult *T. r. amazonicus*; it is unclear if this is the typical juvenile pattern of *T. r. rufus*, or if there is some variation at the BDFFP indicating blending with *T. r. amazonicus* of the lower Amazon basin. Most, if not all adult ♀♀ at the BDFFP lack this color, suggesting it may be limited to FCJ, and perhaps only ♂♂.

FPF    With body and/or ss cov molt, replacing
       buff-tipped ss covs with adult-like finely
       patterned ss covs. Look for scattered
       adult-like feathers on the upperparts to
       determine sex as in adults. Skull relatively
       unossified.

Less and most med covs replaced or missing with two
juvenile med covs and a few outer juvenile gr covs not
yet replaced. Some terts and inner ss missing, but may
have resulted from adventitious loss.

FCF    Molt limits among the ss covs and/or ss. Skull variable.

SPB    Replacing juvenile flight feathers with adult flight feathers. May be difficult in ♀♀ to distinguish from DPB.
       Skull relatively ossified.

A mix of brownish formative body feathers (left); p1–p5 and corresponding pp covs new and definitive basic,
p6 missing, p7–p10 and corresponding pp covs juvenile, s1 growing, s2–s6 juvenile and notably brown,
s7 growing, s8 and s9 new and definitive basic.

SCB    Like DCB, but with one or more retained juv ss. Skull relatively ossified.

DCB      Without molt limits among ss covs and/or ss. Skull relatively ossified.

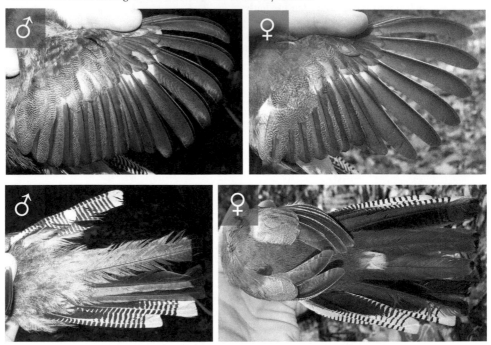

DPB      Replacing adult flight feathers with adult flight feathers. May be difficult in ♀♀ to distinguish from SPB. Skull relatively ossified. TPB should be possible if adult feathers are replacing a mix of adult and juvenile flight feathers.

SAB      Like DCB, but with one or more retained adult ss. Skull relatively ossified.

---

## LITERATURE CITED

Collar, N. [online]. 1999. Violaceous Trogon (*Trogon violaceus*). In J. del Hoyo, A. Elliott, J. Sargatal, D. A. Christie, and E. de Juana (editors), Handbook of the birds of the world alive. Lynx Edicions, Barcelona, Spain. <http://www.hbw.com>.

Collar, N. [online]. 2001. Green-backed Trogon (*Trogon viridis*). In J. del Hoyo, A. Elliott, J. Sargatal, D. A. Christie, and E. de Juana (editors), Handbook of the birds of the world alive. Lynx Edicions, Barcelona, Spain. <http://www.hbw.com>.

Dickey, D. R., and A. J. Van Rossem. 1938. The birds of El Salvador. Field Museum of Natural History, Zoological Series, Chicago, IL.

Junge, G. C. A., and G. F. Mees. 1961. The avifauna of Trinidad and Tobago. E. J. Brill, Leiden, Netherlands.

Pyle, P. 1997. Identification guide to North American birds, Part I. Slate Creek Press, Bolinas, CA.

Ryder, T. B., and J. D. Wolfe. 2009. The current state of knowledge on molt and plumage sequences in selected Neotropical bird families: a review. Ornitología Neotropical 20:1–18.

# CHAPTER FOURTEEN

# Alcedinidae (Kingfishers)

# Species in South America: 6

# Species recorded at BDFFP: 5

# Species captured at BDFFP: 2

The aptly named kingfishers are an entertaining group of birds, as they use their long and disproportionately large and heavy bills to dive into rivers, ponds, lakes, and marshes after minnows and aquatic invertebrates, although some, particularly Old World taxa, have adapted to drier lands and eat a variety of insects, reptiles, and small mammals. Many are colorful, ranging from electric blues and greens, to reds, oranges, browns, and striking black and whites. Many are sexually dichromatic and, in New World species (subfamily Cerylinae), females will have an extra color band across the chest. Kingfishers have zygodactylous feet on short legs, with two toes pointed forward and two pointed backward (Clement 2009). Diversity peaks in Australasia, whereas only six species are present in the New World. Kingfishers have 10 pp, 13–15 ss, and 12 rects (Pyle 1997).

Kingfishers may follow a Complex Basic Strategy, although a Simple Basic Strategy has been suggested as a possible alternative (P. Pyle, pers. comm.). The three subfamilies (sometimes each recognized as full families) utilize slightly different molting sequences (Howell 2010, Radley et al. 2011, Pyle et al. 2016), although here we focus on the New World group. In *Megaceryle alcyon* pp molt has two waves, one starting at p1 progressing sequentially outward, and a second starting at p7 progressing sequentially inward and outward simultaneously (Pyle 1995, 1997) or perhaps alternatively starting at p4 (Stone 1896). Similar patterns should be looked at for other *Chloroceryle*, and we noted a single starting node of p6 or p4 in both C. *inda*

and C. *aenea*, progressing bidirectionally, without a corresponding p1 node. Secondary replacement in *Chloroceryle* appears to have multiple nodes, including s1 and proceeding distally, as well as one or two nodes among the inner five secondaries. DPB molts may regularly be incomplete retaining some secondaries and/or primary coverts, especially in *Megaceryle*, but are probably more often complete in *Chloroceryle* (Pyle 1995, 1997). For incomplete replacement patterns, it may be difficult to determine whether retained feathers were juvenile or adult, making it challenging to distinguish between SCB and SAB (Pyle 1995, 1997; Howell 2010). At least in *Chloroceryle*, it appears that retained juvenile secondaries are longer than replaced feathers in SCB, whereas retained feathers the same length as those replaced should be SAB. The juvenile plumage is often similar to that of adults, but less brilliant and, in species with bands or "belts" across the front, these are typically reduced in intensity and size in females and faintly present in males. Preformative molts in M. *alcyon* and C. *americana* have been reported to be incomplete and eccentric, and relatively extensive, perhaps following the DPB sequence then arresting (Pyle 1995, 1997; Wolfe et al. 2009). However, an alternative interpretation could be that because the juvenile plumage is structurally sound, kingfishers may instead have an absent or limited FPF, not including flight feathers, and what has been interpreted as the FPF is instead the SPB (P. Pyle, pers. comm.). This may be more consistent with molting strategies of some Old World kingfishers (Radley et al. 2011, Pyle et al. 2016, but see Cramp 1985). In an examination of 39 C. *inda* and 21 C. *aenea* specimens (LSUMZ), we found only one obviously juvenile bird in molt (among 15 with juvenile-like chest patterns): a young male *C. inda* with a relatively ossified skull (75%) and moderate wear of the retained flight feathers that was more consistent with a set of 1-year-old feathers being replaced during the SPB than relatively fresh juvenile feathers as would have been expected in an FPF. Only one of the other 14 specimens was reported to have body molt, but this may have been finishing the FPJ. As such, we provisionally consider *Chloroceryle* as following a Simple Basic Strategy.

In general, incubation is shared by the sexes and both probably develop similar brood patches, but cloacal protuberances are poorly developed in males (Pyle 1997). Skull ossification is rarely complete (Pyle 1997, LSUMZ), but like in other near-passerines, the extent of ossification may be useful for aging.

## SPECIES ACCOUNTS

### *Chloroceryle inda* (monotypic)

Green-and-rufous Kingfisher • Martim-pescador-da-mata

Band Size: G

\# Individuals Captured: 42

| | | |
|---|---|---|
| Wing | ♂ | 87.0–98.0 mm (93.3 ± 2.8 mm; n = 18) |
| | ♀ | 91.0–101.5 mm (95.4 ± 2.5 mm; n = 20) |
| Tail | ♂ | 53.0–67.0 mm (61.7 ± 3.6 mm; n = 14) |
| | ♀ | 56.0–64.5 mm (61.7 ± 2.3 mm; n = 18) |
| Mass | ♂ | 43.0–55.0 g (49.4 ± 3.6 g; n = 18) |
| | ♀ | 45.0–61.0 g (53.5 ± 4.8 g; n = 20) |

Similar species: From other *Chloroceryle* by large size and rufous belly.

Skull: Probably does not complete, but percentage may be useful in determining age (Pyle 1997).

Brood patch: May occur in both sexes. Timing uncertain, but may occur during late wet and early dry season, with a juvenile caught in Oct.

Sex: ♂ lacks green band across chest.

Molt: Group 1 (or rarely Group 2)/Simple Basic Strategy(?); FPF probably absent (or limited?), DPBs complete (or rarely incomplete). DPBs may regularly be complete, but rarely could retain 1 or 2 pp and/or ss. Timing of molt uncertain with one bird each seen finishing molt in Jul, Aug, and Nov, but 20 other birds captured were not molting including during the span of Jul–Nov, which may be explained by DPBs progressing rapidly and/or substantially decreased capture probability during molt.

(LSUMZ 172814.)

FCJ     Adult-like in many respects, but note reduced chest band in ♀♀ chest, lacking white, and some light green spotting across the chest in ♂♂. Also perhaps with reduced spotting in ss covs. Rects average broader than DCB. Skull is variably ossified.

(LSUMZ 50966.)

(LSUMZ 119400.)

| SPB | With flight feather and body molt. Retained ss may be slightly longer and browner than in DPB. Skull is relatively ossified. |
|---|---|

s1–s3 new, s4 retained juvenile, s5 molting, s6–s13 new, p10 molting, p1–p9 and corresponding pp covs new.

| DCB | Without molt limits among the ss, pp, or pp covs. Rects average less broad than FCJ. ♂♂ lack any green fleeking in chest, whereas ♀♀ are boldly banded with green and white. Skull is relatively ossified. Note that SCB and SAB may be possible where incomplete molts result in retained juvenile or adult-like feathers, respectively. |
|---|---|

| DPB | With mix of retained and replaced adult-like feathers, even in length across especially the ss. Skull is relatively ossified. |
|---|---|

p6 molting (symmetrical), all other pp and ss retained. (LSUMZ 172814.)

*Chloroceryle aenea aenea*

American Pygmy Kingfisher • Martinho

Band Size: D

# Individuals Captured: 27

| Wing | ♂ | 52.5–57.0 mm (54.4 ± 1.7 mm; n = 7) |
| | ♀ | 52.0–59.0 mm (55.2 ± 1.9 mm; n = 11) |
| Tail | ♂ | 30.0–35.0 mm (32.1 ± 1.9 mm; n = 5) |
| | ♀ | 28.0–35.0 mm (32.7 ± 2.1 mm; n = 10) |
| Mass | ♂ | 11.0–15.0 g (12.8 ± 1.3 g; n = 10) |
| | ♀ | 12.0–16.8 g (14.2 ± 1.3 g; n = 14) |
| Bill | ♂ | 23.7–24.4 mm (24.1 ± 0.5 mm; n = 2) |
| | ♀ | 24.6–26.0 mm (25.3 ± 1.0 mm; n = 2) |
| Tarsus | ♂ | 9.5 mm (n = 1) |
| | ♀ | 8.7–11.0 mm (9.9 ± 1.6 mm; n = 2) |

Similar species: From other *Chloroceryle* by small size and rufous and white belly.

Skull: Probably does not complete, but percentage may be useful in determining age (Pyle 1997).

Brood patch: May occur in both sexes. Timing unknown, but three juveniles captured in Jun.

Sex: ♂ lacks green band across chest.

Molt: Group 1 (or rarely Group 2)/Simple Basic Strategy(?); FPF probably absent (or limited?), DPBs complete (or rarely incomplete). DPBs may regularly be complete, but rarely could retain 1 or 2 pp and/or ss. One bird each was seen finishing molt in Aug, Sep, and Oct, and all other captures not seen molting, including 10 birds captured during these same 3 months. Suggests a rapid molt, variable timing, and/or substantially decreased capture probability during molt.

FCJ  Like adult, but with paler underparts. ♀ with an incomplete breast band. ♂ with streaks on belly (Wolfe et al. 2009). Rects average broader than DCB. Skull relatively unossified.

SPB  With flight feather and body molt. Retained ss may be slightly longer and browner than replaced ss. Skull is relatively unossified.

DCB     Without molt limits among the ss, pp, or pp covs. Rects average less broad than FCJ. Skull is relatively ossified. Note that SCB and SAB may be possible where incomplete molts result in retained juvenile or adult-like feathers, respectively.

DPB     With mix of retained and replaced adult-like feathers, even in length across especially the ss. Skull is relatively ossified.

p8–p10 retained and relatively fresh, p7 molting, p5–p6 new, p1–p4 old, s1 new, s2–s7 retained, s8 missing, s9 molting, s10–s11 new, s12 missing, s13 new. (LSUMZ 114731.)

## LITERATURE CITED

Clement, P. 2009. Kingfishers. Pp. 164–166 in T. Harris (editor), National Geographic complete birds of the world. National Geographic Society, Washington, DC.

Cramp, S. 1985. Birds of the Western Palearctic (Vol. 4). Oxford University Press, Oxford, UK.

Howell, S. N. G. 2010. Molt in North American birds. Houghton Mifflin Harcourt, Boston, MA.

Pyle, P. 1995. Incomplete flight feather molt and age in certain North American non-passerines. North American Bird Bander 15:15–26.

Pyle, P. 1997. Identification guide to North American birds, Part I. Slate Creek Press, Bolinas, CA.

Pyle, P., K. Tranquillo, K. Kayano, and N. Arcilla. 2016. Molt patterns, age criteria, and molt-breeding dynamics in American Samoan landbirds. Wilson Journal of Ornithology 128:56–69.

Radley, P., A. L. Crary, J. Bradley, C. Carter, and P. Pyle. 2011. Molt patterns, biometrics, and age and gender classification of landbirds on Saipan, Northern Mariana Islands. Wilson Journal of Ornithology 123:588–594

Stone, W. 1896. The molting of birds with special reference to the plumages of smaller land birds of eastern North America. Proceedings of the Academy of Natural Sciences of Philadelphia 48:108–167.

Wolfe, J. D., P. Pyle, and C. J. Ralph. 2009. Breeding seasons, molt patterns, and gender and age criteria for selected northeastern Costa Rican resident landbirds. Wilson Journal of Ornithology 121:556–567.

# CHAPTER FIFTEEN

# Momotidae (Motmots)

# Species in South America: 8

# Species recorded at BDFFP: 1

# Species captured at BDFFP: 1

Motmots are colorful birds limited to Neotropical woodlands and are known for their iridescent blue and green facial patterns and racquet-shaped tails, formed by the breaking of weakly attached barbs. Their strong bills with serrated edges are capable of grabbing a variety of insects, arthropods, and small vertebrates, as well as some fruit. They have syndactylous toes (the outer two are partially fused), presumably providing an advantage for digging burrows in which they nest (Chartier 2009). Although monochromatic, males in at least one species (*Eumomota superciliosa*) average larger and slightly longer tailed for their body size, and have tail wires that average 10% longer, but these characteristics overlap between sexes (Murphy 2008). Motmots have 10 pp, 11 ss, and 10–12 rects (Murie 1872, Wetmore 1968).

Molt in motmots is not well understood, but it is believed to follow a Complex Basic Strategy. The preformative molt can be limited to partial in some species (Dickey and Van Rossem 1938, Ryder and Wolfe 2009), but can be extremely cryptic, as the adult body plumage and ss covs are relatively soft, thus difficult to distinguish from juvenile feathers. Juvenile plumages of motmots are like that of the adult, but duller, and in some species with more brownish tones to flight feathers. Molt appears to proceed distally from p1, proximally from s1, and bidirectionally from s10.

Both sexes incubate and should therefore develop brood patches. Surprisingly for such a large nonpasserine, skulls may completely ossify in at least some species.

## SPECIES ACCOUNT

*Momotus momota momota*

Amazonian Motmot • Udu-de-coroa-azul

Band Size: L, H

# Individuals Captured: 248

| | |
|---|---|
| Wing | 131.0–152.0 mm (141.5 ± 4.1 mm; n = 124) |
| Tail | 219.0–275.0 mm (247.3 ± 13.8 mm; n = 101) |
| Mass | 95.0–175.0 g (133.7 ± 14.2 g; n = 206) |
| Bill | 25.3–32.4 mm (28.9 ± 2.0 mm; n = 14) |
| Tarsus | 30.5–37.0 mm (34.6 ± 1.5 mm; n = 15) |

Similar species: None.

Skull: Apparently completes (n = 24), but small windows may persist that are hard to see.

Brood patch: Apparently both sexes incubate, but not known if ♂♂ develop less extensive brood patches than ♀♀. BPs observed in Dec and also late Mar through Jun. One fledgling observed in Feb and one FPF observed in Aug, suggesting a mainly wet season breeding schedule.

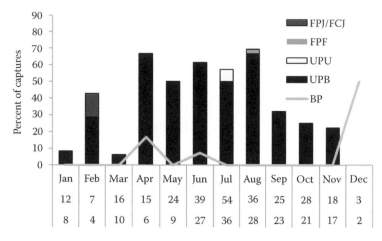

| | Jan | Feb | Mar | Apr | May | Jun | Jul | Aug | Sep | Oct | Nov | Dec |
|---|---|---|---|---|---|---|---|---|---|---|---|---|
| | 12 | 7 | 16 | 15 | 24 | 39 | 54 | 36 | 25 | 28 | 18 | 3 |
| | 8 | 4 | 10 | 6 | 9 | 27 | 36 | 28 | 23 | 21 | 17 | 2 |

Sex: ♀ = ♂; racket length and color apparently do not vary by sex (Stiles 2009).

Molt: Group 3/Complex Basic Strategy; FPF partial, DPBs complete. FPF may include 5 to all(?) gr covs, and perhaps one or more med covs can occasionally be retained; resulting molt limits can be cryptic to see. Some DPB molts may rarely be incomplete; look for retained feathers among s3–s7. Note that freshly molted tail feathers lack rackets in all cycles. Molt appears to peak Apr–Aug, but some birds appear to continue into at least Nov.

FCJ      Plumage like adult, but perhaps duller with more bluish and less bright green tones on the ss covs and flight feathers. Rects browner, especially toward the bases. Skull is unossified.

FPF      With body molt. Skull is unossified.

FCF      With molt limits among the gr covs (or perhaps between the gr covs and pp covs) and with retained juvenile rects that are brownish toward the base. Otherwise very much like DCB. Skull is unossified or ossified.

All rects retained juvenile (left); outer 3 gr covs retained (right).

SPB      Wing molt replacing juvenile and bluish flight feathers (caution: this may be hard to determine from DPB and more study is needed; age as UPB when uncertain). Skull is ossified.

DCB      Without molt limits among the gr covs or between the gr covs and pp covs (caution: this can be difficult to see). Skull is ossified.

DPB      With wing molt replacing adult-like and greenish flight feathers (especially ss) with greenish flight feathers and greenish rects with greenish rects. Skull is ossified.

p8–p10 old, p7 missing, p1–p6 new; s1,s2 growing, s3–s6 old, s7 growing, s8–s11 new.

## LITERATURE CITED

Chartier, A. 2009. Motmots. Pp. 168 in T. Harris (editor), National Geographic complete birds of the world. National Geographic Society, Washington, DC.

Dickey, D. R., and A. J. Van Rossem. 1938. The birds of El Salvador. Field Museum of Natural History, Zoological Series, Chicago, IL.

Murie, J. 1872. On the motmots and their affinities. Ibis 14:383–412.

Murphy, T. G. 2008. Lack of assortative mating for tail, body size, or condition in the elaborate monomorphic Turquoise-browed Motmot (*Eumomota superciliosa*). Auk 125:11–19.

Ryder, T. B., and J. D. Wolfe. 2009. The current state of knowledge on molt and plumage sequences in selected Neotropical bird families: a review. Ornitología Neotropical 20:1–18.

Stiles, F. G. 2009. A review of the genus *Momotus* (Coraciiformes: Momotidae) in northern South America and adjacent areas. Ornitología Colombiana 8:29–75.

Wetmore, A. 1968. Momotidae in the birds of the Republic of Panama, part II. Smithsonian Miscellaneous Collections 150:437–455.

# CHAPTER SIXTEEN

# Galbulidae (Jacamars)

# Species in South America: 18

# Species recorded at BDFFP: 4

# Species captured at BDFFP: 3

Jacamars are a colorful group of birds only found in Neotropical forests and woodlands, often garbed in metallic green with varying amounts of rufous or white. They are impressive sally feeders, dashing out from branches where they otherwise inconspicuously sit to catch flying insects (Chartier 2009). Their feet are zygodactylous with toes 2 and 3 pointed forward and toes 1 and 4 pointed backward, like in woodpeckers and other Piciformes. Jacamars have 10 pp (the 10th reduced), 10–11? ss, and 12 rects (r6 often reduced).

Very little is known or has been described about jacamar molt strategies, although they probably follow a Complex Basic Strategy, lacking prealternate molts. Like puffbirds, jacamars may have two nodes of primary flight feather replacement, both of which proceed distally (Haffer 1968), although we have not noted this in our banding records or specimen examinations. Juvenile plumages are similar to adults, being metallic and brightly colored, but the plumage is even more loosely textured and it probably takes several weeks or months for their bills to reach the adult size, which can aid in aging (Figure 16.1). The extent of preformative molts remains uncertain, but may regularly be incomplete based on our banding records and a review of specimens as a consistent pattern emerges across taxa (at least *Jacamerops* and *Galbula*); up to five inner primaries, up to six inner secondaries, and up to all rects can be replaced. Alternatively, this could be viewed as a suspension limit before preformative molts (and definitive prebasic molts) complete; however, given the consistency and frequency with which this pattern emerges, we, for now, treat this as an incomplete preformative molt. It is possible during the DPB that one or more ss and/or rects can be retained occasionally, allowing for more specific aging codes (i.e., SCB and SAB) than we suggest in the following species accounts.

Jacamars nest in burrows and, like other cavity and burrow nesters at the BDFFP, jacamars may often nest during the late wet and early dry seasons. In other studies of jacamars, birds have been observed breeding during a dry season (Skutch 1937, 1963; Tobias and Seddon 2003). Many jacamars are sexually dichromatic, with males having a white patch in the throat or chest, and, in many cases, this dichromatism is apparent in the juvenile plumage. Apparently both sexes incubate, often the female at night and both sexes taking turns during the day (Skutch 1937, 1963; Tobias et al. 2002). At the BDFFP only females have been observed with developed brood patches; it is possible that males develop less extensive brood patches given their less intensive incubation schedule. Skulls may regularly not complete with at least small windows (<3 mm) persisting for life.

**Figure 16.1.** The juvenile (foreground) has a more loosely textured body plumage than the nonjuvenile (background). Also note differences in plumage color intensity and bill length. *Galbula albirostris* is shown here, but similar juvenile criteria have been observed in other Galbulidae.

## SPECIES ACCOUNTS

### *Galbula albirostris albirostris*

Yellow-billed Jacamar • Ariramba-de-bico-amarelo

Band Size: E
# Individuals Captured: 428

| Wing | ♂ | 63.0–76.0 mm |
| | | (70.3 ± 2.3 mm; n = 135) |
| | ♀ | 63.0–73.5 mm |
| | | (68.4 ± 2.1 mm; n = 106) |
| Tail | ♂ | 56.0–73.0 mm |
| | | (66.2 ± 2.9 mm; n = 138) |
| | ♀ | 56.0–77.0 mm |
| | | (64.0 ± 3.3 mm; n = 101) |
| Mass | ♂ | 15.0–22.0 g |
| | | (18.0 ± 1.2 g; n = 199) |
| | ♀ | 14.0–22.0 g |
| | | (17.8 ± 1.5 g; n = 141) |
| Bill | ♂ | 26.6–33.7 mm |
| | | (30.3 ± 2.1 mm; n = 24) |
| | ♀ | 25.3–30.0 mm |
| | | (28.5 ± 1.4 mm; n = 16) |
| Tarsus | ♂ | 12.2–15.8 mm |
| | | (14.3 ± 0.9 mm; n = 19) |
| | ♀ | 12.4–15.2 mm |
| | | (13.8 ± 1.0 mm; n = 13) |

Similar species: None at the BDFFP.

Skull: Usually does not complete with small (<5 mm) openings (n = 107).

Brood patch: From the banding data, it breeds mainly during the late wet and early dry season, from Apr–Aug, and perhaps into Nov, but there exists at least one Dec record (Oniki and Willis 1982). Only ♀♀ have been observed with BPs.

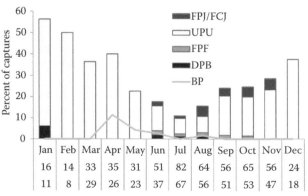

| | Jan | Feb | Mar | Apr | May | Jun | Jul | Aug | Sep | Oct | Nov | Dec |
|---|---|---|---|---|---|---|---|---|---|---|---|---|
| | 16 | 14 | 33 | 35 | 31 | 51 | 82 | 64 | 56 | 65 | 56 | 24 |
| | 11 | 8 | 29 | 26 | 23 | 37 | 67 | 56 | 51 | 53 | 47 | 18 |

Sex: ♂ has a white patch in the throat, which the ♀ lacks.

Molt: Group 3/Complex Basic Strategy; FPF partial-incomplete, DPBs complete. The FPF includes the replacement of most to all gr covs (rarely up to 5 outer gr covs retained), 0–5 inner pp, 3–5 (or occasionally more?) inner ss, and 0–12 rects. Molt peaks from Dec–Apr.

FCJ      Previously undescribed. Adult-like in many respects, but duller orange below, more loosely textured body feathers and ss covs, and often with relatively short bill. Pp covs and flight feathers relatively dull green with slightly reduced metallic sheen. Sex criteria as in subsequent plumages. Skull is relatively unossified.

FPF      Relatively fresh, but dull and nonmetallic juvenile body, inner pp, inner ss, and/or rects replaced by bright, metallic adult-like feathers. Skull is relatively unossified.

p1–p3 new, p4 in molt, p5–p10 retained juvenile.

FCF      With molt limits among the gr covs, pp, ss, and/or rects. Skull is variably ossified.

p1–p4 new, p5–p10 retained juvenile; s8–s11 new, s1–s7 retained juvenile. Note the difference in contrast between replaced and retained feathers compared to FPF; this is an artifact of lighting.

SPB      Mix of retained juvenile and adult-like (formative) flight feathers being replaced by adult-like feathers. Skull relatively ossified.

DCB    Without molt limits among the gr covs, ss, pp, and rects. Skull relatively ossified.

DPB    Adult-like body, tail, and flight feathers being replaced by adult-like feathers. Skull is relatively ossified.

## *Galbula dea dea*

Paradise Jacamar • Ariramba-de-paraíso

Band Size: Unknown

# Individuals Captured: 10

Wing    81.5–91.0 mm (87.4 ± 2.9 mm; n = 9)

Tail    149.0–160.0 mm (150.8 ± 6.5 mm; n = 7)

Mass    26.0–32.0 g (29.5 ± 2.3 g; n = 10)

Similar species: None.

Skull: Does not fully ossify as in *G. albirostris* (n = 8, BDFFP; n = 11, LSUMZ).

Brood patch: Unknown if both sexes develop BPs. Timing unknown.

Sex: ♂ = ♀.

Molt: Group 3/Complex Basic Strategy; FPF partial(?)-incomplete, DPBs complete. Based on molt limits in three specimens of *G. d. amazonus* from Bolivia (LSUMZ), the FPF at least includes the replacement of all gr covs (rarely up to 5 outer gr covs retained?), 0–4 inner pp (or rarely more?), 2–6 inner ss, and 8–12 rects (or rarely fewer?). Timing unknown, but one bird captured in wing molt in Nov and a specimen (LSUMZ) captured in wing molt in Apr.

FCJ    Adult-like, but less glossy/metallic blue and with bill relatively short. Tail also perhaps relatively short and mostly or entirely lacking whitish tips to outer rects. Skull is relatively unossified.

FPF    Replacing relatively fresh juvenile body feathers, inner ss, and inner pp with adult-like feathers. Skull is relatively unossified.

FCF    With molt limits among the gr covs, pp, ss, and/or rects. Skull is variably ossified.

SPB    Mix of retained juvenile and adult-like (formative) flight feathers being replaced by adult-like feathers. Skull is relatively ossified.

DCB    Without molt limits among the gr covs, ss, pp, and rects. Skull is relatively ossified.

DPB    Adult-like body, tail, and flight feathers being replaced by adult-like feathers. Skull is relatively ossified.

*Jacamerops aureus ridgwayi*

Great Jacamar • Jacamaraçu

| Wing | ♂ | 106.0–118.0 mm |
|---|---|---|
| | | (111.9 ± 4.3 mm; n = 8) |
| | ♀ | 111.0–114.0 mm |
| | | (112.0 ± 2.0 mm; n = 3) |
| Tail | ♂ | 121.0–137.0 mm |
| | | (129.6 ± 6.1 mm; n = 7) |
| | ♀ | 122.0–127.0 mm |
| | | (124.7 ± 2.5 mm; n = 3) |
| Mass | ♂ | 59.0–74.5 g |
| | | (66.4 ± 5.1 g; n = 8) |
| | ♀ | 51.5–73.0 g |
| | | (62.4 ± 8.8 g; n = 4) |
| Bill | ♂ | 39.5 mm (n = 1) |
| | ♀ | No data at the BDFFP |

Similar species: Larger than other Galbulidae. Superficially similar to large kingfishers, like *Chloroceryle inda*, but without spots in wing and tail and a more curved bill.

Skull: Usually does not fully ossify (n = 4, BDFFP; n = 15, LSUMZ), with DCB retaining small (<3 mm) openings.

Brood patch: Unknown if both sexes develop BPs. One BP was observed in a ♀ in Sep.

Sex: ♂ has a white patch in the throat, which the ♀ lacks.

Molt: Group 3/Complex Basic Strategy; FPF partial(?)-incomplete, DPBs complete. Based on one FCF capture and four FPF/FCF of 15 specimens examined from Peru, Panama, and Costa Rica (LSUMZ), the FPF includes the replacement of all gr covs, 2–5 inner pp (or sometimes none?), 3–6 inner ss, and 0–12 rects. One bird with wing molt was captured in Oct, but otherwise the timing is not known.

FCJ      Adult-like, but less glossy/metallic blue and copper with bill and perhaps tail relatively short. Skull is relatively unossified.

FPF      With body molt, inner ss molt, and inner pp molt replacing relatively fresh juvenile, but dull metallic feathers. Skull is relatively unossified.

FCF      With molt limits among the pp, ss, and/or rects. Skull is variably ossified.

p1–p5 new, p6–p10 retained juvenile; s8–s11 new, s1–s7 retained juvenile.

| SPB | With mix of retained juvenile and adult-like flight feathers being replaced by adult-like feathers. Skull is relatively ossified. |
| DCB | Without molt limits among the ss, pp, or rects. Skull is relatively ossified. |
| DPB | Adult-like body, tail, and flight feathers being replaced by adult-like feathers. Skull is relatively ossified. |

## LITERATURE CITED

Chartier, A. 2009. Jacamars. Pp. 185–186 in T. Harris (editor), National Geographic complete birds of the world. National Geographic Society, Washington, DC.

Haffer, J. 1968. Über die Flügel und Schwanzmauser columbianischer Piciformes. Journal für Ornithologie 109:157–171 (in German).

Oniki, Y., and E. O. Willis. 1982. Breeding records of birds from Manaus, Brazil, part II, Apodidae to Furnariidae. Revista Brasileira de Biologia 42:745–752.

Skutch, A. F. 1937. Life-history of the Black-chinned Jacamar. Auk 54:135–148.

Skutch, A. F. 1963. Life history of the Rufous-tailed Jacamar Galbula ruficauda in Costa Rica. Ibis 105:354–368.

Tobias, J., T. Züchner, and T. A. de Melo Júnior. 2002. Family Galbulidae (Typical Antbirds). Pp. 74–101 in J. del Hoyo, A. Elliott, and D. A. Christie (editors), Handbook of the birds of the world. Volume 7: Jacamars to Woodpeckers. Lynx Editions, Barcelona, Spain.

Tobias, J. A., and N. Seddon. 2003. Breeding, foraging, and vocal behavior of the White-throated Jacamar (Brachygalba albogularis). Wilson Bulletin 115:237–240.

# CHAPTER SEVENTEEN

# Bucconidae (Puffbirds)

# Species in South America: 35

# Species recorded at BDFFP: 8

# Species captured at BDFFP: 6

Puffbirds are a Neotropical group of birds with large heads, long rictal bristles, small legs, zygodactylous feet, and puffy plumage. With strong bills armed with a formidable hook, often with the tip of the maxilla forked such that the mandible fits cleanly within this hook, puffbirds sally out from branches to snag large invertebrates and small vertebrates (Chartier 2009). Puffbirds have 10 pp, 10(?)–12 ss, and 12 rects.

Puffbirds likely follow a Complex Basic Strategy, but little literature exists on puffbird molts. In our experience, the juvenile plumage is often nearly identical to the definitive basic plumage, except in some cases where reduced feather markings or soft part colors are distinct. Preformative molts are probably typically partial (P. Pyle, pers. comm.) resulting in a mix of juvenile and formative ss covs. We have also found two cases (a *Bucco capensis* and a *Malacoptila fusca*) in which inner pp were replaced apparently during the FPF. Dickey and van Rossem (1938) also suggested that *Notharchus hyperrhynchus* "birds of the year" have an incomplete preformative molt, retaining only one or two pp, but this sounds more consistent with a second prebasic molt. During definitive prebasic molts, at least some (or all?) puffbirds replace pp and ss in multiple molt series (points of molt initiation). Silveira and Marini (2012) described that *Nystalus chacuru* replaced pp in two waves, starting at p1 distally to p8 and p10 proximally to p9, but in some of our species at the BDFFP, an additional series may start at or near p5 and proceed distally. Variation in this theme may be present among successful replacement cycles, individuals, and species. For example, Dickey and van Rossem (1938) describe one N. *hyperrhynchus* in molt starting at p2 and proceeding distally, followed by p1, and also p10 proceeding proximally. Ss replacement is likely even more complicated with three initiation points reported among s12, s9, and/or s7 proceeding distally, and s5 or s4 and s1 or s2 proceeding proximally (Silveira and Marini 2012). Definitive prebasic molts may regularly be incomplete as in a *Staffelmauser* pattern, but it seems possible that smaller species may often or regularly replace all pp and ss. Ryder and Wolfe (2009) suggest the possibility of a prealternate molt in *Monasa*, which is based on ss covs that appear variably glossy; more likely this is the result of suspended or protracted definitive prebasic molts or partial preformative molts (P. Pyle, pers. comm.).

Many puffbirds nest in burrows dug into termite mounds in trees, whereas others burrow into the ground. Apparently both sexes share in incubation and both are likely to develop brood patches (Hilty 2003). Skulls apparently completely ossify, but some individuals probably retain small windows.

# SPECIES ACCOUNTS

*Notharchus tectus tectus*

Pied Puffbird • Macuru-pintado

Band Size: G (CEMAVE 2013)

\# Individuals Captured: 5

| | |
|---|---|
| Wing | 66.0–71.0 mm (69.1 ± 2.1 mm; n = 5) |
| Tail | 45.0–52.0 mm (48.3 ± 3.3 mm; n = 4) |
| Mass | 22.0–31.0 g (26.4 ± 3.2 g; n = 5) |
| Bill | 16.0 mm (n = 1) |
| Tarsus | 14.1 mm (n = 1) |

Similar species: Much smaller than *N. macrorhynchos* and with black forehead.

Skull: Skull apparently completes, probably during the FCF.

Brood patch: Both sexes probably incubate (Hilty 2003) and develop BPs. Timing unknown.

Sex: ♀ = ♂.

Molt: Group 3 (or sometimes Group 5?)/Complex Basic Strategy; FPF partial, DPBs complete (or occasionally incomplete?). Based on an examination of three specimens from LSUMZ, the FPF appears to include body feathers and most to all ss covs with outer med or gr covs retained. Timing unknown.

FCJ Rects paler, especially on the underside. Ss covs with pale fringing. Skull is unossified.

(LSUMZ 52986.)

FPF Pale blackish juvenile plumage being replaced by black adult-like plumage on the body and ss covs. Skull is unossified.

Skull 100% ossified with "trace" body molt (adventitious or protracted FPF?). (LSUMZ 52986.)

FCF Molt limits among the med and gr covs with retained feathers paler black and with buffy fringing. Rects as in FCJ. Skull is unossified or ossified.

SPB Juvenile-like flight feathers being replaced by adult-like flight feathers. Skull is ossified.

DCB    Without molt limits among the glossy black ss covs or pp. Rects adult-like and relatively blackish on the underside. Skull is ossified.

(LSUMZ 136721.)                    (LSUMZ 52986.)

DPB    Adult-like flight feathers replacing adult-like flight feathers. Skull is ossified.

SAB    With one or more retained adult-like pp or ss. Ss covs and rects as in DCB. Skull is ossified.

## *Bucco tamatia tamatia*

Spotted Puffbird • Rapazinho-carijó

Band Size: G

\# Individuals Captured: 30

| | |
|---|---|
| Wing | 73.0–80.0 mm (76.2 ± 2.2 mm; n = 18) |
| Tail | 56.0–69.0 mm (60.8 ± 3.6 mm; n = 18) |
| Mass | 30.9–40.0 g (34.9 ± 2.4 g; n = 27) |
| Bill | 19.3–21.2 mm (20.3 ± 1.0 mm; n = 3) |
| Tarsus | 19.4–19.9 mm (19.6 ± 0.3 mm; n = 3) |

Similar species: None.

Skull: May complete or retain small (<3 mm) windows (n = 9).

Brood patch: Both sexes probably incubate (Hilty 2003) and develop BPs. Seasonal timing not understood as no captures with BPs have been observed at the BDFFP.

Sex: ♀ = ♂.

Molt: Group 5 (or rarely Group 3?)/Complex Basic Strategy; FPF partial, DPBs incomplete-complete(?). One DPB bird had replaced p1–p3 distally, p8 and p10, with p4–p7 and p9 still retained, with s3, s7–s9, and s11 replaced and s1–s2, s4–6, and s10 retained, consistent with a *Staffelmauser*-like replacement pattern in other puffbirds, thus may result in one or more retained flight feathers as in other puffbirds. One SCB with a retained rect did not have apparent molt limits among the pp or ss, suggesting that some DPB (perhaps especially SPB) may be complete. Flight feather replacement has been observed between Jul and Feb.

FCJ    With reduced buffy spotting on upperparts, but otherwise very similar to the adult. Iris brownish. Skull unossified.

FPF    With body molt; replaced feathers with more distinct buffy markings than retained feathers. Iris brownish or perhaps brownish-red. Skull relatively unossified.

FCF   Look for molt limits within the gr or med covs, with retained feathers being dull brown and with less distinct buffy markings. The skull completely or nearly completely ossifies and the iris probably changes from brown to bright red during FCF.

Two outer gr covs retained.

SPB   With body molt and flight feather molt, replacing relatively worn juvenile flight feathers and rects with adult-like feathers. May be difficult to distinguish from DPB in many cases (use UPB). Iris red (or sometimes brownish-red?). Skull relatively ossified.

SCB   With one or more wing or tail feathers retained, being relatively worn. May be difficult to distinguish from SAB (use DCB). Iris red. Skull relatively ossified.

Right r5 retained and extremely worn.

DCB   Without molt limits among ss covs, pp, ss, or rects. Iris red. Skull relatively ossified.

Although limits are not obvious here, pp covs 6–7 may be retained (they may be SAB). DCB could be used in puffbirds for cases where FCF and FCJ can be excluded, *Staffelmauser*-like molt limits are not apparent, or where DPBs result in a complete replacement, which may be rare.

DPB    Adult-like flight feathers replacing relatively unworn adult-like flight feathers. May be difficult to distinguish from SPB in many cases (use UPB). TPB may be possible where molts limits from the SCB are still apparent during the next cycle of flight feather replacement. Iris red. Skull relatively ossified.

SAB    One or more feathers or blocks of feathers retained and adult-like among the ss and/ or pp. Iris red. Skull relatively ossified.

Molt limits within the pp covs (with 6 and 7 and possibly 9 and 10 retained), pp (with at least p6 and maybe p6–p8 replaced); it seems likely given the pp cov replacement pattern that additional pp are replaced.

## *Bucco capensis* (monotypic)

Collared Puffbird • Rapazinho-de-colar

| | |
|---|---|
| Wing | 77.0–85.5 mm (80.9 ± 2.1 mm; n = 31) |
| Tail | 59.5–68.0 mm (63.7 ± 2.2 mm; n = 25) |
| Mass | 47.0–62.0 g (52.5 ± 3.7 g; n = 37) |
| Bill | 22.4–23.5 mm (22.9 ± 0.5 mm; n = 3) |
| Tarsus | 18.8–23.0 mm (21.1 ± 2.1 mm; n = 3) |

Similar species: None.

Skull: May complete, or with small (<3 mm) windows (n = 8).

Brood patch: Both sexes probably incubate (Hilty 2003) and develop BPs. Seasonal pattern not well understood; only two have been observed, in Feb and May, and juveniles were observed in Jul and Aug, one in each month.

Sex: ♀ = ♂.

Molt: Group 3 (or Group 5?)/Complex Basic Strategy; FPF partial-incomplete, DPBs incomplete-complete(?). One FPF was caught with p1 and p2 replaced, and may be the exception among puffbirds where this molt is usually partial (see also *Malacoptila fusca*). In such cases, we would expect replacement of flight feathers should be less extensive than in DPBs, involving only a variable number of inner pp and/or terts. DPBs seasonal timing of molt not well understood.

FCJ    Skull unossified, but otherwise very similar to the adult, probably including eye color. Look for heavier barring and more rusty orange in upperparts and crown, a dusky ridge along length of maxilla, and reduced orange markings on the outer webs of the outer gr covs.

FPF    Skull relatively unossified with body molt and sometimes wing molt. Maxilla with dusky ridge.

p1–p2 molting, p3–p10 juvenile; s1–s7 juvenile, s8–s9 molting; ss covs appear juvenile and flight feathers are relatively fresh, suggesting this is not an SPB.

FCF    Molt limits among the gr covs or between the gr covs and pp covs. In some (all?) birds, ss, pp, and/or pp covs also with molt limits. Look for retained ss with unbarred and narrow orange leading edge to ss. Beware of pseudolimits between outer and inner gr covs, and between inner and outer ss. Maxilla may lose dusky ridge before SPB. Skull variably ossified.

SPB    With body molt and flight feather molt, replacing relatively worn juvenile flight feathers and rects with adult-like feathers. Retained ss should be relatively unbarred and narrowly orange-margined compared to replaced ss. Ridge of maxilla is relatively orange. Skull relatively ossified.

SCB    With one or more wing or tail feathers retained, being relatively worn. Ridge of maxilla mostly orange. Skull relatively ossified.

DCB    Without molt limits among ss covs, pp, ss, or rects. Ridge of maxilla mostly orange. Skull relatively ossified.

DPB    Adult-like flight feathers replacing relatively unworn adult-like flight feathers. May be difficult to distinguish from SPB in many cases (use UPB). TPB may be possible where molts limits from the SCB are still apparent during the next cycle of flight feather replacement. Skull is ossified.

SAB One or more feathers or blocks of feathers retained and adult-like among the ss, pp, and/or pp covs. Beware of pseudolimits between outer and inner gr covs, and between inner and outer ss. Ridge of maxilla mostly orange. Skull relatively ossified.

p3–p4 and p8–p10 retained and adult-like; s1 and s3–s5 retained and adult-like (left); maxilla mostly orange (right).

## Malacoptila fusca (monotypic)

White-chested Puffbird • Barbudo-pardo

Band Size: H, G

# Individuals Captured: 240

| | |
|---|---|
| Wing | 80.0–95.0 mm (87.7 ± 2.7 mm; n = 155) |
| Tail | 59.0–75.0 mm (65.9 ± 3.3 mm; n = 141) |
| Mass | 35.5–54.6 g (44.2 ± 3.4 g; n = 213) |
| Bill | 18.0–22.8 mm (20.1 ± 1.2 mm; n = 15) |
| Tarsus | 22.2–24.5 mm (22.8 ± 0.7 mm; n = 15) |

Similar species: None at the BDFFP.

Skull: Completes or may retain small windows (<3 mm) into DCB (n = 43).

Brood patch: Both sexes probably incubate (Hilty 2003) and develop BPs. Timing uncertain, but perhaps during the late wet and early dry seasons; one BP was observed in May, and two juveniles have been captured in Jul and Aug.

Sex: ♀ = ♂.

Molt: Group 5/Complex Basic Strategy; FPF partial-incomplete; DPBs incomplete. The FPF usually includes all ss covs except the outer 2–8 gr covs, but one apparent FCF with a brown iris replaced all gr covs and also p1–p5. DPBs appear to follow a *Staffelmauser*-like pattern as in other puffbirds, with pp molt beginning at two or more nodes. Timing of molt uncertain, and possibly year-round, although probably peaks following breeding.

FCJ      Plumage very similar to subsequent plumages, but with reduced buffy markings on upperparts. Iris brown. Skull relatively unossified.

FPF      With body molt replacing relatively unmarked upperpart feathers with more distinctly marked feathers. Iris brownish or perhaps brownish-red. Skull relatively unossified.

FCF      Look for molt limits among the gr covs with the outer 2 to 8 retained, and a single generation of pp and ss or with a variable number of inner pp replaced from a single series (always starting at p1?). Iris probably changes from brown to bright red during FCF; look especially for irises intermediate in color. Skull variably ossified.

Iris intermediately reddish-brown (left); outer 5 gr covs and all ss and pp retained and juvenile (right).

An unusual bird a distinctly brown iris, p1–p5 replaced, and relatively worn retained outer pp.

SPB     With body molt and flight feather molt, replacing relatively worn juvenile flight feathers and rects with adult-like feathers. May be difficult to distinguish from DPB in many cases (use UPB). Iris red (or sometimes brownish-red?). Skull relatively unossified.

SCB     Mix of retained juvenile and adult-like flight feathers. May be difficult to distinguish from SAB (use DCB). Iris deep red. Skull relatively ossified.

DPB     Adult-like flight feathers being replaced by adult-like feathers. Iris deep red. Skull relatively ossified.

SAB     One or more feathers or blocks of feathers retained and adult-like among the ss and/or pp. Iris deep red. Skull relatively ossified.

Two versions with two generations of adult-like remiges: (left) p6–p10, p3, p1, s2, s4–5, and s6–s8 retained; (right) p8–p10, p1(?), s6–s9, and 4 outer gr covs retained.

## Nonnula rubecula tapanahoniensis

Rusty-breasted Nunlet • Macuru

Band Size: G, F

# Individuals Captured: 13

Wing    65.0–75.0 mm (67.9 ± 3.3 mm; n = 8)

Tail    58.0–61.0 mm (59.5 ± 2.1 mm; n = 2)

Mass    17.5–21.0 g (19.4 ± 1.1 g; n = 9)

Bill    17.7 mm (n = 1)

Tarsus  16.5 mm (n = 1)

Similar species: None at the BDFFP.

Skull: May complete, or with small (<3 mm) windows (n = 6).

Brood patch: Both sexes probably incubate (Hilty 2003) and develop BPs. Seasonal pattern not understood as no captures with BPs have been observed at the BDFFP.

Sex: ♀ = ♂.

Molt: Group 5 or Group 3/Complex Basic Strategy; FPF partial, DPBs incomplete(?)-complete(?). Probably like other puffbirds, it appears that most ss covs are replaced during the FPF with 1–8(?) gr covs retained, but not the remiges, at least usually (see also discussion in family, *Bucco capensis*, and *Malacoptila fusca* accounts). DPBs poorly known, and may be like other puffbirds, but less likely to retain pp and ss. One capture was initiating molt in Sep.

FCJ     Plumage very similar to adult, but duller, upperparts fringed rufescent, and bill with a pale tip (Rasmussen et al. 2013). Skull is unossified.

FPF     Replacing juvenile body feathers with adult-like feathers. Skull relatively unossified.

<table>
<tr><td>FCF</td><td>Look for molt limits among the gr covs, or perhaps already between replaced inner and retained outer pp. Skull variably ossified.</td></tr>
</table>

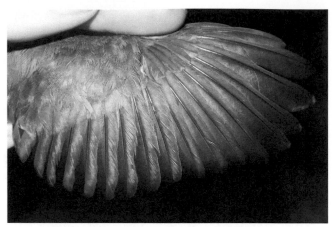

Distinctly brown rachis on each remige suggest these are juvenile and outer 7 or 8 gr covs may be retained.

SPB    With body molt and flight feather molt, replacing relatively worn juvenile flight feathers and rects with adult-like feathers. May be difficult to distinguish from DPB in many cases (use UPB). Skull relatively ossified.

SCB    Mix of retained juvenile and adult-like flight feathers. May be difficult to distinguish from SAB (use DCB). Skull relatively ossified.

DPB    Adult-like flight feathers being replaced by adult-like feathers. Skull relatively ossified.

SAB    One or more feathers or blocks of feathers retained and adult-like among the ss and/or pp. Skull relatively ossified.

## *Monasa atra* (monotypic)

Band Size: H, L

Black Nunbird • Chora-chuva-de-asa-branca

\# Individuals Captured: 39

Wing     118.0–134.5 mm (128.8 ± 4.3 mm; n = 24)

Tail     112.0–128.0 mm (121.5 ± 4.8 mm; n = 20)

Mass     70.0–110.0 g (88.2 ± 9.4 g; n = 34)

Bill     21.8–29.8 mm (27.1 ± 3.2 mm; n = 6)

Tarsus   21.7–26.4 mm (25.0 ± 1.6 mm; n = 7)

Similar species: None at the BDFFP.

Skull: May complete, or with small (<3 mm) windows (n = 4).

Brood patch: Both sexes probably incubate (Hilty 2003) and develop BPs. Seasonal pattern not understood as no captures with BPs have been observed at the BDFFP, but one FPJ was captured in Oct.

Sex: ♀ = ♂.

Molt: Group 5/Complex Basic Strategy; FPF partial, DPBs incomplete. FPF probably like other puffbirds, retaining one or more outer gr covs, but also other ss covs. DPB replacement in a *Staffelmauser*-like pattern as in other puffbirds (see family account). SPB may average more extensive than subsequent DPBs. Timing uncertain, but occurs at least Nov–Feb.

FCJ    Dull grayish-brown body plumage and dull orange bill (Ryder and Wolfe 2009). Sometimes white shoulder feathers tipped brown. Iris relatively dull reddish-brown. Skull unossified.

FPF      With body molt and perhaps wing molt. Bill dull orange. Iris relatively dull reddish-brown. Skull relatively unossified.

FCF      Look for molt limits with one or more outer gr covs, but also retained coverts elsewhere among the ss covs. Remiges and pp covs without molt limits. Bill orange to reddish-orange. Iris variably dull or brighter reddish-brown. Skull variably ossified.

Iris dull brown (left); outer 4 gr covs and several other ss retained, and with an a1–a2 molt limit (right).

SPB      With body molt and flight feather molt, replacing relatively worn juvenile flight feathers and rects with adult-like feathers. May be difficult to distinguish from DPB in some cases (use UPB). Iris relatively glossy and variably reddish-brown. Bill reddish-orange. Skull relatively ossified.

SCB      Glossy black pp and/or ss contrasting against one or more retained brownish juvenile feathers. Bill reddish-orange. Iris variably glossy reddish-brown to reddish. Skull relatively ossified.

Iris glossy, but still brownish (left); s8, p9, and pp cov 8 retained as are various ss covs (right).

DPB      Adult-like body and flight feathers being replaced by adult-like feathers. Bill reddish-orange. Iris glossy and reddish. Skull relatively ossified.

SAB      One or more feathers or blocks of feathers retained and adult-like among the ss and/or pp; the contrast between these feathers may be minimal, but differences in wear should be apparent. Bill reddish-orange. Iris glossy and reddish. Skull relatively ossified.

Iris glossy and relatively reddish-brown (left); note difference in wear between retained and replaced pp and ss: p7–p8, s1, s3–s4, and s6–s8 retained (right).

## LITERATURE CITED

CEMAVE. 2013. Lista das espécies de aves brasileiras com tamanhos de anilha recomendados. Centro Nacional de Pesquisa e Conservação de Aves Silvestres, Cabedelo, Brasil (in Portuguese).

Chartier, A. 2009. Puffbirds. Pp. 187–188 in T. Harris (editor), National Geographic complete birds of the world. National Geographic Society, Washington, DC.

Dickey, D. R., and A. J. van Rossem. 1938. The birds of El Salvador. Field Museum of Natural History, Zoological Series, Chicago, IL.

Hilty, S. L. 2003. Birds of Venezuela (2nd ed.). Princeton University Press, Princeton, NJ.

Rasmussen, P. C., N. Collar, and G. Kirwan. [online]. 2013. Rusty-breasted Nunlet (Nonnula rubecula). In J. del Hoyo, A. Elliott, J. Sargatal, D. A. Christie, and E. de Juana (editors), Handbook of the birds of the world alive. Lynx Edicions, Barcelona, Spain. <http://www.hbw.com>.

Ryder, T. B., and J. D. Wolfe. 2009. The current state of knowledge on molt and plumage sequences in selected Neotropical bird families: a review. Ornitología Neotropical 20:1–18.

Silveira, M. B., and M. Â. Marini. 2012. Timing, duration, and intensity of molt in birds of a Neotropical savanna in Brazil. Condor 114:435–448.

# CHAPTER EIGHTEEN

# Capitonidae (New World Barbets)

# Species in South America: 14

# Species recorded at BDFFP: 1

# Species captured at BDFFP: 1

Placed in the order Piciformes with the woodpeckers, the New World barbets were once taxonomically lumped with the Old World barbets at the family-level, but the barbets are now treated as three distinct families (Capitonidae, Megalaimidae, and Lybiidae). In fact, recent evidence suggests that Capitonidae are more closely related to the toucans (Ramphastidae), another exclusively New World family (e.g., Burton 1984, Barker and Lanyon 2000, Johannson and Ericson 2003, Moyle 2004), and sometimes New World barbets are lumped together with Semnornithidae, the toucan-barbets, into the Ramphastidae (AOU 1998, Harris 2009). The Capitonidae probably warrant their own family status, forming a superfamily with Semnornithidae and Ramphastidae. Barbets and toucans differ in their arrangement of caudal vertebrae and sleeping posture (Short and Horne 2001) and by their cranial morphology (Höfling 1991, 1998). Barbets often have enlarged rictal bristles, which are lacking in

toucans (Dittmann and Cardiff 2009). Genetic data also support the distinction between toucans and barbets (Moyle 2004, reviewed by Remsen et al. 2016). New World barbets have 10 pp, 9 ss, and 10 rects.

Molt in barbets is poorly known, but should be expected to follow a Complex Basic Strategy. The juvenile plumage is weak, like in passerines, often lacks the gaudy colors of adults, and may get replaced quickly by an adult-like formative plumage, although the extent of this molt in most species is not known, but may regularly be partial as in Ramphastidae. Definitive prebasic molts are expected to be complete (Dittmann and Cardiff 2009). Barbets typically exhibit marked sexual dichromatism, and this may be apparent in the formative plumage in many species.

Many species apparently breed with the onset of the rainy season. Incubation is usually or always shared by both sexes, and probably both develop brood patches. The extent of skull ossification is not known and may not be complete. Beware of strong musculature supporting their large bill obscuring the skull, as in closely related woodpeckers.

# SPECIES ACCOUNTS

*Capito niger niger*

Black-spotted Barbet • Capitão-de-bigode-carijó

Band Size: H, J (CEMAVE 2013)

# Individuals Captured: 1

| Wing | ♂ | 80 mm (n = 1) |
| | | 80–86 mm |
| | | (Haverschmidt 1968) |
| | ♀ | Unknown at the BDFFP |
| Tail | ♂ | 50 mm (n = 1) |
| | ♀ | Unknown at the BDFFP |
| Mass | ♂ | 53 g (n = 1) |
| | | 52–60 g (Haverschmidt 1968) |
| | ♀ | Unknown at the BDFFP |
| | | 50–54 g (Haverschmidt 1968) |

Similar species: None at the BDFFP.

(Courtesy of Cameron Rutt.)

Skull: Probably completely ossifies based on a review of specimens (LSUMZ). Timing is unknown, but we would expect the skull to ossify fully during the FCF.

Brood patch: Both sexes probably incubate and develop BPs. Seasonal pattern not understood as no captures with BPs have been observed at the BDFFP, but may be a wet-season breeder (Short and Horne 2001).

Sex: ♂♂ have solid black auriculars and no spotting on chest (spotting confined to lower flanks). ♀♀ have streaked auriculars, scaly wing covs, and heavy spotting on chest and flanks.

Molt: Group 3/Complex Basic Strategy; FPF partial(?), DPBs complete. FPF based on an examination of *C. auratus auratus* and *C. niger insperatus* specimens that had molt limits among the med and/or gr covs (LSUMZ). Tail molt may proceed centripetally as suggested by specimens of *C. auritus auritus*. Timing unknown.

FCJ     Probably without or with limited bright feathers in crown, throat, and back. Rects pointed. Iris is dull gray. Skull is unossified.

FPF     Replacing juvenile body feathers with adult-like feathers. Look for dull brownish pp covs, relatively little contrast in retained terts, and pointed rects. Note the throat may be adult-like early in this molt. Iris is dull gray, but may become redder with age.

This bird with body molt may not replace additional med or gr covs, or pp, ss (left), and rects (right), resulting in molt limits among the med and gr covs. (Courtesy of Cameron Rutt.)

| | |
|---|---|
| FCF | Plumage adult-like with bright throat, crown, and back markings, but with molt limits among the gr covs. Pp covs relatively dull brownish-black. Rects pointed. Iris is grayish to reddish-brown. Skull is unossified or ossified. |
| SPB | Replacing juvenile flight and tail feathers with adult-like feathers. Iris is reddish-brown. Skull is ossified. |
| DCB | Like FCF, but without molt limits among the gr covs. Pp covs glossy black, especially when fresh. Rects rounded. Skull is ossified. |
| DPB | Replacing adult-like body and flight feathers with adult-like feathers. Look for relatively blackish-brown pp covs, relatively strong contrast in retained terts, and similar shape between retained and replaced rects. Iris is reddish-brown. Skull is ossified. |

---

## LITERATURE CITED

AOU. 1998. Check-list of North American birds (7th ed.). American Ornithologists' Union, Lawrence, KS.

Barker, F. K., and S. M. Lanyon. 2000. The impact of parsimony weighting schemes on inferred relationships among toucans and Neotropical barbets (Aves: Piciformes). Molecular Phylogenetics and Evolution 15:215–234.

Burton, P. J. K. 1984. Anatomy and evolution of the feeding apparatus in the avian orders Coraciiformes and Piciformes. Bulletin of the British Museum of Natural History, Zoology 47:331–443.

CEMAVE. 2013. Lista das espécies de aves brasileiras com tamanhos de anilha recomendados. Centro Nacional de Pesquisa e Conservação de Aves Silvestres, Cabedelo, Brasil (in Portuguese).

Dittmann, D. L., and S. W. Cardiff. 2009. Barbets and Toucans. Pp. 177–179 in T. Harris (editor), National Geographic complete birds of the world. National Geographic Society, Washington, DC.

Harris, T. 2009. National Geographic complete birds of the world. National Geographic Society, Washington, DC.

Haverschmidt, F. 1968. Birds of Surinam. Oliver and Boyd, Edinburgh, UK.

Höfling, E. 1991. Étude comparative due cräne chez des Ramphastidae (Aves, Piciformes). Bonner zoologische Beiträge 42:55–65 (in German).

Höfling, E. 1998. Comparative cranial anatomy of Ramphastidae and Capitonidae. Proceedings of the 22nd International Ornithological Congress, Ostrich 69:389–390.

Johansson, U. S., and P. G. P. Ericson. 2003. Molecular support for a sister group relationship between Pici and Galbuldae (Piciformes sensu Wetmore 1960). Journal of Avian Biology 34:185–197.

Moyle, R. G. 2004. Phylogenetics of barbets (Aves: Piciformes) based nuclear and mitochondrial DNA sequence data. Molecular Phylogenetics and Evolution 30:187–200.

Remsen, J. V., C. D. Cadena, A. Jaramillo, M. Nores, J. F. Pacheco, M. B. Robbins, T. S. Schulenberg, F. G. Stiles, D. F. Stotz, and K. J. Zimmer. [online]. 2016. A classification of the bird species of South America. American Ornithologists' Union. Version 30 November 2016. <http://www.museum.lsu.edu/~Remsen/SACCBaseline.htm>.

Short, L., and J. F. M. Horne. 2001. Toucans, Barbets, and Honeyguides: Ramphastidae, Capitonidae, and Indicatoridae. Oxford University Press, Oxford, UK.

# CHAPTER NINETEEN

# Ramphastidae (Toucans)

# Species in South America: 34

# Species recorded at BDFFP: 4

# Species captured at BDFFP: 3

The toucans are a New World family of medium to large birds restricted to tropical regions. Taxonomically, they are closely related to the barbets (Capitonidae, Megalaimidae, and Lybiidae) and may be the sister family to the New World barbets (Capitonidae). Toucans are familiar, if not instantly recognizable, by their oversized bills designed to reach fruits on outer branches. Toucans are also characterized by powerful and zygodactylous feet. Only toucans in the genus *Selenidera* are sexually dimorphic (see *Selenidera culik*), although differences in bill and wing lengths between sexes may be common in all toucan species. The extent and timing of skull ossification is unknown. Toucans have 10 pp, 13 ss, and 12 rects.

The molt strategy in toucans has not been studied in depth or previously reviewed to our knowledge. Dickey and Van Rossem (1938) provide a detailed account of some smaller toucans from El Salvador (sometimes called toucanets), and Van Tyne (1929) details the molt of *Ramphastos sulphuratus brevicarinatus*. Both authors suggest that primary and secondary replacement are sequential and complete, starting with p1 and finishing on p10, and starting in the tertials and s1 progressing to about s5 or s6 from both directions. Rectrices may be replaced starting with the third or fourth pair of rectrices and proceeds outward, then finishing with the central rects (Dickey and Van Rossem 1938), although Van Tyne (1929) suggests a centripetal molt starting with r6 and proceeding inward finishing at r1. Both authors also report the preformative molt to be partial, but Dickey and Van Rossem (1938) indicate that rectrix and tertial replacement occurs later in at least some species, suggesting an interrupted or protracted incomplete FPF molt, or more likely an early onset to a protracted SPB as some adults (DPB) apparently also show this molt. Evidence of a partial preformative molt in *R. sulfuratus* is also provided by Jones and Griffiths (2011) where birds are reported to replace all body plumage, but no remiges and rectrices, followed by single annual prebasic molt in all subsequent years. Interestingly, Jones and Griffiths (2011) also reported centripetal rectrix molt during the preformative molt in *R. sulfuratus*. Given this assessment by Jones and Griffiths (2011) and Van Tyne (1929) of *R. sulfuratus* molt, we might expect *Ramphastos* and perhaps other genera of Ramphastidae at our study site to exhibit similar patterns, which so far appears to be the case in a few species at the BDFFP.

Ramphastidae are apparently sexually dimorphic in bill length with males averaging about 10% longer bills (Short and Horne 2002). Both sexes incubate, although females more than males, and subsequently brood patches may be more extensive in females. Based on an examination of specimens from *Ramphastos*, *Selenidera*, and *Pteroglossus*, the skull probably typically ossifies (LSUMZ), although this may be difficult to see in live birds.

# SPECIES ACCOUNTS

*Ramphastos vitellinus vitellinus*

Band Size: N, P (CEMAVE 2013)

Channel-billed Toucan • Tucano-de-bico-preto

# Individuals Captured: 1

| | |
|---|---|
| Wing | 185.0 mm (n = 1) |
| | 186 mm (n = 1, Trinidad and Tobago; Junge and Mees 1961) |
| Tail | 132.0 mm (n = 1) |
| | 137 mm (n = 1, Trinidad and Tobago; Junge and Mees 1961) |
| Mass | Unknown at BDFFP |
| | 315 g (n = 1, Trinidad and Tobago; Junge and Mees 1961) |
| | 285–455 g (Short et al. 2014) |

Similar species: From *Ramphastos tucanus tucanus* by smaller size (*R. t. tucanus*: >515 g, Short and Horne 2002; >189 mm wing and >148 mm tail, Jones and Griffiths 2011), yellow wash to throat, lack of yellow ridge along maxilla, and bluish (not yellowish) base to mandible.

Skull: May nearly or completely ossify (n = 1 BDFFP, n = 2 LSUMZ), probably during the FCF.

Brood patch: Both species probably incubate and develop brood patches, although more extensively in ♀♀ than in ♂♂.

Sex: ♂ > ♀, averaging by about 5% (Short et al. 2014).

(Courtesy of Aida Rodrigues.)

Molt: Group 3/Complex Basic Strategy; FPF partial-incomplete(?), DPBs probably complete. FPF based on examination of *R. v. vitellinus* and *R. v. ariel* from the eastern Amazon (n = 3, LSUMZ; Guyana and Pará, Brazil), and likely includes body and most (or all?) ss coverts, 1–3 terts (always?), perhaps sometimes multiple rects. Timing unknown.

| | |
|---|---|
| FCJ | Similar to adult, but less boldly patterned. Black replaced with dark browns. |
| FPF | Replacing dull juvenile body feathers, ss covs, and perhaps rects with adult-like brighter feathers. |
| FCF | Look for molt limits among the gr covs, terts, and/or rects. |

One inner gr cov replaced and two or three outer med covs retained. Note that the pp and ss appear relatively bright and glossy, but are retained from the juvenile plumage. (LSUMZ 132308.)

| | |
|---|---|
| SPB | Replacing juvenile-like (relatively dull brownish-black) pp and ss with adult-like pp and ss. |
| DCB | Without molt limits among the gr covs, terts, or rects. |
| DPB | Replacing adult-like (relatively blackish) pp and ss with adult-like pp and ss. |

## Selenidera piperivora (monotypic)

Band Size: Unknown

Guianan Toucanet • Açaçari-negro

# Individuals Captured: 7

Wing    110.0–122.0 mm
              (116.6 ± 3.6 mm; n = 7)

Tail      89.0–102.0 mm (96.6 ± 5.0 mm; n = 5)

Mass    126.0–146.0 g (135.5 ± 6.5 g; n = 6)

Similar species: From *Pteroglossus viridis* lacking yellow underparts and presence of yellow streak behind eye, among other plumage differences.

Skull: Probably ossifies during the FCF.

Brood patch: Both species probably incubate and develop brood patches, although more extensively in ♀♀ than in ♂♂.

Sex: ♂ with black throat and chest. ♀ with gray throat and chest.

(Courtesy of Lindsay Wieland.)

Molt: Group 3/Complex Basic Strategy; FPF partial-incomplete, DPBs complete. FPF includes body and most ss covs, 1–3 terts (always?), and sometimes at least one pair of rects.

Notes: Previously known as *S. culik* (see Pacheco and Whitney 2006 for taxonomic justification).

FCJ       Iris paler orange and plumage pattern duller than subsequent plumages.

FPF       Replacing dull juvenile body feathers, ss covs, and perhaps rects with adult-like brighter feathers.

FCF       Look for molt limits among the gr covs, terts, and/or rects.

SPB      Replacing juvenile-like (relatively dull green) pp and ss with adult-like pp and ss.

DCB      Without molt limits among the gr covs, terts, or rects.

DPB      Replacing adult-like (relatively bright green) pp and ss with adult-like pp and ss.


## Pteroglossus viridis (monotypic)

Band Size: L (CEMAVE 2013)

Green Aracari • Açaçari-miudinho

# Individuals Captured: 7

Wing    103.0–117.0 mm (108.8 ± 5.4 mm; n = 5)

Tail      93.0–106.0 mm (98.5 ± 6.7 mm; n = 3)

Mass    100.0–121.0 g (113.2 ± 8.6 g; n = 5)

Similar species: From *Selenidera piperivora* by yellow underparts and lack of yellow streak behind eye, among other plumage differences.

Skull: Probably ossifies during the FCF.

Brood patch: Both species probably incubate and develop brood patches, although more extensively in ♀♀ than in ♂♂.

Sex: ♂ with black crown, nape, and underparts. ♀ with gray crown and underparts and maroon nape.

Molt: Group 3/Complex Basic Strategy; FPF partial-incomplete, DPBs complete. FPF includes body and most ss covs, 1–3 terts (always?), and at least one pair of rects (always?). One capture initiating molt in Oct.

FCJ       Plumage pattern similar to subsequent plumages, but more dull.

FPF       Replacing dull juvenile body feathers, ss covs, and perhaps rects with adult-like brighter feathers.

FCF      Look for molt limits among the gr covs, terts, and/or rects.

r5 left and r1 retained (left); 4 outer gr covs and 1 outer med cov retained, and interestingly p1–p2 and s11 have been replaced (right). Is this the onset of a suspended SPB or part of the FPF? (Courtesy of Luiza Figuera and Pedro Martins.)

SPB      Replacing juvenile-like (relatively dull green) pp and ss with adult-like pp and ss.

DCB      Without molt limits among the gr covs, terts, or rects.

DPB      Replacing adult-like (relatively bright green) pp and ss with adult-like pp and ss.

## LITERATURE CITED

CEMAVE. 2013. Lista das espécies de aves brasileiras com tamanhos de anilha recomendados. Centro Nacional de Pesquisa e Conservação de Aves Silvestres, Cabedelo, Brasil (in Portuguese).

Dickey, D. R., and A. J. Van Rossem. 1938. The birds of El Salvador. Field Museum of Natural History, Zoological Series, Chicago, IL.

Jones, R., and C. S. Griffiths. [online]. 2011. Keel-billed Toucan (*Ramphastos sulfuratus*). In T. S. Schulenberg (editor), Neotropical birds online. Cornell Lab of Ornithology, Ithaca, NY. <http://neotropical.birds .cornell.edu>.

Junge, G. C. A., and G. F. Mees. 1961. The avifauna of Trinidad and Tobago. E. J. Brill, Neider, Netherlands.

Pacheco, J. F., and B. M. Whitney. 2006. Mandatory changes to the scientific names of three Neotropical birds. Bulletin of the British Ornithologists' Club 126:242–244.

Short, L. L., and J. F. M. Horne. 2002. Ramphastidae (Toucans). Pp. 220–273 in J. del Hoyo, A. Elliott, and J. Sargatal (editors), Handbook of the birds of the world. Lynx Edicions, Barcelona, Spain.

Short, L. L., E. de Juana, C. J. Sharpe, and G. M. Kirwan. [online]. 2014. Channel-billed Toucan (*Ramphastos vitellinus*). In J. del Hoyo, A. Elliott, J. Sargatal, D. A. Christie, and E. de Juana (editors), Handbook of the birds of the world alive. Lynx Edicions, Barcelona, Spain. <http://www.hbw.com>.

Van Tyne, J. 1929. The life history of the toucan *Ramphastos brevicarinatus*. University of Michigan, Ann Arbor, MI.

# CHAPTER TWENTY

# Picidae (Woodpeckers, Piculets, and Allies)

# Species in South America: 83

# Species recorded at BDFFP: 11

# Species captured at BDFFP: 5

Woodpeckers, flickers, and piculets, but not sapsuckers, are South American representatives of the family Picidae, all of which are characterized by chisel-like bills and long tongues used to extract insect prey. All woodpeckers, except piculets, also have stiffened tails that are used to brace themselves against trees and snags when foraging. The tiny piculets instead forage along thin branches much like nuthatches. The generalized and ancestral toe arrangement among Picidae is zygodactylous, but this arrangement has been modified in several different ways among lineages (Winkler et al. 1995). Only piculets, wrynecks, and flickers retain the classic zygodactyl form. Other medium-sized woodpeckers have a reduced hallux (which is even absent in some species, like Three-toed Woodpeckers [*Picoides dorsalis*]), and the largest species (e.g., *Campephilus*) have an elongated hallux and even rely on their tarsus for supporting their weight when clinging and climbing trunks. Many Picidae are sexually dichromatic where males often exhibit more red (or sometimes yellow) in the head or stronger facial markings (Winkler et al. 1995). Woodpeckers and piculets have 10 pp, 9 (*Picumnus*)–12 ss, and 10 visible rects (the outer rect is vestigial).

Picidae molt is fairly well studied in North America, but has been almost completely neglected throughout the Neotropics (but see Pyle et al. 2004, 2015; Guallar et al. 2009). Picidae follow a Complex Basic Strategy and the preformative molt is variable among species, although regularly incomplete and sometimes partial.

The preformative molt often follows an unusual replacement pattern that includes primaries and rectrices, but not primary coverts or secondaries (Pyle and Howell 1995, Pyle 1997, Howell 2010, Seigel et al. 2016). In such cases, the primaries can begin molting before leaving the nest and before body feathers begin molting. In some individuals of two *Melanerpes* species, the preformative molt arrests, retaining up to five outer primaries. In two other *Melanerpes* species, however, the preformative molt is partial and does not include any flight feathers (Pyle and Howell 1995, Pyle 1997). Preformative molts in *Picumnus*, however, may only include crown feathers (Winkler et al. 1995). During second and subsequent definitive prebasic molts, most North American woodpeckers retain some secondaries and primary coverts, but these molts may complete in *Picumnus* (Winkler et al. 1995). In many woodpeckers, the juvenile plumage is female-like, although in some cases juvenile females, juvenile males, and older females may occasionally or even regularly exhibit adult-male like characteristics such as small amounts of red in the crown. A simple three-step process for aging woodpeckers is as follows:

1. Are there molt limits in the pp covs?
    a. If no, then FCJ or DCB.
    b. If yes, then go to step 2.
2. How many generations of pp covs are there?
3. Is the oldest generation juvenile?
    a. If no, then count generations starting with 2. Take the total number of generations and call the bird "after" the penultimate generation. For example,

if three generations are present, the bird is an SAB.

b. If yes, then count generations starting with 1. Use the number of generations to determine the cycle. For example, if three generations are present, then the bird is a TCB.

Note that additional confirmation of age can be achieved by examining the ss, and sometimes pp, for molt limits, but beware that flight feather replacement is typically more extensive than pp cov replacement.

Determining the age of woodpeckers and piculets by examining the skull for ossification is generally not possible because of the presence of tongue musculature that obscures areas of ossification. According to Pyle (1997), North American woodpecker skulls remain unossified throughout their life, and this appears true in several species of *Picumnus* and Neotropical woodpeckers based on LSUMZ specimens, although variation in ossification extent may be related to age. Eye color can be a dependable way to distinguish juveniles from older individuals in at least some woodpecker species (possibly piculets as well).

## SPECIES ACCOUNTS

*Picumnus exilis undulatus*                Band Size: D, E

Golden-spangled Piculet • Pica-pau-anão-de-pintas-amarelas        # Individuals Captured: 4

Wing        47.0–49.0 mm (47.8 ± 1.0 mm; n = 4)

Tail        23.0–24.5 mm (23.6 ± 0.8; n = 4)

Mass        8.0–9.7 g (8.6 ± 0.8 g; n = 4)

Similar species: No other *Picumnus* is known from the BDFFP.

Brood patch: Probably both sexes incubate like in woodpeckers (Winkler et al. 1995).

Sex: ♂ from ♀ by presence of red in crown.

Molt: Group 3(?)/Complex Basic Strategy; FPF limited(?), DPBs complete(?). FPF probably only includes crown feathers as in other *Picumnus* (Winkler et al. 1995). Timing unknown.

FCJ        Like adult, but duller overall. Olive or brownish crown is streaked (not spotted) and both sexes lack red (Winkler et al. 1995).

FPF        Streaked juvenile crown feathers being replaced with spotted adult-like feathers. Most notable in ♂♂ by first incoming red feathers.

FCF        Crown like that of adult, but body and flight feathers duller overall and relatively worn.

SPB        Replacing juvenile-like body feathers (and adult-like crown feathers) and flight feathers with adult-like body and flight feathers.

DCB        All body and flight feathers bright, distinctly marked, and relatively fresh.

DPB        Replacing adult-like body and flight feathers with adult-like body and flight feathers.

## *Veniliornis cassini* (monotypic)

Golden-collared Woodpecker • Pica-pau-de-colar-dourado

| | |
|---|---|
| Wing | 90.0–97.0 mm (93.2 ± 2.4 mm; n = 6) |
| Tail | 52.0–62.0 mm (55.3 ± 4.6 mm; n = 4) |
| Mass | 29.0–36.0 g (33.1 ± 2.8 g; n = 9) |

Similar species: Larger than *Picumnus* and smaller than *Piculus*. *Piculus chrysochlorus* has an olive face and yellowish-olive chest and belly, and *Piculus flavigula* has a yellow face and scalloped (not barred) underparts.

Brood patch: Both sexes incubate and develop BPs. One ♂ caught in Jul with a BP.

Sex: ♂ from ♀ by presence of red in crown.

Molt: Group 5/Complex Basic Strategy; FPF partial(?), DPBs probably incomplete as in other woodpeckers. FPF probably includes several inner less, med, and gr covs, but perhaps not pp or rects based on a single capture with fault-bar alignment across the ss and pp, and distinctly worn pp and rects. Timing unknown, although one bird was captured beginning molt in Oct and another was ending molt in Apr.

| | |
|---|---|
| FCJ | Duller nape and darker face than subsequent plumages. Ss covs relatively streaked. Both sexes with red on the crown (Winkler et al. 1995). |
| FPF | Adult-like, replacing body feathers and pp, but not pp covs, or ss. |
| FCF | Without molt limits among the pp covs, pp, and ss; these are relatively dull olive against the brighter gr covs and sometimes one or more terts. |

All rects appear retained from the juvenile plumage(?) as they are faded and worn (left); pp and ss with pale growth-bar alignment suggesting they are of the same generation, and perhaps 3 inner gr covs, 1 med cov, and several less covs are replaced.

| | |
|---|---|
| SPB | Replacing body, flight feathers, and pp covs. Retained juvenile ss and pp covs relatively worn compared to replaced pp and ss. |
| SCB | Adult-like with a block of retained inner juvenile pp covs and sometimes one or more ss. |
| TPB | With a mix of retained juvenile and adult-like pp covs (and sometimes ss) being replaced by a third generation of pp covs (and ss). |
| TCB | With three generations of pp covs, the oldest being juvenile. Perhaps also with three generations of ss, but more likely two adult-like generations. |
| SAB | With two generations of pp covs, the oldest being definitive basic. |
| DPB | One or more generations of retained nonjuvenile ss and pp covs being replaced by adult-like pp covs and ss. Look for molt limits among the retained pp covs, ss, and/or pp. Replacing body and flight feathers. |
| DCB | With two or more generations of pp covs, ss, and/or pp of undetermined age. |

*Celeus undatus undatus*

Waved Woodpecker • Pica-pau-barrado

Band Size: H (CEMAVE 2013)

# Individuals Captured: 3

| | | |
|---|---|---|
| Wing | ♂ | 120.0–124.0 mm (122.0 ± 2.8 mm; n = 2) |
| | ♀ | 125.0 mm (n = 1) |
| Tail | ♂ | 68.0–77.0 mm (72.5 ± 6.4 mm; n = 2) |
| | ♀ | 75.0 mm (n = 1) |
| Mass | ♂ | 72.0 g (n = 1) |
| | | 61–73 g (Winkler and Christie 2013) |
| | ♀ | 80.0 g (n = 1) |
| | | 58–68 g (Winkler and Christie 2013) |

Similar species: From other *Celeus* by coarse barring on rump, back, flight feathers, and rects. *C. torquatus* much larger with contrastingly pale head.

Brood patch: Both sexes probably develop BPs as in other woodpeckers. Timing unknown, but based on timing of molt, may breed during the wet season.

Sex: ♂ from ♀ by presence of red in malar.

Molt: Group 5/Complex Basic Strategy; FPF probably incomplete as in *C. elegans*, DPBs incomplete. FPF probably includes most ss covs except outer gr covs, 0–3 terts, and all pp. Timing unknown, but all three birds captured were nearly completing molt in Jul and Sep, so may molt from late wet season to mid to late dry season.

FCJ     Like adult, but duller with less barring on upperparts (Winkler et al. 1995).

FPF     Adult-like, replacing body feathers, some ss covs, and pp, but not pp covs, or ss (or sometimes 1–3 terts?).

FCF     Without molt limits among the pp covs, pp, and ss (except sometimes terts); probably usually with molt limits among the gr covs.

SPB     Replacing body, flight feathers, and outer pp covs. Retained juvenile ss and pp covs relatively worn compared to replaced pp and ss.

SCB     Adult-like with a block of retained inner juvenile pp covs and sometimes one or more ss.

TPB     With a mix of retained juvenile and adult-like pp covs (and sometimes ss) being replaced by a third generation of pp covs (and ss).

TCB     With three generations of pp covs, the oldest being juvenile. Perhaps also with three generations of ss, but more likely two adult-like generations.

SAB     With two generations of pp covs, the oldest being definitive basic.

DPB     Retained ss and pp covs relatively worn compared to retained pp. Look for molt limits among the retained pp covs, ss, and/or pp. Replacing body and flight feathers.

DCB     With two or more generations of pp covs, ss, and/or pp of undetermined age.

*Celeus elegans elegans*

Band Size: L, H

Chestnut Woodpecker • Pica-pau-chocolate

# Individuals Captured: 17

| | | |
|---|---|---|
| Wing | ♂ | 148.0–163.0 mm (157.3 ± 6.5 mm; n = 4) |
| | ♀ | 149.0–171.0 mm (161.5 ± 10.0 mm; n = 4) |
| Tail | ♂ | 88.0–101.5 mm (94.9 ± 5.5 mm; n = 4) |
| | ♀ | 86.0–114.0 mm (98.8 ± 9.9 mm; n = 6) |
| Mass | ♂ | 125.0–153.0 g (140.4 ± 12.6 g; n = 5) |
| | ♀ | 121.0–145.0 g (131.5 ± 7.4 g; n = 9) |

Similar species: Only *Celeus* with a distinct golden-yellow crown.

Brood patch: Both sexes probably develop BPs as in other woodpeckers. The only BP was observed in Nov in a ♀, and based on molt we might expect a late dry to wet season breeding schedule.

Sex: ♂ from ♀ by presence of red in malar.

Molt: Group 5/Complex Basic Strategy; FPF incomplete, DPBs incomplete. FPF in one bird included only one gr cov, but may be more or less extensive among individuals. Timing unknown, but based on three birds captured in molt, may molt from about Apr or May to Nov.

FCJ    Like adult, but darker faced and darker mottling below (Winkler et al. 1995). Iris brown.

FPF    Adult-like, replacing body feathers, some ss covs, and pp, but not pp covs, or ss. Iris brownish.

p1–p3 replaced, p4 missing, p5–p10 juvenile, all ss juvenile, mostly juvenile ss covs with a few less covs replaced (left); iris distinctly brown.

FCF    Pp covs, pp, and ss uniform in wear and without molt limits within these tracts (except for perhaps within the terts). Molt limits among the gr covs or between the med and gr covs. Iris brownish-red or deep red.

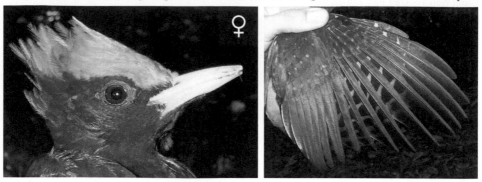

Iris brownish-red (left); less and med covs replaced, inner 2 gr covs replaced, all ss and pp covs juvenile, and all pp replaced (right).

SPB    Replacing body, flight feathers, and pp covs. Retained juvenile ss and pp covs relatively worn compared to replaced pp and ss. Iris reddish to deep red.

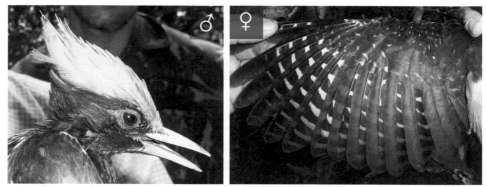

Iris red (left); s1–s5 and s7–s10 new, s6 molting, p1–p6 new, p7 molting, p8–p10 old, pp covs 1–5 old, pp covs 6 new, pp cov 7 molting, pp covs 8–10 old (right).

SCB    Adult-like with a block of retained inner juvenile pp covs and sometimes one or more ss. Iris deep red.

TPB    With a mix of retained juvenile and adult-like pp covs (and sometimes ss) being replaced by a third generation of pp covs (and ss). Iris deep red.

TCB    With three generations of pp covs, the oldest being juvenile. Perhaps also with two or rarely three generations of ss. Iris deep red.

SAB    With two generations of pp covs, the oldest being definitive basic. Iris deep red.

DPB    One or more generations of retained non-juvenile ss and pp covs being replaced by adult-like pp covs and ss. Replacing body and flight feathers. Iris deep red.

p1–p7 new, p8 growing, p9–p10 old, inner 3 pp covs replaced, all ss replaced except s7.

DCB    With two or more generations of pp covs, ss, and/or pp of undetermined age. Iris deep red.

*Campephilus rubricollis rubricollis*

Red-necked Woodpecker • Pica-pau-de-barriga-vermelha

Band Size: M (CEMAVE 2013)

# Individuals Captured: 4

Wing      171.0–179.0 mm (175.0 ± 5.7 mm; n = 2)

Tail       104.0–112.0 mm (108.0 ± 5.7 mm; n = 2)

Mass      195.0–220.0 g (209.0 ± 12.8 g; n = 3)

Similar species: From *Dryocopus* by lack of white stripe down neck onto back and by beige chest and belly.

Brood patch: Unknown if both sexes develop BPs. Timing unknown, but based on timing of molt, may breed during the wet season.

Sex: ♂ from ♀ by lack of a white cheek stripe.

Molt: Group 5/Complex Basic Strategy; FPF probably partial as in other *Campephilus* and woodpeckers (see family overview), DPBs incomplete. Timing unknown, but based on one bird captured in wing molt and one bird with tail molt (one capture was not assessed for molt) may molt from about May or Jun to Nov or Dec.

FCJ      Duller than adults, being more brownish. ♂ has a white cheek stripe with some red mixed and ♀ like adult, but with black extending into the forehead and crown (Winkler et al. 1995).

FPF      Adult-like, replacing body feathers, some ss covs, and pp, but not pp covs, or ss.

FCF      Pp covs, pp, and ss uniform in wear and without molt limits within these tracts. Perhaps with molt limits among the gr covs or between the med and gr covs.

SPB      Replacing body, flight feathers, and pp covs. Retained juvenile ss and pp covs relatively worn compared to replaced pp and ss.

SCB      Adult-like with a block of retained inner juvenile pp covs and sometimes one or more ss.

TPB      With a mix of retained juvenile and adult-like pp covs (and sometimes ss) being replaced by a third generation of pp covs (and ss).

TCB      With three generations of pp covs, the oldest being juvenile. Perhaps also with two or rarely three generations of ss.

SAB      With two generations of pp covs, the oldest being definitive basic.

s5 appears retained and definitive basic, as do 2 inner pp covs, and pp covs 5–6.

DPB      Retained ss and pp covs relatively worn compared to retained pp. Look for molt limits among the retained pp covs, ss, and/or pp. Replacing body and flight feathers.

DCB      With two or more generations of pp covs, ss, and/or pp of undetermined age.

## LITERATURE CITED

CEMAVE. 2013. Lista das espécies de aves brasileiras com tamanhos de anilha recomendados. Centro Nacional de Pesquisa e Conservação de Aves Silvestres, Cabedelo, Brasil (in Portuguese).

Guallar, S., E. Santana, S. Contreras, H. Verdugo, and A. Gallés. 2009. Paseriformes del Occidente de México: Morfometría, datación y sexado. Monografies del Museu de Ciències Naturals 5 (in Spanish).

Howell, S. N. G. 2010. Molt in North American Birds. Houghton Mifflin Harcourt, Boston, MA.

Pyle, P. 1997. Identification guide to North American birds, Part I. Slate Creek Press, Bolinas, CA.

Pyle, P., A. Engilis, and D. A. Kelt. 2015. Manual for ageing and sexing birds of Bosque Fray Jorge National Park and Northcentral Chile, with notes on range and breeding seasonality. Special Publication of the Occasional Papers of the Museum of Natural Science, Louisiana State University, Baton Rouge, LA.

Pyle, P., and S. N. G. Howell. 1995. Flight-feather molt patterns and age in North American woodpeckers. Journal of Field Ornithology 66:564–581.

Pyle, P., A. McAndrews, P. Veléz, R. L. Wilkerson, R. B. Siegel, and D. F. DeSante. 2004. Molt patterns and age and sex determination of selected southeastern Cuban landbirds. Journal of Field Ornithology 75:136–145.

Seigel, R. B., M. W. Tingley, R. L. Wilkesron, C. A. Howell, M. Johnson, and P. Pyle. 2016. Age structure of Black-backed Woodpecker populations in burned forests. Auk 133:69–78.

Winkler, H., and D. Christie. [online]. 2013. Waved Woodpecker (*Celeus undatus*). In J. del Hoyo, A. Elliott, J. Sargatal, D. A. Christie, and E. de Juana (editors), Handbook of the birds of the world alive. Lynx Edicions, Barcelona, Spain. <http://www.hbw.com>.

Winkler, H., D. A. Christie, and D. Nurney. 1995. A guide to the Woodpeckers, Piculets, and Wrynecks of the world. Pica Press, Sussex, UK.

# CHAPTER TWENTY-ONE

# Falconidae (Falcons)

# Species in South America: 26

# Species recorded at BDFFP: 9

# Species captured at BDFFP: 4

Falcons have long been believed to be closely allied to hawks given similarities of a hooked beak, a cere, and legs and feet with strong claws (except caracaras; Pyle 2008, Dittmann and Cardiff 2009). Recent accumulating genetic evidence points to the contrary, placing falcons as closely related to parrots (Psittacidae) and Passeriformes (Ericson et al. 2006, Hackett et al. 2008, Suh et al. 2011). Interestingly, this close relationship between falcons and parrots is supported by a unique shared sequence of primary and secondary molt, beginning at p4–p6 and s5, and progressing in both directions with each of these tracts (Pyle 2013). Within the Falconidae, there are three major groups: the forest-falcons (Herpetotherinae), true falcons, and caracaras (the latter two placed in Falconinae; Remsen et al. 2016). Falcons have 10 pp, 12–13 ss, and 12 rects (Pyle 2008).

Falcons follow the Complex Basic Strategy, with wing molts initiating at p4–p5, s5, or both synchronously and proceeding in both directions simultaneously, with the last feathers to be replaced including p10, p1, s1, and among s9–s13, but some falcons also have a node at the second or innermost tertial and replace proximally (Bond 1936, Miller 1941, Stresemann 1958, Willoughby 1966, Pyle 2008). Preformative molts can range from absent/limited to partial and rarely incomplete (Pyle 2005). Prebasic molts are usually complete, but may sometimes retain a few less covs or body feathers. At least in North American species, the prebasic molt often begins during breeding and may often suspend during the chick feeding period, creating suspension limits among feathers of the same generation (Pyle 2008, Howell 2010). Forest-falcons (*Micrastur* spp.) are the most regularly captured falcon at Neotropical forest banding stations, but their molt has been largely unstudied (but see Pyle 2013). It appears that preformative molts may regularly be absent to limited, and that definitive prebasic molts are complete (Pyle 2013), but certainly much more study is needed.

In general, falcons are sexually dimorphic with females being larger than males, and sometimes substantially so. Some species are also sexually dichromatic, but many are not. In most species, incubation is largely or entirely done by the females (Dittmann and Cardiff 2009), but in others both sexes incubate and can develop brood patches. Females will also develop distended cloacae (Pyle 2008).

## SPECIES ACCOUNTS

*Micrastur ruficollis concentricus*
Barred Forest-Falcon • Falcão-caburé

Band Size: M, N (CEMAVE 2013); lock-on recommended
# Individuals Captured: 20

| Wing | 151.0–185.0 mm (170.4 ± 9.1 mm; n = 10) |
|---|---|
| | ♂ 157–178 mm (mean: 169 mm; n = 23; Schwartz 1972) |
| | ♀ 170–186 mm (mean: 177 mm; n = 17; Schwartz 1972) |
| Tail | 151.0–170.0 mm (158.3 ± 7.6 mm; n = 10) |
| | ♂ 140–181 mm (mean: 157 mm; n = 23; Schwartz 1972) |
| | ♀ 151–174 mm (mean: 162 mm; n = 17; Schwartz 1972) |
| Mass | 150.0–250.0 g (198.6 ± 29.4 g; n = 16) |
| | ♂ 156–185 g (Bierregaard and Kirwan 2013) |
| | ♀ 196–269 g (Bierregaard and Kirwan 2013) |

Similar species: Smallest of four *Micrastur* at the BDFFP and most similar to *M. gilvicollis*. Iris color is variable, but probably typically gray in adults compared to white in *M. gilvicollis*. The facial skin is typically yellowish in *M. ruficollis* and orange in *M. gilvicollis*. The number of white tail bars (tips excluded) may not be reliable, ranging from 3 to 6 (usually 4[?] or up to 8[?]), usually with at least one obscured by the folded wings when perched (Howell and Webb 1995). The wing:tail ratio may be especially useful and is between 0.99 and 1.15 (Schwartz 1972), compared to between 1.16 and 1.35 in *M. gilvicollis* (Schwartz 1972, Whittaker 2002). Immature (FCJ?) *M. r. concentricus* apparently often (but not always?) has white spotting or barring on the upper tail coverts, which is lacking in immature *M. gilvicollis* (Schwartz 1972).

Brood patch: Unknown if both sexes develop BPs. Timing unknown, but based on limited observations of molt, possibly late wet to early dry season, consistent with a description in WikiAves and a Jun record noted by Bierregaard and Kirwan (2015).

Sex: ♀ > ♂.

Molt: Group 3/Complex Basic Strategy; FPF limited(?)-partial(?), DPBs complete. The FPF may involve similar replacement patterns as in *M. gilvicollis*. Based on five birds, molts at least between May and Aug.

| FCJ | With a whitish crescent behind the auriculars, buffy underwing covs and/or in underparts (or whitish), and brownish above (Ridgway 1875, Hilty 2003, Schulenberg et al. 2007). Iris brownish. As the formative feathers may be similar to juvenile feathers, it may be safest to label juvenile-like birds with undetermined body plumage limits as FCU. |
|---|---|

| FPF | Probably replacing only body or perhaps also ss covs. |
|---|---|
| FCF | Look for two generations of body feathers, perhaps with the newer feathers appearing more grayish. It may be safest to label juvenile-like birds with undetermined body plumage limits as FCU. |

SPB     Replacing brownish flight feathers with gray flight feathers. Iris may range from brownish to gray.

DCB     White underparts and variably barred below with gray back and head. Iris gray or brownish. Rufous morph is probably rare (Hilty 2003, Schulenberg et al. 2007).

DPB     Replacing brownish-gray flight feathers with gray flight feathers. Iris gray or brownish.

## *Micrastur gilvicollis* (monotypic)

Lined Forest-Falcon • Falcão-mateiro

Band Size: N, P (CEMAVE 2013); lock-on recommended

# Individuals Captured: 104

Wing    164.0–213.0 mm (184.1 ± 7.6 mm; n = 57)
        ♂ 165–195 mm (mean: 182 mm; n = 27; Schwartz 1972)
        ♀ 170–198 mm (mean: 186 mm; n = 24; Schwartz 1972)

Tail    130.0–161.0 mm (146.8 ± 6.3 mm; n = 52)
        ♂ 135–155 mm (mean: 146 mm; n = 27; Schwartz 1972)
        ♀ 130–161 mm (mean: 148 mm; n = 24; Schwartz 1972)

Mass    170.0–262.0 g (212.9 ± 19.7 g; n = 74)

Tarsus  69.1–71.9 mm (70.5 ± 2.0 mm; n = 2)

Similar species: Adults distinctive by white iris and orange facial skin, but these characters are probably variable in younger birds. The number of white tail bars (tips excluded) may not be reliable, ranging from 1 to 3 (or sometimes 4), usually with at least one obscured by the folded wings when perched (Howell and Webb 1995). The wing:tail ratio may be especially useful for distinguishing from M. *ruficollis* for all ages, being >1.16 (Schwartz 1972, Whittaker 2002). Immature M. r. *concentricus* apparently often (but not always) have white spotting or barring on the upper tail coverts, which is lacking in immature M. *gilvicollis* (Schwartz 1972).

Brood patch: Unknown if both sexes develop BPs. Timing unknown, but based on limited observations of molt, possibly late dry to early wet season.

Sex: ♀ > ♂.

Molt: Group 3 (or sometimes Group 5?)/ Complex Basic Strategy; FPF limited(?)-partial, DPBs incomplete(?)-complete. FPF includes some body, less covs, and med covs. May be found molting year-round, but probably peaks Apr to Sep.

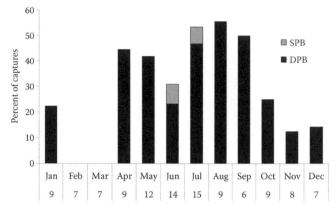

| | Jan | Feb | Mar | Apr | May | Jun | Jul | Aug | Sep | Oct | Nov | Dec |
|---|---|---|---|---|---|---|---|---|---|---|---|---|
| | 9 | 7 | 7 | 9 | 12 | 14 | 15 | 9 | 6 | 9 | 8 | 7 |

FCJ     With a whitish crescent behind the auriculars, ochraceous underwing covs and in underparts, and grayish-brown above, much like *M. ruficollis* (Hilty 2003, Schulenberg et al. 2007; see also "Similar species"). Iris brownish. It may be safest to label juvenile-like birds with undetermined body plumage limits as FCU.

FPF     Brownish-gray body feathers and less and med covs replaced by brighter gray feathers. Iris apparently remains dusky at least through the first few weeks of the preformative molt.

FCF     Look for molt limits among the body feathers and sometimes (always?) among the less and med covs. The older feathers should appear more worn and distinctly brownish-gray compared to the gray formative feathers. Iris may regularly transition from brownish to white, and look for mixed white and grayish patterns in the iris. Flight feathers will become progressively brown and relatively worn, and their rachii will fade more rapidly than in DCB. It may be safest to label juvenile-like birds with undetermined body plumage limits as FCU.

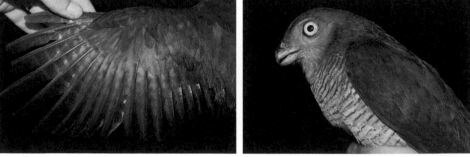

Not in active molt with about 50% of the ss covs replaced (left) and a mostly white iris with grayish flecking around the outer edge (right).

SPB    Replacing grayish-brown flight feathers with gray flight feathers. Iris should be whitish, but may still have evidence of brownish or grayish tones.

p9–p10 retained juvenile, p8 molting, p6–p7 new, p5 molting, p1–p4 retained juvenile, s1–s3 retained juvenile, s4 new (adventitious?), s5–s11 retained juvenile, s12 new.

DCB    White underparts and barred below with gray back and head. Iris white, but some birds with brownish-white irises may be distinguished as SCB (Hilty 2003, Schulenberg et al. 2007).

DPB    Replacing brownish-gray flight feathers with gray flight feathers. Iris white.

p1 appears new, s2–s3 molting, s4–s7 old, s8–s9 new, s10 old, s11 new; this appears to be an unusual molt sequence among falcons, not starting a s5 or p4–p5. Is this an anomaly or do some Micrastur regularly diverge from the expected pattern? Furthermore, s1 appears a generation older, perhaps suggesting a prior incomplete DPB.

SAB    With two generations of adult-like ss (and/or sometimes pp?). Iris white. It is not clear if this unusual plumage is a molt suspension or the result of a truly incomplete DPB.

Suspension or molt limits? All pp apparently new, s3–s5, s10 new; s1–s2, s6–s9, and s11–s12 old. Alternatively this may have resulted from an incomplete DPB, thus SAB should be used.

FALCONIDAE (FALCONS)                                                            147

*Micrastur mirandollei* (monotypic)

Slaty-backed Forest-Falcon • Tanatau

Band Size: Unknown; lock-on recommended

# Individuals Captured: 2

| | |
|---|---|
| Wing | 240.0–275.0 mm (257.5 ± 17.5 mm; n = 2)<br>224–233 mm (Chubb et al. 1916) |
| Tail | 155.0–172.0 mm (163.5 ± 8.5 mm; n = 2)<br>184 mm (Chubb et al. 1916) |
| Mass | 500 g (n = 1)<br>♂ 420 g, ♀ 500–556 g (Bierregaard and Kirwan 2013) |

Similar species: Larger and lack of barring underneath separates this species from *M. ruficollis* and *M. gilvicollis*. Also lacks partial collar of even larger *M. semitorquatus*.

Brood patch: Unknown if both sexes develop BPs. Timing unknown.

Sex: ♀ > ♂.

Molt: Group 3 (or sometimes Group 5?)/Complex Basic Strategy; FPF limited(?)-partial(?), DPBs incomplete(?)-complete. Timing unknown.

| | |
|---|---|
| FCJ | Like adult, but with scaling below (Hilty 2003, Schulenberg et al. 2007). It may be safest to label juvenile-like birds with undetermined body plumage limits as FCU. |
| FPF | Probably replacing only body or perhaps also ss covs. |
| FCF | Look for two generations of body feathers, perhaps with the newer feathers appearing more grayish. It may be safest to label juvenile-like birds with undetermined body plumage limits as FCU. |
| SPB | Unknown how to distinguish from DPB unless scaled underparts is evident. |
| DCB | White underparts and unscaled below with gray back and head. Without molt limits in the wing or tail. Iris gray (Hilty 2003, Schulenberg et al. 2007). |

(Courtesy of Angelica Hernández Palma.)     (Courtesy of Angelica Hernández Palma.)

| | |
|---|---|
| DPB | Replacing brownish-gray flight feathers with gray flight feathers without any scaling evident in underparts. Iris gray. |
| SAB | With two generations of adult-like ss (and/or sometimes pp?). Iris gray. However, it is not clear whether some DPBs are incomplete or temporarily suspend. See *M. gilvicollis*. |

*Micrastur semitorquatus semitorquatus*

Collared Forest-Falcon • Falcão-relógio

Wing    201.0 mm (n = 1)
        264–277 mm (Guyana; Chubb et al. 1916)
        246 mm (Paraguay; Dickey and Van Rossem 1938)

Tail    205.0 mm (n = 1)
        233 mm (Guyana; Chubb et al. 1916)
        248 mm (Paraguay; Dickey and Van Rossem 1938)

Mass    No data from BDFFP
        ♂ 535 g, ♀ 700 g (Venezuela; Hilty 2003)
        ♂ 479–646 g, ♀ 660–940 g (Bierregaard et al. 2015)

Similar species: Largest forest-falcon with extremely long tail (tail > wing; Ridgway 1875). Adults have partial white collar. Immatures are like some *Accipiters* in plumage, being heavily barred and spotted brown and buff, but have bare facial skin.

Brood patch: Only ♀♀ incubate and develop BPs. Timing unknown, but early to mid dry season in Guatemala (Thorstrom 2012).

Sex: ♀ > ♂.

Molt: Group 3 (or sometimes Group 5?)/Complex Basic Strategy; FPF limited, DPBs incomplete(?)-complete.

FCJ     Variable with brownish above variable spotted with buff and white to buff below variably barred or scaled brownish (Hilty 2003, Schulenberg et al. 2007). Cere and facial skin greenish-yellow.

FPF     Replacing body feathers with feathers similar in color and pattern to juvenile feathers.

FCF     Like FCJ, but with a mix of newer and older body feathers.

Given the wear on the wing and tail, it seems this is not a fresh FCJ; additionally, upper back feathers appear fresher and replaced.

| | |
|---|---|
| SPB | Replacing brown flight feathers and barred underparts with grayish-black flight feathers and unbarred underparts (which can be white, buff, or black; see DCB). |
| DCB | Three color morphs. Light morph: back blackish with white nuchal collar. Tawny morph: like light phase, but white replaced with buff. Dark phase: mostly sooty black without nuchal collar and often with faint white barring on rump, flanks, and underparts (Hilty 2003, Schulenberg et al. 2007). Cere and facial skin yellow. |
| DPB | Replacing blackish-gray flight feathers with blackish flight feathers. Underparts without barring. |
| SAB | With two generations of adult-like ss (and/or sometimes pp?). It is not clear whether, however, some DPBs are incomplete or temporarily suspend. See M. gilvicollis. |

## LITERATURE CITED

Bierregaard, R. O., and G. M. Kirwan. [online]. 2013. Slaty-backed Forest-Falcon (Micrastur mirandollei). In J. del Hoyo, A. Elliott, J. Sargatal, D. A. Christie, and E. de Juana (editors), Handbook of the birds of the world alive. Lynx Edicions, Barcelona, Spain. <http://www.hbw.com>.

Bierregaard, R. O., and G. M. Kirwan. [online]. 2015. Barred Forest-Falcon (Micrastur ruficollis). In J. del Hoyo, A. Elliott, J. Sargatal, D. A. Christie, and E. de Juana (editors), Handbook of the birds of the world alive. Lynx Edicions, Barcelona, Spain. <http://www.hbw.com>.

Bierregaard, R. O., G. M. Kirwan, and G. M. Boesman. 2015. Collared Forest-Falcon (Micrastur semitorquatus). In J. del Hoyo, A. Elliott, J. Sargatal, D. A. Christie, and E. de Juana (editors), Handbook of the birds of the world alive. Lynx Edicions, Barcelona, Spain.

Bond, J. 1936. Resident birds of the Bay Islands of Spanish Honduras. Proceedings of the Academy of Natural Sciences of Philadelphia 88:353–364.

CEMAVE. 2013. Lista das espécies de aves brasileiras com tamanhos de anilha recomendados. Centro Nacional de Pesquisa e Conservação de Aves Silvestres, Cabedelo, Brasil (in Portuguese).

Chubb, C., H. Grönvold, F. V. McConnell, H. F. Milne, P. Slud, and Bale & Danielsson. 1916. The birds of British Guiana: based on the collection of Frederick Vavasour McConnell. Bernard Quaritch, London, UK.

Dickey, D. R., and A. J. Van Rossem. 1938. The birds of El Salvador. Field Museum of Natural History, Zoological Series, Chicago, IL.

Dittmann, D. L., and S. W. Cardiff. 2009. Falcons. Pp. 73–75 in T. Harris (editor), National Geographic complete birds of the world. National Geographic Society, Washington, DC.

Ericson, P. G. P., C. L. Anderson, T. Britton, A. Elzanowski, U. S. Johansson, M. Kallerrsjo, J. I. Ohlson, T. J. Parsons, D. Zuccon, and G. Mayr.

2006. Diversification of Neoaves: integration of molecular sequence data and fossils. Biology Letters 2:543–547.

Hackett, S. J., R. T. Kimball, S. Reddy, R. C. K. Bowie, E. L. Braun, M. J. Braun, J. L. Chojnowski, W. A. Cox, K.-L. Han, J. Harshman, C. J. Huddleston, B. D. Marks, K. J. Miglia, W. S. Moore, F. H. Sheldon, D. W. Steadman, C. C. Witt, and T. Yuri. 2008. A phylogenomic study of birds reveals their evolutionary history. Science 320:1763–1768.

Hilty, S. L. 2003. Birds of Venezuela (2nd ed.). Princeton University Press, Princeton, NJ.

Howell, S. N. G. 2010. Molt in North American birds. Houghton Mifflin Harcourt, Boston, MA.

Howell, S. N. G., and S. Webb. 1995. A guide to the birds of Mexico and northern Central America. Oxford University Press, Oxford, UK.

Miller, A. H. 1941. The significance of molt centers among secondary remiges in the Falconiformes. Condor 43:113–115.

Pyle, P. 2005. Remigial molt patterns in North American Falconiformes as related to age, sex, breeding status, and life-history strategies. Condor 107:823–834.

Pyle, P. 2008. Identification guide to North American birds, Part II. Slate Creek Press, Bolinas, CA.

Pyle, P. 2013. Evolutionary implications of synapomorphic wing-molt sequences among falcons (Falconiformes) and parrots (Psittaciformes). Condor 115:593–602.

Remsen, J. V., C. D. Cadena, A. Jaramillo, M. Nores, J. F. Pacheco, M. B. Robbins, T. S. Schulenberg, F. G. Stiles, D. F. Stotz, and K. J. Zimmer. [online]. 2016. A classification of the bird species of South America. American Ornithologists' Union. Version 30 November 2016. <http://www.museum.lsu.edu/~Remsen/SACCBaseline.htm>.

Ridgway, R. 1875. Studies of the American Falconidae: monograph of the genus Micrastur. Proceedings of the Academy of Natural Sciences of Philadelphia 27:470–502.

Schulenberg, T. S., D. F. Stotz, D. F. Lane, J. P. O'Neill, and T. A. Parker. 2007. Birds of Peru. Princeton University Press, Princeton, NJ.

Schwartz, P. 1972. *Micrastur gilvicollis*, a valid species sympatric with M. *ruficollis* in Amazonia. Condor 74:399–415.

Stresemann, V. 1958. Sind die Falconidae ihrer Mauserweise nach eine einheitliche Gruppe? Journal für Ornithologie 99:81–88 (in German).

Suh, A., M. Paus, M. Kiefmann, G. Churakov, F. A. Franke, J. Brosius, J. O. Kriegs, and J. Schmitz. 2011. Mesozoic retroposons reveal parrots as the closest living relatives of passerine birds. Nature Communications 2:443–448.

Thorstrom, R. K. 2012. Collared Forest-Falcon. Pp. 250–264 in D. F. Whitacre (editor), Neotropical birds of prey: biology and ecology of a Forest Raptor community. Cornell University Press, Ithaca, NY.

Whittaker, A. 2002. A new species of forest-falcon (Falconidae: *Micrastur*) from southeastern Amazonia and the Atlantic rainforests of Brazil. Wilson Bulletin 114:421–445.

Willoughby, E. J. 1966. Wing and tail molt of the Sparrow Hawk. Auk 83:201–206.

PART THREE

# Passerines, the Suboscines

# CHAPTER TWENTY-TWO

# Thamnophilidae (Antbirds)

# Species in South America: 229

# Species Recorded at BDFFP: 28

# Species Captured at BDFFP: 22

Antbirds are in an insectivorous family that includes many forest understory species and subcanopy species. Most are well suited for living in tropical habitats, including forest interior and second growth, as well as various specialized microhabitats like gaps and vine tangles. The most unifying feature of the family is their bill shape, which has a terminal hook and a pinched base. Many species are sexually dichromatic with males being grayish to black with variably white wingbars, wing spots, tail spots, concealed intrascapular patch, or other black-and-white patterning. The females are often brownish, rufous-orange, or tawny, often similar in pattern to the male, except that blacks are replaced by brown colors contrasting with brighter orange or buffy barring. Although sexual dichromatism is common in the family, obligate ant-followers often either minimize this (e.g., *Gymnopithys*) or are not sexually dichromatic (e.g., *Pithys*).

All antbirds appear to follow the Complex Basic Strategy (Ryder and Wolfe 2009, Wolfe et al. 2009, Johnson and Wolfe 2014). The first preformative molt can be partial to complete and may be related to the amount of time self-sufficient young remain within their parent's territory (Wolfe et al. 2009, Johnson and Wolfe 2014). For species that follow their parents for some time after fledging, young males often retain a female-like plumage through a partial preformative molt such that young birds can be distinguished by their molt limits, but their sex may not be known. However, juveniles of species that probably disperse away from their parents at an early age appear to have formative plumages similar

to that of the adult plumage of the same sex. Several species, including many obligate ant-following Thamnophilids (e.g., *Pithys*, *Gymnopithys*, *Phaenostictus*, and *Rhegmatorhina*), exhibit complete preformative molts resulting in an adult-like plumage (Willis 1969, 1973; Johnson and Wolfe 2014). Interestingly, *Percnostola rufifrons* has two inserted molts within the first cycle: a partial auxiliary preformative molt followed by a complete preformative molt (Johnson and Wolfe 2014). The variation in preformative molt extents across this large family presents an interesting opportunity to explore the role of phylogeny, habitat use, and solar exposure (see also Johnson and Wolfe 2014), as patterns have emerged in North American passerines (Pyle 1998).

The skull completes in adults, probably at about six months of age, but more study is needed to pin down the exact timing, and this may vary within and across species by several months or more. Both sexes often share in incubation and develop brood patches, but brood patches usually appear more developed and are more frequently seen in females, although in at least some species brood patches are equal in size between males and females (Schwartz 2008). In most species, the timing of breeding often corresponds with the dry season, but there are exceptions, as well as some asynchrony in these populations with some individuals breeding at any time of the year, and sometimes when less than one year of age (during the formative plumage), as in many temperate Passeriformes. Molt often occurs after the breeding season, but can again be highly asynchronous within a species. Molt-breeding overlap is generally frequent, but occurs more at the population level than the individual level (Johnson et al. 2012, see also Pyle et al. 2016). In at least *Pithys*, but perhaps also other slow-molting species, molt can suspend for breeding (see also Pyle et al. 2016).

## SPECIES ACCOUNTS

### Cymbilaimus lineatus lineatus

Fasciated Antshrike • Papa-formiga-barrado

Band Size: G, F

# Individuals Captured: 24

| Wing | ♂ | 71.0–77.0 mm (72.8 ± 2.0 mm; n = 9) |
|------|---|-------------------------------------|
| | ♀ | 70.5–77.0 mm (73.4 ± 2.5 mm; n = 8) |
| Tail | ♂ | 65.0–72.0 mm (68.2 ± 2.9 mm; n = 6) |
| | ♀ | 68.0–79.0 mm (71.0 ± 4.6; n = 5) |
| Mass | ♂ | 29.0–39.0 g (35.0 ± 2.9 g; n = 13) |
| | ♀ | 30.0–37.5 g (32.9 ± 2.2 g; n = 9) |
| Bill | ♂ | 13.9 g (n = 1) |
| | ♀ | 14.5 g (n = 1) |
| Tarsus | ♂ | 29.5 g (n = 1) |
| | ♀ | 28.9 g (n = 1) |

Similar species: Fine black and white barring throughout body of ♂ is unique at BDFFP. ♀ most like ♀ of *Frederickena viridis*, but smaller with barred black and white back and wings.

Skull: Probably completes (n = 5), probably during FPF or perhaps also FCF.

Brood patch: Observed in three ♀♀ and one ♂, so both sexes likely share in incubation as in other antbirds. Occurs at least in Aug and probably mainly during the dry season.

Sex: ♂ black with white barring and more solid black cap. ♀ dark brown above with buff barring, buffy below with dark barring, and bright rufous-orange crown.

Molt: Group 3 (or sometimes Group 4?)/Complex Basic Strategy; FPF probably incomplete or eccentric (Ryder and Wolfe 2009) or sometimes complete(?), DPBs complete. The variation in extent of FPF is not well known. American Museum of Natural History specimens of this subspecies (n = 4) and C. l. *intermedius* from Amazonas and Pará (n = 10) suggest that the FPF molt may occasionally complete (skull <100%, n = 2), with others retaining 2 or 3 inner ss (among s4–s7) and/or p9–p10 (n = 2). The timing of molt is not well understood, but should follow breeding, perhaps mainly during the late dry through the wet season.

FCJ      ♂ ≠ ♀. ♂ patterned like ♀, but perhaps blacker and with narrower pale markings on outer webbing of pp and ss. ♀ generally similar to adult ♀, but may have black barring on rusty-brown crown as in C. l. *intermedius*. Iris probably brownish. Skull is unossified.

FPF      ♂ ≠ ♀. Adult-like feathers (including flight feathers) replacing juvenile plumage. Iris brownish to reddish. Skull is unossified, but perhaps sometimes ossified especially if FPF goes to completion or near-completion.

(AMNH 120942.)

FCF  ♂ ≠ ♀. Both sexes like DCB, but look for molt limits among pp, terts, and/or ss. Iris dull brownish-orange to becoming bright red. "Subadult" (perhaps referring to FCF?) ♂♂ may have broader white barring on crown and body (Zimmer and Isler 2003). Skull is variably ossified, but probably usually completely ossified.

p1–p8 new, p9–p10 juvenile; s1–s2, s7–s9 new, s3–s6 juvenile.

FAJ  ♂ ≠ ♀. Without molt limits among the pp, terts, or ss. Iris red. Skull is ossified.

SPB  ♂ ≠ ♀. With mix of juvenile and adult-like feathers being replaced by adult-like feathers. Iris red. Skull is ossified.

UPB  ♂ ≠ ♀. With adult-like feathers being replaced by adult-like feathers. Iris red. Skull is ossified.

p1–p2 new, p3 missing, p4–p10 old and adult-like; s1–s8 old and adult-like, s9 new.

## *Frederickena viridis* (monotypic)

Black-throated Antshrike • Borralhara-do-norte

Band Size: H

# Individuals Captured: 125

| | | |
|---|---|---|
| Wing | ♂ | 89.9–98.0 mm (92.9 ± 2.3 mm; n = 35) |
| | ♀ | 87.0–98.0 mm (92.6 ± 2.9 mm; n = 32) |
| Tail | ♂ | 67.0–80.0 mm (74.8 ± 3.0 mm; n = 33) |
| | ♀ | 66.0–81.0 mm (76.4 ± 3.4 mm; n = 22) |
| Mass | ♂ | 58.7–77.5 g (67.2 ± 4.0 g; n = 48) |

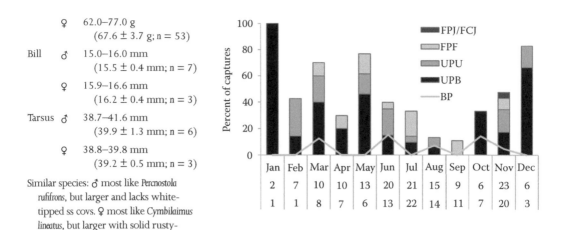

♀ 62.0–77.0 g
(67.6 ± 3.7 g; n = 53)

Bill ♂ 15.0–16.0 mm
(15.5 ± 0.4 mm; n = 7)

♀ 15.9–16.6 mm
(16.2 ± 0.4 mm; n = 3)

Tarsus ♂ 38.7–41.6 mm
(39.9 ± 1.3 mm; n = 6)

♀ 38.8–39.8 mm
(39.2 ± 0.5 mm; n = 3)

Similar species: ♂ most like *Percnostola rufifrons*, but larger and lacks white-tipped ss covs. ♀ most like *Cymbilaimus lineatus*, but larger with solid rusty-brown wings and upperparts.

Skull: Completes, probably during or perhaps rarely just following the FPF (n = 22).

Brood patch: Develops in both sexes, although more extensive in ♀♀. May occur year-round, but perhaps most regularly during the early to mid dry season (about Jun–Oct).

Sex: ♂ unmarked charcoal gray underparts, back, and wing with faintly barred gray tail. ♀ finely barred black and white underparts, tail, and face; crown, back, and wings are solid rusty-brown.

Molt: Group 4/Complex Basic Strategy; FPF complete, DPBs complete. Molt probably peaks from Dec–May and is least frequent from Aug–Oct. Molt is protracted.

FCJ ♂ = ♀. Like adult ♀, but the iris is dull brown and the mandible has extensive yellow. The remiges are unmarked brown and the chest is less boldly patterned than in definitive ♀♀ and barred black, gray, and white. There seems to be variation in the strength of barring on FCJ and ♀ FAJ rects that may be age/sex related, but more study is needed. Skull is unossified.

FPF ♂ ≠ ♀. ♂ is mixed gray and brown. Dull brown iris, yellowish bill, and unossified skull are especially useful for separating FPF ♀♀ from DPB ♀♀, but during the FPF, the iris becomes red, the bill becomes black, and the skull becomes ossified, so adult-like females late in molt may be best aged as UPU.

In two stages of FPF: just beginning (left two images) and nearly complete (right two images).

FAJ ♂ ≠ ♀. Without molt limits among the pp, terts, or ss. Skull is ossified.

UPB ♂ ≠ ♀. With adult-like feathers being replaced by adult-like feathers. Iris red. Skull is ossified.

p1–p2 new, p3 molting, p4–p10, and all ss retained and adult-like (left); p1–p7 new, p8 molting, p9–p10 retained and adult-like, s1–s2 and s8–s9 new, s5 molting (adventitious?), and s3–s4 and s6–s7 retained and adult-like (right).

## *Thamnophilus murinus murinus*

Mouse-colored Antshrike • Choca-murina

Band Size: E

\# Individuals Captured: 263

| | | |
|---|---|---|
| Wing | ♂ | 54.0–65.0 mm (60.4 ± 1.8 mm; n = 86) |
| | ♀ | 56.0–64.0 mm (60.1 ± 1.8 mm; n = 69) |
| Tail | ♂ | 47.0–56.0 mm (51.0 ± 2.0 mm; n = 73) |
| | ♀ | 48.0–59.0 mm (52.6 ± 2.1 mm; n = 65) |
| Mass | ♂ | 15.1–22.0 g (17.5 ± 1.1 g; n = 126) |
| | ♀ | 15.0–21.5 g (18.0 ± 1.4 g; n = 94) |
| Bill | ♂ | 10.2–12.9 mm (11.6 ± 0.6 mm; n = 19) |
| | ♀ | 10.7–12.4 mm (11.4 ± 0.5 mm; n = 15) |
| Tarsus | ♂ | 20.0–23.8 mm (21.8 ± 1.0 mm; n = 19) |
| | ♀ | 20.9–23.2 mm (22.2 ± 0.6 mm; n = 17) |

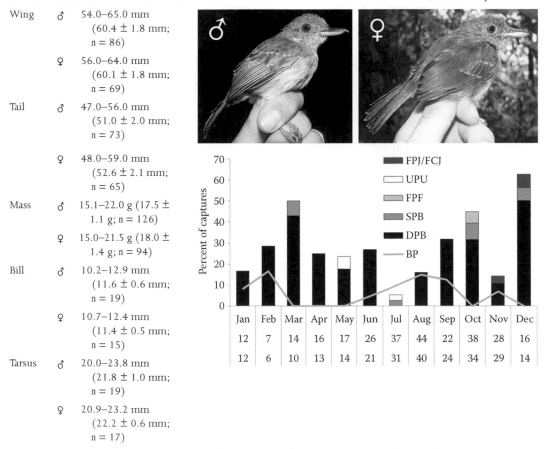

Similar species: ♂ from other medium-sized, all-gray antbirds (e.g., *Thamnomanes caesius*) by grayish iris, brownish wings, and faint white tips to ss covs. ♀ from other brown antbirds by contrastingly brighter rufous crown, grayish iris, and overall lack of pattern.

Skull: Completes during the FCF (n = 57).

Brood patch: Develops in both sexes, although more extensive in ♀♀. Seen from Jun–Feb and probably peaks in Aug–Sep; most juveniles are seen during the late dry season (Oct–Dec), but there is a record from Apr.

Sex: ♂ is grayish overall with brownish-edged gray pp and ss. ♀ is all brown with contrastingly brighter chestnut crown.

Molt: Group 3/Complex Basic Strategy; FPF partial-incomplete, DPBs complete. FPF includes body feathers, a variable amount of rects (from none to all?), and usually all ss covs except pp covs. The alula is not replaced, but 1 or 2 alula covs are (always?). The proportion of individuals in DPB molt is lowest in Jun–Jul, but otherwise may occur year-round.

FCJ      ♂ = ♀. Both sexes like DCB ♀ (at least in *T. m. canipennis*, LSUMZ), but duller and darker. ♂ remiges may average more grayish, whereas ♀ remiges may average more brownish. Pp covs are dull dark brown with paler dull brown edging. The outer webbing of p9 and p10 appears to be consistently broader than in DCB, but more study is needed. The iris may be dull brownish-gray to gray. Rects are relatively pointed. Skull is unossified.

FPF      ♂ ≠ ♀. Adult-like body feathers replacing juvenile plumage. Skull is unossified.

FCF      ♂ ≠ ♀. Retained rects are longer and more pointed than replaced rects, although all rects may be replaced. The outer webs of p9–p10 may be relatively broad. Molt limits between the pp and gr covs as well as between the a1 or a2 and the greater alula. ♂ like DCB, but with more brownish pp covs, pp, and ss. Replaced med and gr covs average more brownish-gray often with slightly larger buffy tips. The central belly is usually buffy-gray. ♀ is similar to DCB ♀, but with duller brown edging to pp covs, pp, and ss. The iris is gray or brownish-gray. Skull is variably unossified to ossified.

In both images, all gr covs are replaced, as is the alula covert (right), and the alula covert and lesser alula (left).

SPB      ♂ ≠ ♀. Replacing juv flight feathers with adult flight feathers. May be difficult in ♀♀ to distinguish from DPB (use UPB). Skull is ossified.

r1 and r3 replaced, r2 and r4–r6 retained.

p7–p10 and associated pp covs retained juvenile, p6 molting, p1–p5 new, s1–s2 new, s3 molting, s4–s6 retained juvenile, s7 missing, s8–s9 new (left). p6–p10 and associated pp covs retained juvenile, p5 missing, p4 molting, p1–p3 new, s1 new, s2–s6 retained juvenile, s7 new, s8 missing, s9 new (right).

DCB      ♂ ≠ ♀. Without molt limits among the med, gr, and pp covs. The outer webs of p9–p10 may average narrower than FCJ/FCF. ♂ has brownish-gray edged gray pp and ss. Med and gr covs are bright olive-gray with a small white spot at the tip. Iris is gray. ♀ has contrastingly brighter chestnut crown. Pp covs are rich brown with bright brown edging. Pp and ss are rich brown. The iris is gray or brownish-gray. Skull is ossified.

DPB      ♂ ≠ ♀. Replacing adult-like body and flight feathers with adult-like feathers. May be difficult in ♀♀ to distinguish from SPB (use UPB). Skull is ossified.

p4–p10 and associated pp covs retained definitive basic, p3 molting, p1–p2 new, s1–s6 retained definitive basic, s7–s9 new.

*Thamnophilus punctatus punctatus*

Northern Slaty-Antshrike • Choca-bate-cabo

| Wing | ♂ | No data from the BDFFP |
| | | 64–69 mm (66.2 ± 1.7 mm; n = 10; Isler et al. 1997) |
| | ♀ | 63.0–66.5 mm (65.3 ± 2.0 mm; n = 3) |
| Tail | ♂ | No data from the BDFFP |
| | | 52–56 mm (53.9 ± 1.4 mm; n = 10; Isler et al. 1997) |
| | ♀ | 53.0–57.0 mm (55 ± 2.0 mm; n = 3) |
| Mass | ♂ | No data from the BDFFP |
| | ♀ | 16.9–23.7 mm (19.8 ± 3.5 mm; n = 3) |

Similar species: Similar in size to *T. murinus*, but with bold white spots on outer webbing of less, med, and gr covs in ♂♂ and ♀♀.

Skull: Probably completes during FPF or FCF (n = 1).

Brood patch: Probably develops in both sexes as in *T. atrinucha* (Skutch 1934, Johnson 1953), although probably more extensive in ♀♀. Timing unknown.

Sex: ♂ is dark gray overall with contrasting black crown, wing, and tail. Ss covs, terts, and rects also with bold white spots. ♀ is similar to ♂ in pattern, but brown and rufous-brown instead of dark gray and black. White spotting on the covs are mostly on the outer web and are bordered by black, which blends into an otherwise brown feather. The iris is reddish-brown.

Molt: Group 3/Complex Basic Strategy; FPF partial-incomplete, DPBs complete. The FPF was erroneously reported as complete by Johnson and Wolfe (2014), but with more study (A. Diaz, pers. comm.) appears to be similar to *T. murinus* (see earlier) and *T. atrinucha* (Wolfe et al. 2009), including body feathers, a variable amount of rects (from none to all?), and usually all ss covs except pp covs, but also 0–3 terts. Timing unknown, but an FPF ♀ with body molt captured in Jul was nearly completing wing molt when she was recaptured in Dec.

FCJ      ♂ ≠ ♀. ♀-like, but with reduced blacks in ss covs, reduced rufous in crown, and brown iris. ♂♂ with grayish remiges and pp covs, which are more brownish in ♀♀. Skull is unossified.

FPF      ♂ ≠ ♀. Adult-like body and flight feathers replacing juvenile plumage. Iris is brown, but may become brownish-red upon completion. Skull is unossified.

FCF      ♂ ≠ ♀. Retained rects are longer and more pointed than replaced rects, although all rects may occasionally be replaced. Molt limits between the pp and gr covs as well as between the a1 or a2 and the greater alula are distinct in ♂♂, but subtler in ♀♀. ♀ is similar to DCB ♀, but with duller brown edging to pp covs, pp, and ss. In both sexes, replaced terts have a broader and brighter leading edge than retained juvenile terts and ss. The iris is brownish-red to red. Skull variably unossified to ossified.

SPB      ♂ ≠ ♀. Replacing juv flight feathers with adult flight feathers. May be difficult in ♀♀ to distinguish from DPB (use UPB). Skull is ossified. Iris is red or still brownish. Skull is ossified.

Iris reddish-brown (left); molting p10 and s6 not yet replaced (right).

DCB     ♂ ≠ ♀. Without molt limits between the gr covs and pp covs, or among the alulas. ♂ has blackish pp and ss. Pp covs in ♀♀ are rich brown with bright brown edging. Pp and ss are rich brown. Iris is red. Skull is ossified.

(Courtesy of Lindsay Wieland.)

DPB     ♂ ≠ ♀. Replacing adult-like body and flight feathers with adult-like feathers. May be difficult in ♀♀ to distinguish from SPB (use UPB). Iris is red. Skull is ossified.

## *Thamnomanes ardesiacus obidensis*

Band Size: E

Dusky-throated Antshrike • Cirapuru-de-garganta-preta        # Individuals Captured: 1040

| | | |
|---|---|---|
| Wing | ♂ | 63.0–78.8 mm (72.7 ± 2.0 mm; n = 318) |
| | ♀ | 66.0–78.0 mm (72.4 ± 2.1 mm; n = 220) |
| Tail | ♂ | 45.0–58.0 mm (51.4 ± 2.0 mm; n = 293) |
| | ♀ | 45.0–57.0 mm (52.3 ± 1.7 mm; n = 201) |
| Mass | ♂ | 15.0–22.5 g (18.1 ± 1.1 g; n = 506) |
| | ♀ | 15.0–22.0 g (18.3 ± 1.3 g; n = 343) |
| Bill | ♂ | 10.0–12.4 mm (11.2 ± 0.5 mm; n = 52) |
| | ♀ | 7.8–12.0 mm (11.2 ± 0.8 mm; n = 32) |
| Tarsus | ♂ | 20.0–23.1 mm (21.7 ± 0.6 mm; n = 52) |
| | ♀ | 20.3–23.4 mm (21.9 ± 0.7 mm; n = 31) |

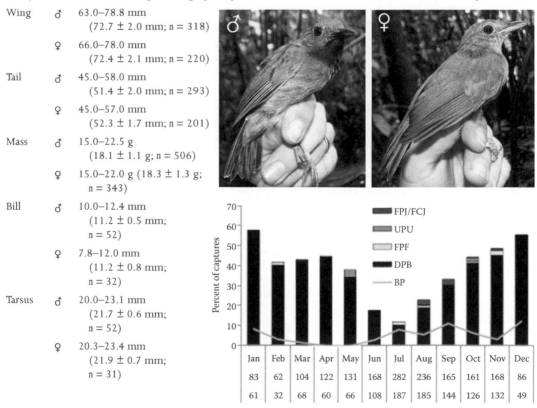

| | Jan | Feb | Mar | Apr | May | Jun | Jul | Aug | Sep | Oct | Nov | Dec |
|---|---|---|---|---|---|---|---|---|---|---|---|---|
| | 83 | 62 | 104 | 122 | 131 | 168 | 282 | 236 | 165 | 161 | 168 | 86 |
| | 61 | 32 | 68 | 60 | 66 | 108 | 187 | 185 | 144 | 126 | 132 | 49 |

Legend: FPJ/FCJ, UPU, FPF, DPB, BP. Percent of captures.

Similar species: Both sexes from other antbirds by lacking wing markings and by measurements (almost no overlap in tail length to separate *Thamnomanes*). ♂ from *T. caesius* by black throat, and presence of a narrow blue eyering. ♀ from *T. caesius* by duller orange underparts. *T. caesius* also has larger white interscapular patch in both sexes.

Skull: Completes during FCF (n = 230).

Brood patch: Develops in both sexes, although more extensive in ♀♀. Observed nearly year-round, except Apr–May, but most frequent in Jul–Dec.

Sex: Distinguishable upon FPJ. ♂ is gray overall with minimal grayish concealed intrascapular patch and variably small whitish tips on the rects. The throat is black, but variably so to almost absent. ♀ is ochre-brown overall, paler on the underparts with a faint grayish concealed intrascapular patch.

Molt: Group 3/Complex Basic Strategy; FPF partial-(rarely incomplete), DPBs complete. FPF includes less covs, med covs, and 0 to all gr covs (usually inner 3 or 4), rarely 1–3 (usually 0) terts, 0 to 3 rects, and rarely inner 1 or 2 pp. The frequency of molt peaks from Nov–Apr and is least frequent in Jul.

FCJ     ♂ ≠ ♀. ♂ is mottled brown and gray with more gray than brown. Rects, some body feathers, and flight feathers are brownish-gray. ♀ is brown with rust tones to back and ss covs. Rects are brown and pointed. Skull is unossified.

FPF     ♂ ≠ ♀. Replacing juvenile body feathers, ss covs, and sometimes some terts or rects with adult-like feathers. Skull is unossified.

Some less and med covs replaced, gr covs and all flight feathers still retained.

FCF    ♂ ≠ ♀. Sexes similar to adults of same sex, but look for molt limits among the gr covs or rarely between the gr covs and pp covs and/or among the ss and pp. This is much easier to see in ♂♂ than ♀♀. Retained rects are distinctly pointed, but some rects may be replaced and are more rounded. Skull is either unossified or ossified.

Inner 3 gr covs and all other ss covs replaced (both); a1–a2 limit (left) or no alula limit (right).

SPB    ♂ ≠ ♀. Replacing juv flight feathers with adult flight feathers. May be difficult in ♀♀ to distinguish from DPB (use UPB). Skull is ossified.

p8–p10 and associated pp covs retained juvenile, p7 missing, p6 molting, p1–p5 new, s1 new, s2 missing, s3–s6 retained juvenile, s7–s9 new (left). p5–p10 and associated pp covs retained juvenile, p4 missing, p1–p3 new, s1–s7 and s9 retained juvenile, s8 new (center). r1, r4 left, r5 right, and r6 all molting, other rects retained juvenile (right).

DCB    ♂ ≠ ♀. Without molt limits among the gr covs, terts, or rects. Rects are slightly pointed or distinctly rounded. Skull is ossified.

DPB    ♂ ≠ ♀. Replacing adult-like body
       and flight feathers with adult-like
       feathers. May be difficult in ♀♀ to
       distinguish from SPB (use UPB).
       Skull is ossified.

p4–p10 retained definitive basic, p1–p3 new, s1–s7 and s9 retained definitive basic, s8 new.

## *Thamnomanes caesius glaucus*

Cinerous Antshrike • Ipecuá

Band Size: E

# Individuals Captured: 1119

| | | |
|---|---|---|
| Wing | ♂ | 67.0–77.0 mm (71.6 ± 1.9 mm; n = 278) |
| | ♀ | 66.0–76.5 mm (71.1 ± 1.8 mm; n = 248) |
| Tail | ♂ | 55.0–70.5 mm (62.9 ± 2.7 mm; n = 255) |
| | ♀ | 56.0–71.0 mm (63.4 ± 2.6 mm; n = 224) |
| Mass | ♂ | 14.2–21.5 g (17.6 ± 1.2 g; n = 489) |
| | ♀ | 14.0–21.5 g (17.5 ± 1.3 g; n = 392) |
| Bill | ♂ | 9.8–12.5 mm (11.0 ± 0.5 mm; n = 50) |
| | ♀ | 9.4–11.4 mm (10.6 ± 0.4 mm; n = 38) |
| Tarsus | ♂ | 18.7–20.8 mm (19.8 ± 0.6 mm; n = 51) |
| | ♀ | 18.9–22.6 mm (20.2 ± 0.8 mm; n = 38) |

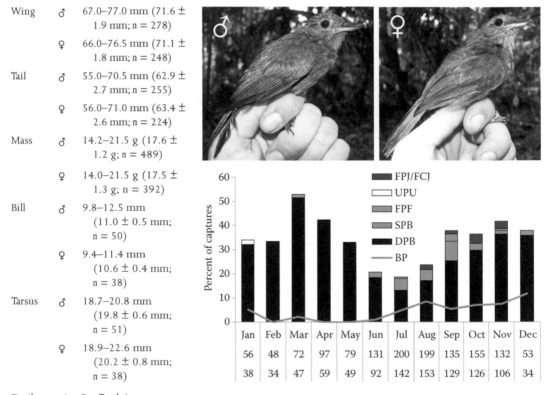

| | Jan | Feb | Mar | Apr | May | Jun | Jul | Aug | Sep | Oct | Nov | Dec |
|---|---|---|---|---|---|---|---|---|---|---|---|---|
| | 56 | 48 | 72 | 97 | 79 | 131 | 200 | 199 | 135 | 155 | 132 | 53 |
| | 38 | 34 | 47 | 59 | 49 | 92 | 142 | 153 | 129 | 126 | 106 | 34 |

Similar species: See *T. ardesiacus*.

Skull: Completes during FCF (n = 228).

Brood patch: Develops in both sexes, although more extensive in ♀♀. Observed from Jul–Jan, peaking in Oct.

Sex: ♂ is gray overall with white concealed intrascapular patch and variably small whitish tips on the rects. ♀ is ochre-brown overall, deep ochre-orange on the underparts with white concealed intrascapular patch.

Molt: Group 3/Complex Basic Strategy; FPF partial-(incomplete), DPB complete. FPF includes the less covs, med covs, and 0 to all gr covs (usually inner 3), and rarely some or all of the terts. Can occur at any time of the year, peaking (>50%) in Mar and least frequent (<20%) in Jul.

**FCJ**   ♂ ≠ ♀. ♂ is mottled brown and gray with grayish-brown remiges and rects. ♀ is brown with rust tones to back and ss covs, and rects are brown and distinctly pointed. Skull is unossified.

**FPF**   ♂ ≠ ♀. Replacing juvenile body feathers, ss covs, and sometimes some terts or rects with adult-like feathers. Skull is unossified.

Some body and less covs replaced, other ss covs, ss, and pp retained juvenile.

FCF     ♂ ≠ ♀. Sexes similar to adults of same sex, but look for molt limits among the gr covs or rarely between the gr covs and pp covs and/or among the terts. This is much easier to see in ♂♂ than ♀♀. Retained rects are distinctly pointed, but some rects may be replaced and are more rounded. Skull is either unossified or ossified.

Inner 3 gr covs and s8 replaced, all other gr covs and flight feathers retained juvenile (left); inner 4 gr covs and s8 replaced, all other gr covs and flight feathers retained juvenile (right).

SPB     ♂ ≠ ♀. Replacing juvenile flight feathers with adult flight feathers. May be difficult in ♀♀ to distinguish from DPB (use UPB). Skull is ossified.

p5–p10 retained juvenile, p4 molting, s1–s3 new, s1 and s7–s9 new, s2–s6 retained juvenile.

DCB     ♂ ≠ ♀. Without molt limits among the gr covs, terts, or rects. Rects are slightly pointed or distinctly rounded. Skull is ossified.

DPB ♂ ≠ ♀. Replacing adult-like body and flight feathers with adult-like feathers. May be difficult in ♀♀ to distinguish from SPB (use UPB). Skull is ossified.

p8–p10 retained definitive basic, p7 molting, p1–p6 new, s1 and s8–s9 new, s2–s7 retained definitive basic.

## *Epinecrophylla gutturalis* (monotypic)

Brown-bellied Antwren • Choquinha-de-barriga-parda

Band Size: D

\# Individuals Captured: 702

| | | |
|---|---|---|
| Wing | ♂ | 45.5–53.5 mm (49.8 ± 1.4 mm; n = 183) |
| | ♀ | 46.0–54.6 mm (49.7 ± 1.4 mm; n = 172) |
| Tail | ♂ | 34.0–46.0 mm (40.2 ± 2.0 mm; n = 160) |
| | ♀ | 34.0–44.0 mm (40.1 ± 2.2 mm; n = 133) |
| Mass | ♂ | 7.0–11.0 g (8.5 ± 0.7 g; n = 299) |
| | ♀ | 7.0–11.0 g (8.8 ± 0.8 g; n = 250) |
| Bill | ♂ | 7.5–9.8 mm (8.7 ± 0.7 mm; n = 21) |
| | ♀ | 7.9–9.4 mm (8.8 ± 0.4 mm; n = 19) |
| Tarsus | ♂ | 16.8–18.9 mm (17.9 ± 0.6 mm; n = 20) |
| | ♀ | 16.2–19.0 mm (17.8 ± 0.7 mm; n = 19) |

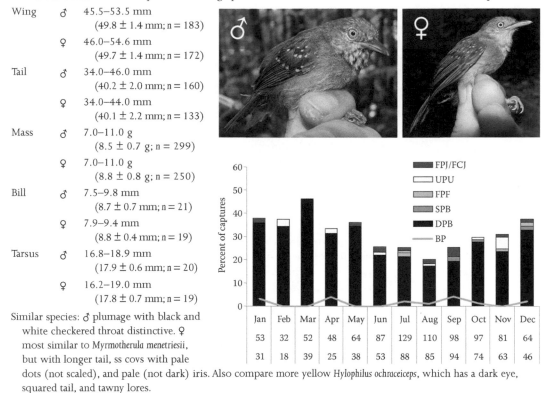

| | Jan | Feb | Mar | Apr | May | Jun | Jul | Aug | Sep | Oct | Nov | Dec |
|---|---|---|---|---|---|---|---|---|---|---|---|---|
| | 53 | 32 | 52 | 48 | 64 | 87 | 129 | 110 | 98 | 97 | 81 | 64 |
| | 31 | 18 | 39 | 25 | 38 | 53 | 88 | 85 | 94 | 74 | 63 | 46 |

Similar species: ♂ plumage with black and white checkered throat distinctive. ♀ most similar to *Myrmotherula menetriesii*, but with longer tail, ss covs with pale dots (not scaled), and pale (not dark) iris. Also compare more yellow *Hylophilus ochraceiceps*, which has a dark eye, squared tail, and tawny lores.

Skull: Nearly (>90%) or more usually completely ossifies during FCF (n = 167).

Brood patch: Develops in both sexes, although more extensive in ♀♀. Observed Jul–Jan and also in Apr, but may breed year-round, peaking in Jul–Sep.

Sex: ♂ has a checkered black and white throat and underparts are gray. ♀ has a plain brown throat concolor with the underparts.

Molt: Group 3/Complex Basic Strategy; FPF partial, DPBs complete. FPF includes body feathers, less covs, med covs, usually no (but occasionally inner 1 or 2) gr covs, and no flight feathers. Molt frequency peaks (>30%) in Dec–May and is least frequent (<20%) in Aug.

FCJ     ♂ = ♀. Chestnut-brown on upperparts, weakly defined buffy spotting in ss covs (although beware of DCB ♀♀ with similar spotting), and pp covs faintly tipped buffy to pale brown with less contrast and weaker edging than the similar, but brighter DCB. Gr covs are brown at the base without an olive wash. It is possible that ♀♀ have browner throats than ♂♂, but more study is needed. The iris is dark gray. Skull is unossified.

FPF     ♂ ≠ ♀. Replacing juvenile body feathers, ss covs, and sometimes some terts or rects with adult-like feathers. Gray iris quickly changes to tawny brown then creamy white, often soon after FPF commences. Look for incoming feathers in the throat to determine sex. Skull is unossified.

FCF     ♂ ≠ ♀. Like DCB, except with molt limits between replaced med covs and retained gr covs and pp covs (or occasionally some inner gr covs are replaced). Molt limits can be difficult to see in ♀♀. Skull is either unossified or ossified.

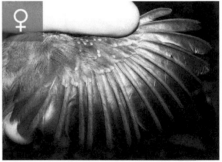

All less and med covs replaced, gr covs, alulas, and flight feathers retained (both images).

SPB     ♂ ≠ ♀. Replacing juvenile flight feathers with adult flight feathers. May be difficult in ♀♀ to distinguish from DPB (use UPB). Skull is ossified, but perhaps sometimes with small windows.

p1 new, p2 molting, p3 missing, p4–p10, associated pp covs retained, and all ss retained juvenile (left); r5–r6 replaced, r1–r4 retained, suggesting a centripetal replacement pattern (right).

DCB     ♂ ≠ ♀. Without molt limits among the gr covs, terts, or rects. Rects are slightly pointed or distinctly rounded. The iris is creamy white in ♂♂ and pale brown in ♀♀. Skull is ossified.

DPB     ♂ ≠ ♀. Replacing juvenile adult-like body and flight feathers with adult-like feathers. May be difficult in ♀♀ to distinguish from SPB (use UPB). Skull is ossified.

## *Isleria guttata* (monotypic)

Rufous-bellied Antwren • Choquinha-de-barriga-ruiva

Band Size: D, E

# Individuals Captured: 281

| | | |
|---|---|---|
| Wing | ♂ | 46.0–54.0 mm (50.3 ± 1.6 mm; n = 82) |
| | ♀ | 47.0–54.0 mm (49.6 ± 1.4 mm; n = 58) |
| Tail | ♂ | 20.0–28.0 mm (24.5 ± 1.6 mm; n = 77) |
| | ♀ | 22.0–28.0 mm (24.5 ± 1.3 mm; n = 48) |
| Mass | ♂ | 8.5–12.0 g (10.2 ± 0.7 g; n = 142) |
| | ♀ | 8.0–13.0 g (10.5 ± 0.8 g; n = 87) |
| Bill | ♂ | 8.3–8.8 mm (8.5 ± 0.3 mm; n = 3) |
| | ♀ | 8.5–8.6 mm (8.6 ± 0.1 mm; n = 2) |
| Tarsus | ♂ | 21.9–22.6 mm (22.3 ± 0.4 mm; n = 3) |
| | ♀ | 22.5 mm (n = 1) |

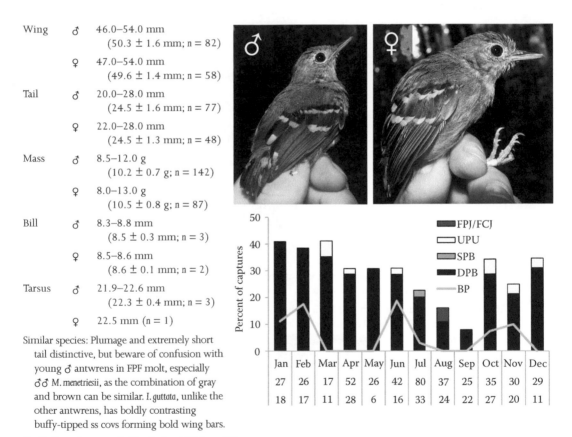

| | Jan | Feb | Mar | Apr | May | Jun | Jul | Aug | Sep | Oct | Nov | Dec |
|---|---|---|---|---|---|---|---|---|---|---|---|---|
| | 27 | 26 | 17 | 52 | 26 | 42 | 80 | 37 | 25 | 35 | 30 | 29 |
| | 18 | 17 | 11 | 28 | 6 | 16 | 33 | 24 | 22 | 27 | 20 | 11 |

Similar species: Plumage and extremely short tail distinctive, but beware of confusion with young ♂ antwrens in FPF molt, especially ♂♂ *M. menetriesii*, as the combination of gray and brown can be similar. *I. guttata*, unlike the other antwrens, has boldly contrasting buffy-tipped ss covs forming bold wing bars.

Skull: Completes, probably during FCF (n = 45).

Brood patch: Develops in both sexes, although more extensive in ♀♀, and probably mainly Jun–Feb, but there are few data available.

Sex: ♂ has gray head, throat, and chest blending to tawny-brown in the belly and rump. ♀ has olive-brown upperparts, head, and throat with a contrastingly ochraceous belly.

Molt: Group 3/Complex Basic Strategy; FPF partial, DPBs complete. FPF includes body, less and med covs, and innermost 3–6 gr covs, but not rects. Most frequent from Jan–Mar and least frequent from Aug–Sep.

FCJ      ♂ = ♀. Brownish med and gr covs with poorly defined spotted tips. Rects relatively pointed. Skull is unossified.

FPF      ♂ ≠ ♀. Replacing juvenile body feathers, ss covs, and sometimes some terts or rects with adult-like feathers. Skull is unossified.

FCF      ♂ ≠ ♀. Like DCB, except with molt limits among the gr covs with the inner replaced gr covs having much more extensive and well-defined terminal spots compared to retained outer gr covs. Skull is either unossified or ossified.

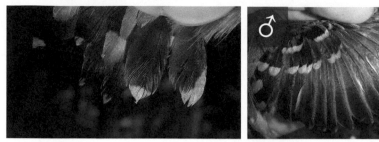

Less, med, and inner 6 gr covs replaced, other gr covs and flight feathers retained juvenile (left); rects juvenile (right).

SPB      ♂ ≠ ♀. Replacing juvenile flight feathers with adult flight feathers. May be difficult in ♀♀ to distinguish from DPB (use UPB). Skull is ossified.

p4–p10 and corresponding pp covs retained juvenile, p3 missing, p1–p2 new, s1–s7 retained juvenile, s8–s9 missing.

DCB      ♂ ≠ ♀. Without molt limits among the gr covs, all distinctly black with broad distinct tawny-orange tips. Rects relatively rounded. Skull is ossified.

DPB      ♂ ≠ ♀. Replacing adult-like body and flight feathers with adult-like feathers. May be difficult in ♀♀ to distinguish from SPB (use UPB). Skull is ossified.

p1 molting, p2 missing, p3–p10 and corresponding pp covs retained definitive basic, s1–s7 retained definitive basic, s8–s9 new (left); p1–p10 and corresponding pp covs new, s1–s3 and s5 retained definitive basic, s4 and s6 molting, s7–s9 new (right).

## Myrmotherula axillaris axillaris

White-flanked Antwren • Choquinha-de-flanco-branco

Band Size: C, D

# Individuals Captured: 536

| | | |
|---|---|---|
| Wing | ♂ | 47.5–55.0 mm (51.2 ± 1.6 mm; n = 162) |
| | ♀ | 46.0–52.5 mm (50.0 ± 1.3 mm; n = 119) |
| Tail | ♂ | 32.0–42.0 mm (37.8 ± 1.9 mm; n = 144) |
| | ♀ | 32.0–42.0 mm (37.3 ± 1.7 mm; n = 102) |
| Mass | ♂ | 6.0–9.5 g (7.6 ± 0.6 g; n = 247) |
| | ♀ | 6.0–9.0 g (7.7 ± 0.6 g; n = 163) |
| Bill | ♂ | 7.3–9.6 mm (8.6 ± 0.5 mm; n = 15) |
| | ♀ | 7.9–9.4 mm (8.7 ± 0.4 mm; n = 18) |
| Tarsus | ♂ | 16.4–18.0 mm (17.3 ± 0.5 mm; n = 15) |
| | ♀ | 16.4–18.1 mm (17.1 ± 0.5 mm; n = 18) |

Percent of captures

Legend: FPJ/FCJ, UPU, SPB, DPB, BP

| | Jan | Feb | Mar | Apr | May | Jun | Jul | Aug | Sep | Oct | Nov | Dec |
|---|---|---|---|---|---|---|---|---|---|---|---|---|
| | 27 | 14 | 36 | 34 | 30 | 37 | 77 | 74 | 54 | 66 | 55 | 15 |
| | 16 | 12 | 20 | 14 | 17 | 27 | 48 | 63 | 47 | 50 | 50 | 9 |

Similar species: All ages and sexes separated from other small antwrens by contrasting white flanks, often concealed under folded wings. Measurements intermediate among antwrens, but plumage characteristics should be diagnostic.

Skull: Completes during FCF (n = 113).

Brood patch: Develops in both sexes, although more extensive in ♀♀. Present year-round at low frequencies.

Sex: ♂ is dark grayish-black overall with distinctive white flanks, which are often concealed by the folded wing. ♀ is tawny-brown overall, slightly paler underneath, and has contrasting white flanks.

Molt: Group 3/Complex Basic Strategy; FPF partial-incomplete, DPBs complete. FPF includes all body feathers, less and med covs, usually 2–8 (occasionally all) gr covs, and none to all rects. Peaks (>30%) Sep–Jan and is least frequent (<10%) in Jun, earlier in the dry season than in other antwrens.

FCJ ♂ ≠ ♀. ♂ is dusky brownish-gray with buffy spots on ss covs and rects. ♀ is generally brown with whitish flanks and similar to definitive plumage ♀♀, but with chestnut-toned upperparts. Rects are distinctly pointed. Skull is unossified.

FPF ♂ ≠ ♀. Replacing juvenile body feathers, ss covs, and sometimes some terts or rects with adult-like feathers. Skull is unossified.

FCF  ♂ ≠ ♀. Molt limits between the replaced med covs and retained pp covs and often among the replaced inner gr covs and the retained outer gr covs, or none or all gr covs are replaced. Replaced rects are shorter and more rounded than any retained rects. Skull is either unossified or ossified.

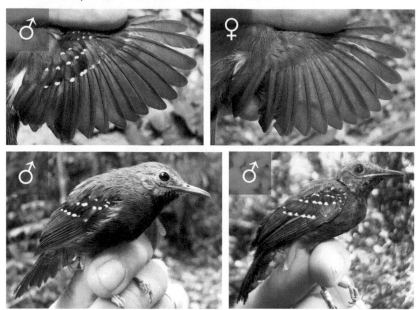

Outer 3 gr covs and all flight feathers retained (upper left); outer 7 gr covs and all flight feathers retained (upper right); variation in extent of black in underparts (lower two images).

SPB  ♂ ≠ ♀. Replacing juv flight feathers with adult flight feathers. May be difficult in ♀♀ to distinguish from DPB. Skull is ossified.

p1–p2 molting, p3–p10, corresponding pp covs, and s1–s3 retained juvenile, other ss adventitiously lost (left); p1 molting, p2 missing, p3–p10 and corresponding pp covs retained juvenile, s1–s7 retained juvenile, s8–s9 missing (center); r1 molting, r2–r6 retained juvenile (right).

DCB  ♂ ≠ ♀. Without molt limits among the gr covs or rects. ♂ pp covs bright grayish black. ♀ pp covs bright brown. Skull is ossified.

DPB  ♂ ≠ ♀. Replacing adult-like body and flight feathers with adult-like feathers. May be difficult in ♀♀ to distinguish from SPB. Skull is ossified.

## Myrmotherula longipennis longipennis

Long-winged Antwren • Choquinha-de-asa-comprida

Band Size: D, C

# Individuals Captured: 766

| | | | |
|---|---|---|---|
| Wing | ♂ | 53.0–63.0 mm | (58.3 ± 1.6 mm; n = 221) |
| | ♀ | 53.0–60.0 mm | (56.6 ± 1.7 mm; n = 69) |
| Tail | ♂ | 30.0–41.0 mm | (35.3 ± 1.8 mm; n = 202) |
| | ♀ | 31.0–39.0 mm | (34.4 ± 1.9 mm; n = 67) |
| Mass | ♂ | 7.0–10.0 g | (8.4 ± 0.6 g; n = 324) |
| | ♀ | 7.0–10.0 g | (8.4 ± 0.7 g; n = 132) |
| Bill | ♂ | 8.4–9.6 mm | (9.0 ± 0.3 mm; n = 30) |
| | ♀ | 8.5–9.4 mm | (8.9 ± 0.3 mm; n = 15) |
| Tarsus | ♂ | 16.0–18.7 mm | (17.4 ± 0.6 mm; n = 32) |
| | ♀ | 16.5–18.7 mm | (17.7 ± 0.6 mm; n = 19) |

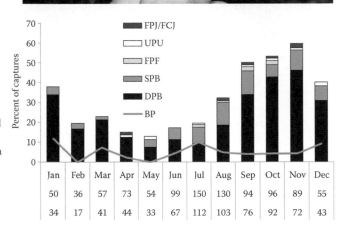

Similar species: Smaller than most other Thamnophilids, but both sexes can be confused with other *Myrmotherula* antwrens. ♂ from *M. menetriesii* by black throat, but beware of some *M. longipennis* that lack or more often have just a trace of black in the throat. From *M. axillaris* by grayer plumage and lack of white flanks. ♀ from other *Myrmotherula* antwrens by white (not brownish) belly and undertail covs. The combination of wing and tail measurements has very little overlap with other *Myrmotherula* antwrens.

Skull: Completes during FCF (n = 167).

Brood patch: Develops in both sexes, although more extensive in ♀♀. Perhaps breeding year-round, but mainly Jun–Jan.

Legend: FPJ/FCJ, UPU, FPF, SPB, DPB, BP

| | Jan | Feb | Mar | Apr | May | Jun | Jul | Aug | Sep | Oct | Nov | Dec |
|---|---|---|---|---|---|---|---|---|---|---|---|---|
| | 50 | 36 | 57 | 73 | 54 | 99 | 150 | 130 | 94 | 96 | 89 | 55 |
| | 34 | 17 | 41 | 44 | 33 | 67 | 112 | 103 | 76 | 92 | 72 | 43 |

Sex: ♂ is gray overall with bold white tips to ss covs. There is considerable variation in the amount of black in the throat (which does not appear to be age-related) with some birds approaching the gray throat of *M. menetriesii*. ♀ is unpatterned tawny-brown overall with a white belly. Sexes not distinguishable until the SPB.

Molt: Group 3/Complex Basic Strategy; FPF partial, DPBs complete. FPF includes all body feathers, less, and med covs, 1–5 gr covs, and most or all rects. Peaks in Sep–Nov and is least frequent Apr–Jul.

FCJ ♂ = ♀. Similar to DCB ♀, but with fewer olive and more chestnut-brown tones overall including edging on remiges. Throat is pale whitish and eyering is buffy. Less covs often distinctly buffy-tipped, but beware of worn definitive feathers that appear similar (other characteristics of juvenile birds like gape, feather texture, aligned growth bars, and skin flaking should help confirm age). Skull is unossified.

FPF ♂ = ♀. Both sexes like DCB ♀♀. Replacing juvenile body feathers, ss covs, and sometimes some terts or rects with adult-like feathers. Skull is unossified.

Some body feathers, less, and med covs replaced, other ss covs and flight feathers retained juvenile (both images).

FCF ♂ = ♀. Both sexes like DCB ♀♀, but with molt limits among the gr covs with the replaced inner gr covs washed olive-brown and contrasting with the retained outer gr covs washed chestnut-brown. Skull is either unossified or ossified.

Inner 2 gr covs replaced, other outer gr covs and flight feathers retained.

SPB ♂ ≠ ♀. Replacing juv flight feathers with adult flight feathers. ♂♂ has a mix of gray and brown feathers. May be difficult in ♀♀ to distinguish from DPB (use UPB). Skull is ossified.

p7–p10 and corresponding pp covs retained juvenile, p6 missing, p1–p5 new, s1 new, s2 molting, s3–s6 retained juvenile, s7–s9 new (left); p7–p10 retained juvenile, p6 missing, p5 molting, p1–p4 new, s1 molting, s2–s7 retained juvenile, s8–s9 new (right).

DCB ♂ ≠ ♀. Lacks molt limits and has rounded rectrices. Rects are relatively rounded. Skull is ossified.

DPB ♂ ≠ ♀. Replacing adult-like body and flight feathers with adult-like feathers. May be difficult in ♀♀ to distinguish from SPB (use UPB). Skull is ossified.

## *Myrmotherula menetriesii cinereiventris*

Gray Antwren • Choquinha-de-garganta-cinza

Band Size: D, C

# Individuals Captured: 454

| | | |
|---|---|---|
| Wing | ♂ | 48.0–57.0 mm (52.5 ± 1.6 mm; n = 127) |
| | ♀ | 48.0–55.0 mm (50.8 ± 1.6 mm; n = 30) |
| Tail | ♂ | 21.6–31.0 mm (27.2 ± 1.7 mm; n = 112) |
| | ♀ | 22.0–30.0 mm (26.8 ± 1.7 mm; n = 39) |
| Mass | ♂ | 6.5–10.0 g (8.2 ± 0.7 g; n = 202) |
| | ♀ | 6.5–10.0 g (8.0 ± 0.6 g; n = 73) |
| Bill | ♂ | 8.3–10.0 mm (9.3 ± 0.4 mm; n = 18) |
| | ♀ | 8.7–10.1 mm (9.4 ± 0.5 mm; n = 9) |
| Tarsus | ♂ | 17.0–19.5 mm (17.9 ± 0.7 mm; n = 17) |
| | ♀ | 16.6–18.7 mm (17.7 ± 0.6 mm; n = 8) |

Similar species: The plainest antwren at the BDFFP. Its short tail separates it from all other antwrens except M. *guttata*. SPB ♂♂ could resemble M. *guttata*, but note the lack of strong buffy-tipped wing bars. M. *menetriesii* lacks the black throat and white flanks or white belly of other small antwrens, but beware of potential confusion with some ♂ M. *longipennis* that have little black on the throat. ♀♀ from other antwrens by uniform tawny brown pattern with no white in flanks or belly.

Skull: Completes during FCF (n = 76).

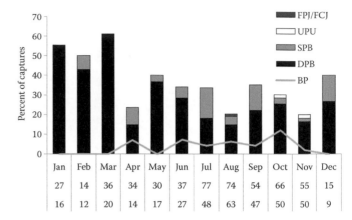

| | Jan | Feb | Mar | Apr | May | Jun | Jul | Aug | Sep | Oct | Nov | Dec |
|---|---|---|---|---|---|---|---|---|---|---|---|---|
| | 27 | 14 | 36 | 34 | 30 | 37 | 77 | 74 | 54 | 66 | 55 | 15 |
| | 16 | 12 | 20 | 14 | 17 | 27 | 48 | 63 | 47 | 50 | 50 | 9 |

Brood patch: Develops in both sexes, although more extensive in ♀♀, mainly from Jun–Oct with one observation each in Nov and Apr.

Sex: ♂ is gray overall with distinct white tips to ss covs. ♀ is unpatterned tawny brown with unmarked olive-brown med covs and gr covs often having slightly buffy-tipped due to wear. Sexes not distinguishable until the SPB.

Molt: Group 3/Complex Basic Strategy; FPF partial-incomplete, DPBs complete. FPF includes all body feathers, less and (usually) med covs, usually 3 (occasionally 1 or 2 more or less) gr covs, and perhaps a variable number of rects, from none to all. Molt irregularly occurs year-round, but peaks from Jan–Mar.

FCJ     ♂ = ♀. Like DCB ♀♀, but with less olive and more chestnut-brown tones. ♂ may have grayer less and med covs than ♀, but this needs confirmation. Less covs often distinctly buffy-tipped, but beware of worn adults looking similar (other characteristics of juvenile birds like gape, feather texture, and aligned growth bars should help confirm age). Rects relatively pointed. Skull is unossified.

FPF     ♂ = ♀. Replacing juvenile body feathers, ss covs, and sometimes some terts or rects with ♀-like feathers. Skull is unossified.

FCF     ♂ = ♀. Both sexes like DCB ♀♀, but with molt limits among the gr covs with bright olive-brown replaced inner gr covs contrasting with the dull chestnut-brown retained outer gr covs. Skull is either unossified or ossified.

Inner 3 gr covs replaced, other outer gr covs and flight feathers retained.

SPB     ♂ ≠ ♀. Replacing juvenile flight feathers with adult flight feathers. ♂♂ have a mix of gray and brown feathers. May be difficult in ♀♀ to distinguish from DPB (use UPB). Skull is ossified.

p10 missing, p9 molting, p1–p8 new, s1–s3 and s7–s9 new, s4–s6 retained juvenile (left); p1 missing, p2–p10 retained juvenile, s1–s7 and s9 retained juvenile, s8 new (right).

DCB     ♂ ≠ ♀. Lacks molt limits and has rounded rectrices. Skull is ossified.

DPB     ♂ ≠ ♀. Replacing adult-like body and flight feathers with adult-like feathers. May be difficult in ♀♀ to distinguish from SPB (use UPB). Skull is ossified.

p5–p10 and corresponding pp covs retained definitive basic, p4 missing, p3 molting, p1–p2 new, s1–s2 missing, s3–s7 retained definitive basic, s8–s9 replaced.

*Hypocnemis cantator cantator*  Band Size: D

Guianan Warbling-Antbird • Cantador-da-guiana  # Individuals Captured: 729

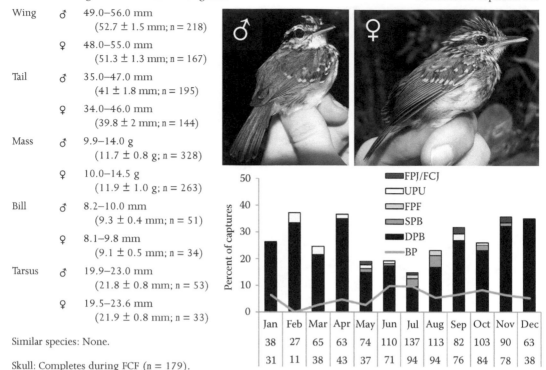

| Wing | ♂ | 49.0–56.0 mm |
| | | (52.7 ± 1.5 mm; n = 218) |
| | ♀ | 48.0–55.0 mm |
| | | (51.3 ± 1.3 mm; n = 167) |
| Tail | ♂ | 35.0–47.0 mm |
| | | (41 ± 1.8 mm; n = 195) |
| | ♀ | 34.0–46.0 mm |
| | | (39.8 ± 2 mm; n = 144) |
| Mass | ♂ | 9.9–14.0 g |
| | | (11.7 ± 0.8 g; n = 328) |
| | ♀ | 10.0–14.5 g |
| | | (11.9 ± 1.0 g; n = 263) |
| Bill | ♂ | 8.2–10.0 mm |
| | | (9.3 ± 0.4 mm; n = 51) |
| | ♀ | 8.1–9.8 mm |
| | | (9.1 ± 0.5 mm; n = 34) |
| Tarsus | ♂ | 19.9–23.0 mm |
| | | (21.8 ± 0.8 mm; n = 53) |
| | ♀ | 19.5–23.6 mm |
| | | (21.9 ± 0.8 mm; n = 33) |

|  | Jan | Feb | Mar | Apr | May | Jun | Jul | Aug | Sep | Oct | Nov | Dec |
|---|---|---|---|---|---|---|---|---|---|---|---|---|
|  | 38 | 27 | 65 | 63 | 74 | 110 | 137 | 113 | 82 | 103 | 90 | 63 |
|  | 31 | 11 | 38 | 43 | 37 | 71 | 94 | 94 | 76 | 84 | 78 | 38 |

Similar species: None.

Skull: Completes during FCF (n = 179).

Brood patch: Develops in both sexes (David and Londoño 2013), although more extensive in ♀♀. Possible year-round, but probably most likely from Jun–Oct and least likely Feb–May.

Sex: Crown of ♂ has bold black and white stripes, whereas the crown has brown and tan stripes in ♀♀. Spots on ss covs tend to be brighter white in ♂ than in ♀, but this is variable with some ♂♂ having only 1 or 2 inner gr covs buffy-tipped and others with all gr covs being buffy-tipped.

Molt: Group 3/Complex Basic Strategy; FPF partial, DPBs complete. FPF includes all body feathers, less and med covs, and inner 0–6 (or rarely more) gr covs, but usually the inner 2–4. FPF probably begins as or before FPJ completes. Peaks from Nov–Apr and is least frequent in Jul.

FCJ      ♂ = ♀. Reminiscent of DCB ♀♀, but without well-defined spots on ss covs, weak head streaking, and tawny underparts with a paler throat. FPF may begin before FPJ completes. Rects relatively pointed. Skull is unossified.

**FPF** ♂ ≠ ♀. Crown feathers are among the first formative feathers to grow in, and these are adult-like in color and useful for sex determination. Replaces juvenile body feathers, ss covs, and sometimes some terts with adult-like feathers. Rects relatively pointed. Skull is unossified.

Body, less, and some med covs being replaced, with a few med covs, gr covs, and all flight feathers retained juvenile.

**FCF** ♂ ≠ ♀. Look for molt limits among the gr covs, or less often between med covs and gr covs. ♂ replaced med and gr covs are black with bold white to slightly buffy terminal spots contrasting against retained gr and pp covs that are dull brown with indistinct buffy tips. ♀ replaced med and gr covs are bright brown with bold buffy terminal spots contrasting against retained gr and pp covs that are dull brown with indistinct buffy tips.

Less, med, and inner 5 gr covs replaced, outer gr covs, alulas, and flight feathers retained juvenile (left); less, med, and inner 3 gr covs replaced, outer gr covs, alulas, and flight feathers retained juvenile (right).

**SPB** ♂ ≠ ♀. Replacing juvenile flight feathers with adult flight feathers. In both sexes, retained pp covs are dull brown. May be difficult to distinguish from DPB when near completion, especially in ♀♀ (use UPB). Skull is ossified.

p6–p10 and corresponding pp covs retained juvenile, p5 missing, p1–p4 new, s1–s7 retained juvenile, s8 new, s9 missing (left); p6–p10 and corresponding pp covs retained juvenile, p4–p5 missing, p1–p3 new, s1 new, s2–s4 retained juvenile, s5–s7new, s8–s9 missing (center); r1 molting, r2 missing, other rects retained juvenile or missing (right).

THAMNOPHILIDAE (ANTBIRDS)

DCB ♂ ≠ ♀. Without molt limits among the gr covs and all less, med, gr, and pp covs are evenly aged and have sharply defined white or buffy spots, but beware of pseudolimits with variable number of buffier-tipped inner gr covs compared whiter-tipped outer gr covs in ♂♂. Skull is ossified.

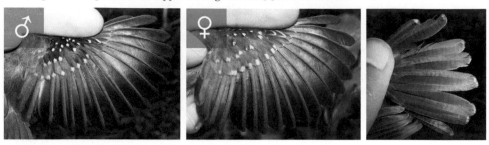

DPB ♂ ≠ ♀. Replacing adult-like body and flight feathers with adult-like feathers. May be difficult to distinguish from SPB (use UPB). Skull is ossified.

## *Cercomacra tyrannina saturatior*

Dusky Antbird • Chororó-escuro

Band Size: E

# Individuals Captured: 35

Wing ♂ 60.0–66.0 mm (62.3 ± 1.5 mm; n = 12)

♀ 57.0–71.0 mm (59.8 ± 4.7 mm; n = 8)

Tail ♂ 50.0–61.0 mm (57.9 ± 3.5 mm; n = 9)

♀ 55.0–68.0 mm (57.9 ± 4.5 mm; n = 9)

Mass ♂ 15.0–19.0 g (17.0 ± 1.0 g; n = 15)

♀ 15.0–20.0 g (16.4 ± 1.4 g; n = 12)

Similar species: Twice as large as antwrens in mass. Note fairly distinct rictal bristles compared to other antbirds. It is almost identical to *C. laeta waimiri* and many birds may not be safely separable in the hand (song is most diagnostic), but *C. laeta waimiri* has not yet been captured knowingly at the BDFFP and has not been detected along ZF-3 despite the type to separate the two species (see Table 22.1). The belly feathers and tips of rects are pale-fringed in definitive ♂ *C. t. saturatior*, but not in *C. l. waimiri*; beware of

(Courtesy of Lindsay Wieland.)

## TABLE 22.1

*Measurements of* Cercomacra tyrannina saturatior *and* C. laeta waimiri *reproduced from Bierregaard et al. (1997).*

| | Wing (mm) | Tail (mm) | Tarsus (mm) | Culmen (mm) |
|---|---|---|---|---|
| | *C. tyrannina saturatior* | | | |
| ♂ | 58.8–66.2 (62.5 ± 1.9, n = 28) | 56.0–63.5 (59.7 ± 2.0, n = 29) | 21.1–23.9 (22.8 ± 0.6, n = 28) | 18.1–20.7 (19.3 ± 0.7, n = 29) |
| ♀ | 55.8–63.6 (59.1 ± 1.9, n = 23) | 55.0–64.9 (58.6 ± 2.7, n = 22) | 20.9–23.2 (22.2 ± 0.6, n = 21) | 17.3–20.1 (18.8 ± 0.7, n = 23) |
| | *C. laeta waimiri* | | | |
| ♂ | 60.0–63.4 (62.0 ± 1.4, n = 4) | 58.3–61.0 (59.3 ± 1.3, n = 4) | 23.6–24.2 (23.9 ± 0.3, n = 4) | 19.0–19.3 (19.1 ± 0.2, n = 4) |
| ♀ | 59.2 (n = 1) | 55.9 (n = 1) | 23.2 (n = 1) | 17.8 (n = 1) |

NOTE: Measurements were taken from specimens and may not be directly comparable to live birds, especially tarsus and culmen, but are indicative of relative differences among sexes and species.

young or worn *C. t. saturatior* that may lack this feature. Ss covs of definitive basic ♀ *C. l. waimiri* are bright rufous-orange and concolor with the underparts and auriculars, unlike *C. t. saturatior*, which have dull brown ss covs and auriculars compared to bright rufous-orange underparts, but this distinction is subtle and may be subject to wear and age (Bierregaard et al. 1997).

Skull: Completes, probably during FCF (n = 7).

Brood patch: Probably occurs regularly in both sexes from Jul–Nov, but also probably occasionally year-round.

Sex: ♂ is all dark gray with white crescents to tips of ss covs. Gr covs may typically be slightly paler gray than the med covs, producing a pseudolimit. ♀ is unmarked tawny-brown. Sexual dichromatism may not be apparent until SPB.

Molt: Not well understood. Group 3/Complex Basic Strategy; FPF partial(?), DCBs complete. FPF appears to include all body feathers, less and med covs, and some inner gr covs, but not pp covs, pp, ss, or rects. Molt may be most regular from Oct–Jul.

| | |
|---|---|
| FCJ | ♂ = ♀. Unknown. Probably ♀-like in appearance, but with grayish wings and rects in ♂♂ and tawny-brownish wings and rects in ♀♀. Skull is unossified. |
| FPF | ♂ = ♀. ♂ with a mix of blackish-gray and retained brown juvenile body feathers. ♀ with incoming brighter tawny-brown ss covs that have brighter orange tips. Skull is unossified. |
| FCF | ♂ = ♀. Both sexes probably similar to DCB except with molt limits in ss covs. Rects are pointed. Skull is either unossified or ossified. |

Inner 4 gr covs and all med and less covs replaced, outer gr covs, all alulas, and all flight feathers retained juvenile (left), as are the rects (right).

SPB    ♂ ≠ ♀. Obvious in ♂♂, with dark grayish-black feathers replacing brown body and flight feathers. Similar to DPB in ♀♀ (use UPB), but look for differences in shape between pointed retained rects and more rounded replaced rects. Skull is ossified.

DCB ♂ ≠ ♀. Without molt limits in gr covs. Rects distinctly rounded. Skull is ossified.

DPB ♂ ≠ ♀. Replacing adult-like feathers with adult like feathers. Obvious in ♂♂, replacing grayish- or brownish-black flight and body feathers with dark grayish-black flight feathers. More similar to SPB in ♀♀ (use UPB), but look for similar shape between retained and replaced rects. Skull is ossified.

p10 retained definitive basic, p9 missing, p1–p8 new, s1 new, s2 molting, s3–s6 retained definitive basic, s7 missing, s8–s9 new. (Courtesy of Lindsay Wieland.)

## *Percnostola rufifrons subcristata*

Band Size: G, F

Black-headed Antbird • Formigueiro-de-cabeça-preta

# Individuals Captured: 928

| | | | |
|---|---|---|---|
| Wing | ♂ | 66.0–78.0 mm (72.3 ± 2.1 mm; n = 193) | |
| | ♀ | 65.5–77.0 mm (70.5 ± 2.1 mm; n = 215) | |
| Tail | ♂ | 51.0–69.0 mm (60.4 ± 2.7 mm; n = 271) | |
| | ♀ | 52.0–68.0 mm (59.5 ± 2.7 mm; n = 279) | |
| Mass | ♂ | 24.4–35.0 g (29.1 ± 1.8 g; n = 343) | |
| | ♀ | 23.5–35.2 g (28.9 ± 2.2 g; n = 342) | |
| Bill | ♂ | 11.3–14.7 mm (12.6 ± 0.6 mm; n = 61) | |
| | ♀ | 11.0–14.1 mm (12.5 ± 0.6 mm; n = 51) | |
| Tarsus | ♂ | 31.1–34.8 mm (33.1 ± 0.9 mm; n = 61) | |
| | ♀ | 30.9–35.4 mm (33.0 ± 1.0 mm; n = 52) | |

Legend: FPJ/FCJ/FPS/FCS, UPU, FPF, DPB, BP

| | Jan | Feb | Mar | Apr | May | Jun | Jul | Aug | Sep | Oct | Nov | Dec |
|---|---|---|---|---|---|---|---|---|---|---|---|---|
| | 49 | 62 | 74 | 72 | 116 | 144 | 193 | 185 | 111 | 97 | 119 | 74 |
| | 34 | 50 | 55 | 50 | 67 | 107 | 152 | 158 | 108 | 88 | 104 | 41 |

Similar species: ♂ from other antbirds with bold white wing markings by red eye and size. ♀ from other antbirds by red eye and uniformly orange throat, breast, and belly.

Skull: Usually completes during FPF (n = 245).

Brood patch: Develops in both sexes, although more extensive in ♀♀. Breeds year-round, peaking Aug–Dec with a second peak in Mar.

Sex: ♂ is charcoal gray with black cap and throat. Ss covs are black with white fringe to tip. The iris is bright red. ♀ is bright tawny orange overall. Ss covs have dark brownish-black bases and bold tawny orange tips. The crown is dusky orange (not black as in *P. r. rufifrons*) and the iris is bright orange to red.

Molt: Group 4/Complex Basic Strategy with auxiliary preformative molt; FPX partial, FPF complete, DPBs complete (Johnson and Wolfe 2014). FPX includes many (or all?) body feathers, most or all less covs, most or rarely all med covs, and occasionally 1–3 inner gr covs, but no remiges. During FPF, auxiliary formative covs are replaced after the juvenile covs, but before the pp complete. It seems that during the FPF, auxiliary formative feathers are generally replaced after juvenile feathers are replaced. Molt (FPF and DPB) regularly occurs at any time of year, peaking from Apr–Jul with a second peak in Dec–Jan. These two peaks do not apparently represent differences in timing between the FPF and DCB molts, but more study is needed. This species regularly loses tail feathers when handled, thus care should be taken when assigning tail molt.

FCJ     ♂ ≈ ♀. Superficially like adult ♀, but duller and lacking bold buffy orange tips to less, med, and gr covs; the tips of ss covs are dull orange. ♂♂ can (always?) have grayish-brown in the tail and back, while ♀♀ (and some ♂♂?) are brown in the tail and back. Pp covs lack small pale-orange spots at tips. Because it is not known if all ♂♂ are readily identifiable by plumage, ♀-like birds (with brown backs and rects) should be sexed as "unknown sex." The iris is brown. The skull is unossified.

THAMNOPHILIDAE (ANTBIRDS)     185

**FPX**    ♂ ≈ ♀. Like adult ♀, but with adult ♀-like ss covs replacing juvenile covs. Iris is brown. Skull is unossified. To our knowledge, at least all ♂♂ (and therefore probably all ♀♀) include this FPX molt, which may overlap in timing with the FPF.

**FCX**    ♂ ≈ ♀. Like FCJ, but with some to all less and med covs and occasionally some gr covs replaced and definitive ♀-like in both sexes and contrasting with the dull retained med and gr covs. Retained juvenile rects may be useful for sexing (see FCJ). The orange edging on replaced covs may be consistently narrower in ♂♂ than ♀♀, but more study is needed. Iris is brown or slightly reddish. Skull is unossified. This plumage may rarely exist as the FPS may regularly overlap the FPF.

**FPF**    ♂ ≠ ♀. In ♂♂, black and white ss covs replacing juvenile and ♀-like ss covs, followed by flight feathers. In ♀♀, this may be difficult or impossible to distinguish from FPX unless excessive numbers of gr covs are molting and/or flight feather replacement commences. During flight feather replacement, the skull becomes ossified and the iris becomes bright orange to bright red.

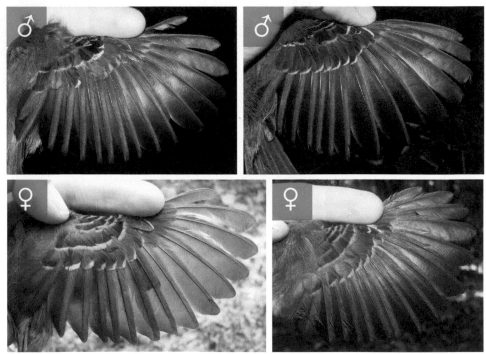

With three generations of ss covs (juvenile: dull brown; auxiliary formative: black with orange tips; formative: black with white tips; upper left); p5–p10 retained juvenile, p4 missing, p1–p3 new, s1–s7 retained juvenile, s8 missing, s9 new (upper right); p4–p10 retained juvenile, p3 missing, p2 molting, p1 new, s1–s6 retained juvenile, s7 missing, s8–s9 new (lower left); p6–p10, p5 molting, p1–p4 new, s1–5 retained juvenile, s6–s9 new (lower right).

FAX ♂ ≠ ♀. Adult-like in all respects, without molt limits. Iris is bright orange to bright red. Skull is ossified.

UPB ♂ ≠ ♀. Replacing adult-like body and flight feathers with adult-like feathers. Iris is bright orange to bright red. Skull is ossified. Can be difficult to determine in females, especially upon completion of molt (age these as UPU).

p8–p10 retained, p7 missing, p1–p6 new, s–s3 new, s4 missing, s5–s6 retained, s7 missing, s8–s9 new.

## *Schistocichla leucostigma leucostigma*

Spot-winged Antbird • Formigueiro-de-asa-pintada

Band Size: F

\# Individuals Captured: 171

| | | |
|---|---|---|
| Wing | ♂ | 60.0–69.0 mm (66.1 ± 1.9 mm; n = 44) |
| | ♀ | 61.0–70.0 mm (65.1 ± 2.1 mm; n = 56) |
| Tail | ♂ | 54.0–64.0 mm (59.1 ± 2.4 mm; n = 29) |
| | ♀ | 54.0–66.0 mm (58.8 ± 3.2 mm; n = 34) |
| Mass | ♂ | 21.0–29.0 g (24.5 ± 1.6 g; n = 54) |
| | ♀ | 21.0–28.3 g (24.3 ± 1.6 g; n = 73) |
| Bill | ♂ | 11.6–13.0 mm (12.2 ± 0.6 mm; n = 4) |
| | ♀ | 10.7–12.9 mm (11.9 ± 0.8 mm; n = 6) |
| Tarsus | ♂ | 27.3–31.4 mm (29.7 ± 1.8 mm; n = 4) |
| | ♀ | 29.0–30.5 mm (29.7 ± 0.6 mm; n = 6) |

Similar species: ♂ potentially confusing only because it superficially resembles several other all dark gray antbirds with white on the ss covs, but it lacks any darker black patterning in the gray and has white spots (not crescents) on the ss covs. ♀ is most similar to *Myrmophylax atrothorax*, but larger with uniformly orange throat and chest (not paler whitish throat) and a gray (not dark brown) iris.

Skull: Probably completes during FCF (n = 31).

Brood patch: Develops in both sexes, although more extensive in ♀♀. Timing not obvious, but perhaps probably mainly breeds during the dry season (Jul–Jan).

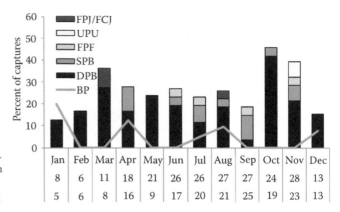

| | Jan | Feb | Mar | Apr | May | Jun | Jul | Aug | Sep | Oct | Nov | Dec |
|---|---|---|---|---|---|---|---|---|---|---|---|---|
| | 8 | 6 | 11 | 18 | 21 | 26 | 26 | 27 | 27 | 24 | 28 | 13 |
| | 5 | 6 | 8 | 16 | 9 | 17 | 20 | 21 | 25 | 19 | 23 | 13 |

Sex: ♂ is all gray with a distinct round white spot on the tip of each less, med, and gr cov and with or without a small white spot on the tip of each pp cov. ♀ has tawny orange underparts with a gray head and orange throat. Each less, med, and gr cov has a large round tawny-orange spot at the tip. Pp covs are brown, usually with a small tawny spot at each tip. The alula is boldly marked with a pale arrow-head.

Molt: Group 3/Complex Basic Strategy; FPF partial, DPB complete. FPF includes body feathers, less and med covs, and inner 3 to all gr covs. Timing not well understood.

FCJ      ♂ = ♀. All dark brown without any markings (at least in *S. l. intensa*), but less, med, and gr covs may have faint pale spots on the tips. The iris is dark brownish-black. Probably in this plumage very briefly.

FPF      ♂ = ♀. Replacing juvenile body feathers, less, med, and/or inner gr covs with adult ♀-like feathers. Skull is unossified.

FCF      ♂ = ♀. Like DCB ♀♀, but slightly paler underneath. Look for molt limits among the gr covs with the replaced inner gr covs having distinct large spots contrasting with the weakly marked retained gr covs, or occasionally molt limits are between the replaced gr covs and retained pp covs. The replaced covs in ♂♂ may have whiter spots than in ♀♀, but more study is needed. The retained alula is weakly marked with a pale tip. The iris is gray. The skull is either ossified or unossified.

Inner 4 gr covs, med, and less covs new, outer gr covs, alulas, and flight feathers retained juvenile.

SPB      ♂ ≠ ♀. Replacing juv flight feathers with adult flight feathers. ♂♂ has a mix of gray and brown feathers. May be difficult in ♀♀ to distinguish from DPB (use UPB); pp covs without buffy spots at tip, but beware of worn DPB with limited or absent buffy tips. The iris is gray. Skull is ossified.

DCB ♂ ≠ ♀. Lacks molt limits in the gr covs. The iris is gray. The skull is ossified.

DPB ♂ ≠ ♀. Replacing adult-like body and flight feathers with adult-like feathers. May be difficult in ♀♀ to distinguish from SPB (use UPB), but pp covs will usually have a terminal buffy spot. The iris is gray. Skull is ossified.

p6–p10 retained definitive basic, p1–p5 new, s1 new, s2–s6 retained definitive basic, s7–s9 new.

## *Drymophila ferruginea ferruginea*

Band Size: F

Ferruginous-backed Antbird • Formigueiro-ferrugem

# Individuals Captured: 301

| | | |
|---|---|---|
| Wing | ♂ | 59.0–67.0 mm (63.2 ± 1.6 mm; n = 86) |
| | ♀ | 57.0–66.0 mm (61.9 ± 2.1 mm; n = 68) |
| Tail | ♂ | 47.0–61.0 mm (53.7 ± 2.8 mm; n = 79) |
| | ♀ | 48.0–60.0 mm (53.5 ± 2.4 mm; n = 59) |
| Mass | ♂ | 21.5–30.0 g (25.0 ± 1.7 g; n = 147) |
| | ♀ | 20.5–30.5 g (24.7 ± 2.2 g; n = 100) |
| Bill | ♂ | 11.1–13.2 mm (12.2 ± 0.6 mm; n = 14) |
| | ♀ | 10.6–12.9 mm (11.7 ± 0.7 mm; n = 10) |
| Tarsus | ♂ | 28.0–30.9 mm (29.4 ± 0.9 mm; n = 15) |
| | ♀ | 27.3–30.9 mm (29.5 ± 1.2 mm; n = 11) |

Similar species: None.

Skull: Completes during FCF (n = 46).

Brood patch: Develops in both sexes, although more extensive in ♀♀. Despite low sample sizes, may breed year-round, but BPs have not been observed in Jan–Feb. A bi-annual peak of breeding is emerging: Mar–Apr and Sep (–Dec?).

Sex: Sexes are generally alike, except ♂ has a black throat and ♀ has a white throat.

Molt: Group 3/Complex Basic Strategy; FPF partial, DPBs complete. FPF

includes all body feathers, rects, less and med covs and gr covs, and 1–2 alula covs, but not pp covs or flight feathers. Molt occurs year-round, but the peak is not clear because of low sample sizes.

FCJ    ♂ = ♀. Generally unmarked brown overall with a black throat and chest. Ss covs are unmarked and uniformly brown. May superficially resemble young of other antbirds, but note long tarsi and bluish-purple bare ocular patch. Skull is unossified.

FPF    ♂ ≠ ♀. Replacing juvenile body feathers, less, med, and/or inner gr covs with adult-like feathers. Sex can be determined by color of incoming throat feathers. Skull is unossified.

Less and some med covs replaced, inner gr covs in molt, outer gr covs, alulas, and flight feathers retained juvenile.

FCF     ♂ ≠ ♀. Molt limits between the unmarked brown retained pp covs and alula, which contrast against the black with bright buffy-tipped replaced gr ss covs. The ocular patch is powder blue. The skull is either ossified or unossified.

One outer gr cov, all alulas, and all flight feathers retained juvenile.

SPB     ♂ ≠ ♀. Replacing juvenile flight feathers with adult flight feathers. ♂♂ have a mix of gray and brown feathers. Readily distinguishable from DPB by unmarked brown pp covs. Skull is ossified.

DCB     ♂ ≠ ♀. Ss covs are without molt limits and are black with a white crescent at the tip (less and pp covs) or buff crescent at the tip (med and gr covs). The ocular patch is powder blue. Skull is ossified.

DPB     ♂ ≠ ♀. Replacing adult flight feathers with adult flight feathers. Readily distinguishable from SPB by blackish pp covs. Skull is ossified.

## *Myrmophylax atrothorax atrothorax*

Band Size: E

Black-throated Antbird • Formigueiro-de-peito-preto

# Individuals Captured: 14

| | |
|---|---|
| Wing | 52.0–57.0 mm (54.8 ± 1.7 mm; n = 8) |
| Tail | 50.0–56.0 mm (54.0 ± 2.3 mm; n = 8) |
| Mass | 14.0–16.0 g (15.1 ± 0.8 g; n = 9) |

Similar species: Smaller than *Schistocichla leucostigma*. ♂ is uniquely patterned with brown cap, black throat, and white spots (not crescents) on ss covs. ♀ most like *S. leucostigma*, but with paler throat than chest (not nearly concolor) and brown tones in crown contrasting with grayish mask.

Skull: Probably completes during FCF (n = 1).

Brood patch: No data available, although both sexes probably incubate as in other Thamnophilidae. Timing unknown.

Sex: ♂ is gray overall with a brown crown, black throat, and a white spot on the tip of each ss cov. ♀ is dull orange-brown on upperparts contrast with grayish mask, whitish throat and brighter orange underparts, and with a buffy spot on the tip of each ss cov.

Molt: Group 3/Complex Basic Strategy; FPF partial, DPBs complete. FPF includes body feathers, less covs, med covs, and a variable number of inner gr covs, but not remiges. Timing unknown.

FCJ     ♂?♀. Unknown, but without spots on ss covs. Skull is unossified.

FPF     ♂?♀. Replacing juvenile body feathers, less, med, and/or inner gr covs with adult ♀-like feathers. Skull is unossified.

FCF     ♂?♀. Has molt limits either between the inner and outer gr covs, and/or between the gr and med covs. Skull is either ossified or unossified.

Inner 5 gr covs, med, and less covs replaced, outer gr covs and all flight feathers retained juvenile.

SPB     ♂ ≠ ♀. Replacing juvenile flight feathers with adult flight feathers and juvenile unspotted gr and pp covs with feathers that have a buffy terminal spot. May be difficult in ♀♀ to distinguish from DPB (use UPB). Skull is ossified.

DCB     ♂ ≠ ♀. Lacks molt limits among the ss covs. Skull is ossified.

DPB     ♂ ≠ ♀. Replacing adult-like body and flight feathers with adult-like feathers. May be difficult in ♀♀ to distinguish from SPB (use UPB). Skull is ossified.

p6–p10 retained definitive basic, p5 missing, p4 molting, p1–p3 retained, s1–s6 retained definitive basic, s7–s9 unknown status.

## Myrmornis torquata torquata

Wing-banded Antbird • Pinto-do-mato-carijó

Band Size: G

# Individuals Captured: 198

| Wing | ♂ | 88.0–98.0 mm (92.3 ± 2.5 mm; n = 47) |
|---|---|---|
| | ♀ | 86.0–99.0 mm (91.7 ± 2.7 mm; n = 45) |
| Tail | ♂ | 35.0–44.0 mm (38.6 ± 2.1 mm; n = 44) |
| | ♀ | 33.0–43.0 mm (37.6 ± 2.4 mm; n = 44) |
| Mass | ♂ | 39.5–50.0 g (44.7 ± 2.6 g; n = 96) |
| | ♀ | 38.0–52.5 g (43.8 ± 2.7 g; n = 82) |

Bill    ♂    15.2–15.9 mm
              (15.6 ± 0.5 mm; n = 2)

         ♀    13.5–15.6 mm
              (14.4 ± 1.1 mm; n = 3)

Tarsus    ♂    29.1–29.9 mm
              (29.5 ± 0.6 mm; n = 2)

         ♀    28.2–28.5 mm
              (28.4 ± 0.2 mm; n = 3)

Similar species: None.

Skull: Completes probably during FCF (n = 41), but little other information available.

Brood patch: Develops in both sexes, although more extensive in ♀♀. Probably peaks from Oct–Mar (–Jun?).

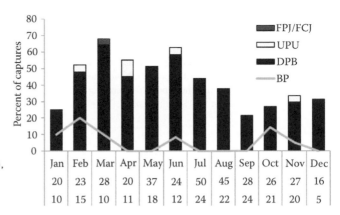

| | Jan | Feb | Mar | Apr | May | Jun | Jul | Aug | Sep | Oct | Nov | Dec |
|---|---|---|---|---|---|---|---|---|---|---|---|---|
| | 20 | 23 | 28 | 20 | 37 | 24 | 50 | 45 | 28 | 26 | 27 | 16 |
| | 10 | 15 | 10 | 11 | 18 | 12 | 24 | 22 | 24 | 21 | 20 | 5 |

Sex: ♂♂ have a black throat and chest, whereas these are orange in ♀♀.

Molt: Group 3(?)/Complex Basic Strategy; FPF partial(?), DPBs complete. FPF may include body feathers, as well as all less, some to all (usually?) med, and some to all (usually?) gr covs retained. Molt is most frequent from Feb–Aug and least frequent in Sep.

FCJ       ♂ ≠ ♀. Described as largely chocolate-brown and gray with a similar pattern as adult (Zimmer and Isler 2003). Skull is unossified.

FPF       ♂ ≠ ♀. Replacing juvenile body feathers, less, med, and/or gr covs. Incoming less covs are distinctly rufous-edged in both sexes. Sex can be determined by color of incoming throat and chest feathers. Skull is unossified.

FCF       ♂ ≠ ♀. Molt limits probably usually occur between the gr covs and pp covs, and although this may be difficult to determine, pp covs are relatively dull and brownish. Occasionally (?) molt limits occur among the outer gr covs or among the med covs. Rects are pointed and average more rufous than adult rects. Less covs in both sexes distinctly rufous edged, which is more useful for aging ♂♂. ♀♀ very much like adult ♀♀, but probably average fewer black and white markings on chest, perhaps usually absent or nearly absent. Skull is either ossified or unossified.

Second outer gr cov and all flight feathers retained juvenile (left); all ss covs new, s2–s4 and s7 new (adventitious replacement?), all other flight feathers retained juvenile (right).

FAJ       ♂ ≠ ♀. Given the difficulty and uncertainty in aging this species, this should be the default aging category for birds with ossified skulls and undetermined molt limits.

SPB       ♂ ≠ ♀. May be extremely difficult to distinguish from DPB, and should be used sparingly until more information is learned about aging this species (use UPB).

DCB ♂ ≠ ♀. Without molt limits among the med or gr covs, or between the gr covs and pp covs. Rects rounded and with fewer rufous tones than in FCJ and FCF. ♂♂ with distinctly gray less covs. ♀♀ with distinct black and white markings in the chest, perhaps averaging more than in FCF. Skull is ossified.

DPB ♂ ≠ ♀. May be extremely difficult to distinguish from SPB, and should be the default category used for birds with wing molt until more information is learned about aging this species (use UPB).

p8–p10 retained definitive basic, p7 molting, p1–p6 new, s1–s7 retained definitive basic, s8 new, s9 missing.

*Pithys albifrons albifrons*

White-plumed Antbird • Papa-formiga-de-topete

Band Size: F

\# Individuals Captured: 2797

Wing      62.5–77.0 mm (69.9 ± 2.0 mm; n = 820)

Tail        34.0–45.0 mm (39.9 ± 1.7 mm; n = 1039)

Mass      16.0–24.9 g (20.1 ± 1.3 g; n = 1798)

Bill          9.7–12.7 mm (11.1 ± 0.5 mm; n = 264)

Tarsus    23.1–28.3 mm (25.7 ± 0.8 mm; n = 261)

Similar species: No other species has a distinctive white crest. The combination of rusty-orange belly and tail, gray back, and orange legs separates all ages from all other species.

Skull: Completes during FPF (n = 699).

Brood patch: Both sexes share in incubation and develop BPs as in other antbirds (Willis 1981). Possible year-round, but most frequent from Sep–Mar.

Sex: ♂ = ♀. Crest length averages larger in ♂♂, but almost completely overlaps ♀♀ (Willis 1981).

Molt: Group 4/Complex Basic Strategy; FPF complete, DPBs complete. As described by Willis (1981), we have twice seen wing molt suspend and then recommence, and there are five additional records from the long-term database. We have also noted several other cases of arrested wing molt (appearing as molt limits among the pp), but did not recapture these birds to determine whether wing molt resumed.

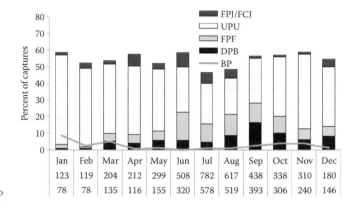

This may be a strategy to deal with unpredictable breeding opportunities (Pyle et al. 2016) and in some cases could lead to incomplete FPF and DPB molts. Tail molt proceeds centripetally (begins with r6 and completes with r1). The adult supercilium and crest are evident and adult-like early in the FPF, long before wing molt completes and often before p3 is replaced. For birds in molt, examine retained wing and tail feathers to accurately determine age. Greater than 40% of birds captured at any time of year have wing molt, but molt peaks Sep–Jan. The progression of wing molt is extremely slow, averaging about 310 days (Johnson 2011).

FCJ     Lacks white crest and supercilium, has dull orange tarsi, a white gape, and extensive whitish-yellow on lower mandible. Ss covs are brownish-gray with chestnut or brownish edging. Pp and ss are brownish-gray and the terts often have chestnut tips. Inner covs and scapulars also often have chestnut tips. Rects are distinctly pointed. The iris is dark sooty brown.

FPF      The adult supercilium and crest are evident and complete early in FPF, long before wing molt completes and often before p3 is replaced. The bill and leg color also become adult-like early in the FPF. Bright gray replaced wing feathers contrast against retained brownish-gray feathers, but beware of worn adult-like feathers that may appear similar to juvenile feathers. Growth bar alignment on inner ss can be useful for aging as FPF. Nearly complete FPF can be similar to nearly complete DPB and may not be reliably aged (use UPU). The skull is unossified (early in molt), and ossifies usually with p3–p7.

p1–p2 new, p3–p10, s1–s9, and gr covs retained juvenile (upper left); p7–p10 retained juvenile, p6 molting, p1–p5 new, s1 new, s2–s6 retained juvenile, s7 missing, s8–s9 new (upper right); body plumage early (lower left) and late (lower center) in molt; r1–r3 right retained juvenile, r4 right missing, r5–r6 new, r1–r2 left retained juvenile r3 left molting, r4–r6 new (lower right).

FAJ      The crest and supercilium are white, tarsi are bright orange, and bill is all black. Flight feathers are bright gray and fade to dull gray with wear. Some birds have small, but distinct, chestnut brown tips in terts, ss, and/or covs (is this age-related?). Rects are rounded. The iris is dark brown. The skull is ossified.

UPB    Replacing adult-like body and gray flight feathers with adult-like feathers. Birds completing DPB are nearly identical to birds completing FPF and are not reliably aged because worn definitive outer pp are extremely similar to worn juvenile outer pp, thus should be aged as UPU. Skull is ossified.

## Gymnopithys rufigula rufigula

Rufous-throated Antbird • Mãe-de-taoca-de-garganta-vermelha

Band Size: G

# Individuals Captured: 1302

| | | |
|---|---|---|
| Wing | ♂ | 69.0–82.0 mm (75.8 ± 2.2 mm; n = 258) |
| | ♀ | 68.0–81.0 mm (74.1 ± 2.2 mm; n = 224) |
| Tail | ♂ | 44.0–58.0 mm (50.8 ± 2.2 mm; n = 261) |
| | ♀ | 43.0–58.0 mm (49.5 ± 2.2 mm; n = 206) |
| Mass | ♂ | 23.0–35.0 g (29.4 ± 2.0 g; n = 564) |
| | ♀ | 23.5–34.6 g (28.7 ± 2.1 g; n = 452) |
| Bill | ♂ | 10.6–13.7 mm (12.6 ± 0.7 mm; n = 73) |
| | ♀ | 10.6–14.3 mm (12.3 ± 0.7 mm; n = 56) |
| Tarsus | ♂ | 29.3–34.9 mm (32.2 ± 0.9 mm; n = 72) |
| | ♀ | 29.6–33.2 mm (31.6 ± 0.9 mm; n = 55) |

Similar species: Only *Drymophila ferruginea* shares bluish bare skin around the eye, but the plumage is otherwise different, thus no species should be confused with *Gymnopithys*.

Skull: Completes during FPF (n = 333).

Brood patch: Develops in both sexes, although more extensive in ♀♀. Possible year-round, but most frequent from Sep–Dec.

Sex: ♂ has a white concealed intrascapular patch, which is tawny-orange in ♀.

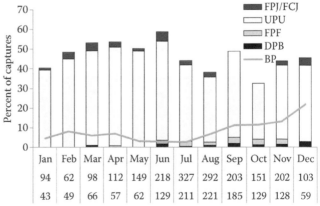

| | Jan | Feb | Mar | Apr | May | Jun | Jul | Aug | Sep | Oct | Nov | Dec |
|---|---|---|---|---|---|---|---|---|---|---|---|---|
| | 94 | 62 | 98 | 112 | 149 | 218 | 327 | 292 | 203 | 151 | 202 | 103 |
| | 43 | 49 | 66 | 57 | 62 | 129 | 211 | 221 | 185 | 129 | 128 | 59 |

Legend: FPJ/FCJ, UPU, FPF, DPB, BP

Molt: Group 4/Complex Basic Strategy; FPF complete, DPBs complete. Molt occurs at any time of year, probably partially due to a slow feather replacement rates like *Pithys*, peaking Feb–Jun.

FCJ     ♂ = ♀. The bare ocular patch is dull powder blue. The chin is reddish-chestnut and the throat is brown. Ss covs are chestnut-brown and tips wear to a pale tawny-brown. Rects are relatively pointed. Gape is white. The concealed sexually dichromatic intrascapular patch is not present, but these are among the first feathers to grow during FPF and can be used to determine sex when present. The iris is dark brown. Skull is unossified.

FPF     ♂ ≠ ♀. Iris transitions from brown to dark red. Skull ossifies during pp molt. Birds completing FPF are nearly identical to birds completing DPB and are not reliably aged because worn definitive outer pp are extremely similar to worn juvenile outer pp.

p4–p10 retained juvenile, p3 molting, p1–p2 new, s1–s7 and s9 retained juvenile, s8 missing.

FAJ     ♂ ≠ ♀. The bare ocular patch is bright powder blue, and the chin and throat are reddish-chestnut. Flight feathers and ss covs are brown with hint of olive and can appear similar to juvenile feathers when worn with pale tips, but with less chestnut color than juvenile feathers of the same wear. Rects are rounded. Iris is dark red. Skull is ossified.

UPB     ♂ ≠ ♀. Replacing adult-like body and flight feathers with adult-like feathers. The bare ocular patch is bright powder blue. Iris is dark red. Skull is ossified.

*Willisornis poecilinotus poecilinotus*

Scale-backed Antbird • Rendadinho

| | | |
|---|---|---|
| Wing | ♂ | 60.0–70.0 mm<br>(64.4 ± 1.8 mm; n = 394) |
| | ♀ | 58.0–70.0 mm<br>(62.8 ± 1.8 mm; n = 498) |
| Tail | ♂ | 36.0–48.0 mm<br>(42.4 ± 1.9 mm; n = 366) |
| | ♀ | 36.0–49.0 mm<br>(41.9 ± 1.9 mm; n = 435) |
| Mass | ♂ | 13.0–21.0 g<br>(16.8 ± 1.1 g; n = 700) |
| | ♀ | 12.5–21.0 g<br>(16.8 ± 1.2 g; n = 749) |
| Bill | ♂ | 9.8–12.2 mm<br>(10.9 ± 0.6 mm; n = 64) |
| | ♀ | 9.7–12.2 mm<br>(10.7 ± 0.6 mm; n = 57) |
| Tarsus | ♂ | 24.9–28.2 mm<br>(26.5 ± 0.8 mm; n = 68) |
| | ♀ | 24.7–27.6 mm<br>(26.3 ± 0.9 mm; n = 52) |

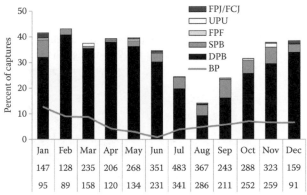

| | Jan | Feb | Mar | Apr | May | Jun | Jul | Aug | Sep | Oct | Nov | Dec |
|---|---|---|---|---|---|---|---|---|---|---|---|---|
| | 147 | 128 | 235 | 206 | 268 | 351 | 483 | 367 | 243 | 288 | 323 | 159 |
| | 95 | 89 | 158 | 120 | 134 | 231 | 341 | 286 | 211 | 252 | 259 | 91 |

Similar species: No other antbird has extensive scaling on back and ss covs.

Skull: Completes during FCF (n = 433).

Brood patch: Develops in both sexes, although more extensive in ♀♀. Breeding can occur year round, but especially from Oct–Mar.

Sex: ♂ is generally all gray with lots of white scaling on the back and blackish ss covs. ♀ has an orange head and a gray throat and underparts; ss covs are blackish with tawny orange tips. Note sexual dichromatism is not apparent until the SPB.

Molt: Group 3/Complex Basic Strategy; FPF partial, DPBs complete. FPF includes all body feathers, less and med covs, and inner 0–4 gr covs. Molt most prevalent from Nov–May and least prevalent from Jul–Sep.

FCJ ♂ = ♀. Like DCB ♀♀, but less, med, and gr covs have dull charcoal gray centers (not glossy), with dull tawny-orange well-defined tips. Pp covs are dull brown with dull tawny-orange well-defined tips. Skull is unossified.

FPF ♂ = ♀. Replacing juvenile body feathers and/or ss covs. Incoming ss covs distinctly blackish with dull tawny-orange tips contrasting against slightly browner and duller tipped juvenile covs. Remnants of white gape usually present. Skull is unossified.

FCF ♂ = ♀. Like DCB ♀♀, but with molt limits among gr covs or less often between med and gr covs. Molt limits can be difficult to see as the retained juvenile gr covs are very similar to replaced med and gr covs, but are dull brownish-black with weaker yellowish-orange tips slightly contrasting against replaced glossy dark black, bolder tawny-orange tipped covs. Retained pp covs are dull brown with poorly defined tawny-orange tips. The skull is either ossified or unossified.

All less and med covs, and 3 inner gr covs replaced, outer gr covs, alulas, and all flight feathers retained juvenile.

SPB ♂ ≠ ♀. Replacing juv flight feathers with adult flight feathers. ♂♂ have a mix of gray and brown feathers. May be difficult in ♀♀ to distinguish from DPB. Skull is ossified.

Scapulars, 1 med cov, and 1 gr cov new and second basic, med and less covs formative, outer gr covs and all flight feathers juvenile (left); mix of second basic and formative ss covs, p7–p10 retained juvenile, p6 molting, p1–p5 new, s1 new, s2–s6 retained juvenile, s7–s9 new (right).

DCB ♂ ≠ ♀. Lacks molt limits among gr covs or between med and gr covs. Only plumage in which ♂♂ are all gray, black, and white.

DPB     ♂ ≠ ♀. Replacing adult-like body and flight feathers with adult-like feathers. May be difficult in ♀♀ to distinguish from SPB. Skull is ossified.

p8–p10 retained definitive basic, p7 molting, p1–p6 new, s1–s3 new, s4 and s6 missing, s5 retained definitive basic, s7–s9 new (left); p8–p10 retained definitive basic, p6–p7 molting, p1–p6 new, s1 new, s2–s5 retained definitive basic, s6–s9 new (right).

## *Hylophylax naevia naevia*

Spot-backed Antbird • Guarda-floresta

Band Size: E

# Individuals Captured: 115

| | | |
|---|---|---|
| Wing | ♂ | 52.0–61.0 mm (57.1 ± 2.0 mm; n = 38) |
| | ♀ | 53.0–60.0 mm (56.2 ± 1.8 mm; n = 29) |
| Tail | ♂ | 33.0–42.0 mm (37.4 ± 2.0 mm; n = 28) |
| | ♀ | 33.0–40.0 mm (36.4 ± 2.0 mm; n = 24) |
| Mass | ♂ | 10.5–15.0 g (12.5 ± 0.9 g; n = 61) |
| | ♀ | 10.0–14.0 g (12.4 ± 1.1 g; n = 34) |
| Bill | ♂ | Unknown |
| | ♀ | 8.6–9.4 mm (9.0 ± 0.4 mm; n = 3) |
| Tarsus | ♂ | 24.2 mm (n = 1) |
| | ♀ | 22.1–23.5 mm (22.8 ± 0.7 mm; n = 3) |

Similar species: None.

Skull: Probably completes during FCF (n = 23).

Brood patch: Probably develops in both sexes, being more extensive in ♀♀. BPs seen in Aug, Sep, and Jan.

Sex: ♂ has a black throat and the necklace is boldly marked black against a white chest. ♀ has a white throat and the black necklace is strewn over a buffy chest.

Molt: Group 3/Complex Basic Strategy; FPF partial, completing within 6 weeks of leaving the nest (Willis 1982), replacing some to all med covs and 0 to 5 (or sometimes more?) inner gr covs; DPBs complete. Wing molt appears to peak from Mar–May.

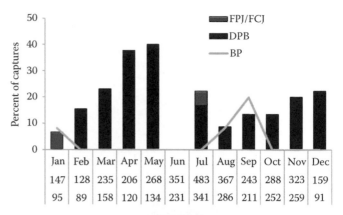

| | Jan | Feb | Mar | Apr | May | Jun | Jul | Aug | Sep | Oct | Nov | Dec |
|---|---|---|---|---|---|---|---|---|---|---|---|---|
| | 147 | 128 | 235 | 206 | 268 | 351 | 483 | 367 | 243 | 288 | 323 | 159 |
| | 95 | 89 | 158 | 120 | 134 | 231 | 341 | 286 | 211 | 252 | 259 | 91 |

FCJ     ♂ = ♀. Like DCB ♀♀, but less, med, and gr covs are dusky black (not jet black) and buff-spotted tips are less distinct. Rects are distinctly pointed. Skull is unossified.

FPF     ♂ ≠ ♀. Replacing juvenile body feathers, less, med, and/or inner gr covs with adult-like feathers. Incoming less and med covs in ♂♂ are black with white tips and in ♀♀ are black with buff tips. Skull is unossified.

Less covs molting, med, gr covs, and all flight feathers retained juvenile.

FCF    ♂ ≠ ♀. Like DCB, but with molt limits among the gr covs, med covs, and/or between the gr covs and med covs. Rects are pointed. The skull is either ossified or unossified.

Less, several inner med covs, and one inner gr cov new, other med and gr covs, and all flight feathers retained juvenile (left); less and med covs new, gr covs and all flight feathers retained juvenile (right).

SPB    ♂ ≠ ♀. Replacing juv flight feathers with adult flight feathers. May be difficult to distinguish from DPB. Skull is ossified.

DCB    ♂ ≠ ♀. Lacks molt limits among gr covs or between med and gr covs. Rects are rounded. Skull is ossified.

(Courtesy of Angelica Hernández.)          (Courtesy of Angelica Hernández.)

DPB    ♂ ≠ ♀. Replacing adult-like body and flight feathers with adult-like feathers. May be difficult to distinguish from SPB. Skull is ossified.

## LITERATURE CITED

Bierregaard, R. O., M. Cohn-Haft, and D. F. Stotz. 1997. Cryptic biodiversity: an overlooked species and new subspecies of antbird (Formicariidae) with a revision of *Cercomacra tyrannina* in northeastern South America. Ornithological Monographs 48:111–128.

Johnson, R. A. 1953. Breeding notes on two Panamanian antbirds. Auk 70:494–496.

Isler, M. L., P. R. Isler, and B. M. Whitney. 1997. Biogeography and systematics of the *Thamnophilus punctatus* (Thamnophilidae) complex. Ornithological Monographs 48:355–381.

Johnson, E. I. 2011. Fragmentation sensitivity and its consequences on demography and host-ectoparasite dynamics in Amazonian birds. Louisiana State University, Baton Rouge, LA.

Johnson, E. I., P. C. Stouffer, and R. O. Bierregaard. 2012. The phenology of molting, breeding and their overlap in central Amazonian birds. Journal of Avian Biology 43:141–154.

Johnson, E. I., and J. D. Wolfe. 2014. Thamnophilidae (antbird) molt strategies in a central Amazonian rainforest. Wilson Journal of Ornithology 126:451–462.

Pyle, P. 1998. Eccentric first-year molts in certain Tyrannid flycatchers. Western Birds 29:29–35.

Pyle, P., K. Tranquillo, K. Kayano, and N. Arcilla. 2016. Molt patterns, age criteria, and molt-breeding dynamics in American Samoan landbirds. Wilson Journal of Ornithology 128:56–69.

Ryder, T. B., and J. D. Wolfe. 2009. The current state of knowledge on molt and plumage sequences in selected Neotropical bird families: A review. Ornitología Neotropical 20:1–18.

Schwartz, B. A. 2008. Sex-specific investment in incubation and the reproductive biology of two tropical antbird species. M.S. thesis, University of Montana, Missoula, MT.

Skutch, A. F. 1934. A nesting of the Slaty Antshrike (*Thamnophilus punctatus*) on Barro Colorado Island. Auk 51:8–16.

Willis, E. O. 1969. On the behavior of five species of *Rhegmatorhina*, ant-following antbirds of the Amazon Basin. Wilson Bulletin 81:363–395.

Willis, E. O. 1973. The behavior of Ocellated Antbirds. Smithsonian Contributions to Zoology 144.

Willis, E. O. 1981. Diversity in adversity: The behaviors of two subordinate antbirds. Arquivos de Zoologia 30:159–234.

Willis, E. O. 1982. The behavior of Scale-backed Antbirds. Wilson Bulletin 94:447–462.

Wolfe, J. D., P. Pyle, and C. J. Ralph. 2009. Breeding seasons, molt patterns, and gender and age criteria for selected northeastern Costa Rican resident landbirds. Wilson Journal of Ornithology 121:556–567.

Zimmer, K., and M. Isler. 2003. Family Thamnophilidae (Typical Antbirds). Pp. 448–681 in J. del Hoyo, A. Elliott, and D. A. Christie (editors), Handbook of the birds of the world. Lynx Editions, Barcelona, Spain.

# CHAPTER TWENTY-THREE

# Conopophagidae (Gnateaters and *Pittasoma* Antpittas)

# Species in South America: 10

# Species recorded at BDFFP: 1

# Species captured at BDFFP: 1

The gnateaters are a small Neotropical family composed of 10 species in two genera: *Conopophaga* and *Pittasoma*. Gnateaters are compact, short-tailed, long-legged, sexually dichromatic insectivores that spend much of their time near the forest floor. Gnateaters have 10 pp, 9 ss, and 12 rects.

Ryder and Wolfe (2009) reviewed *Pittasoma* molt patterns and suggested that they follow a Complex Basic Strategy, and that the preformative molt is partial and does not include primary coverts, flight feathers, and, in some cases, greater coverts.

Molt patterns of *Conopophaga* have not been studied elsewhere to our knowledge; however, at our site, the preformative molt in *Conopophaga aurita* is similar to that described for *Pittasoma* (Ryder and Wolfe 2009) where molt limits were observed between median and greater coverts. Definitive prebasic molts are complete in extent based on the capture of multiple adult-like birds without molt limits. More study is needed to confirm the extent of the preformative molt among *Conopophaga*.

Males exhibit brood patches as well as females as both sexes apparently incubate two-egg clutches (Willis et al. 1983, Alves et al. 2013). Skull ossification appears useful for aging the single species at our study site, *Conopophaga aurita*, where the skull appears to ossify during the formative plumage.

## SPECIES ACCOUNTS

*Conopophaga aurita aurita*                    Band Size: E

Chestnut-bellied Gnateater • Chupa-dente-de-cinta          # Individuals Captured: 170

Wing ♂ 62.5–70.0 mm (65.9 ± 1.5 mm; n = 59)

♀ 61.0–68.0 mm (64.2 ± 2.0 mm; n = 29)

Tail ♂ 24.0–35.0 mm (29.8 ± 1.8 mm; n = 61)

♀ 27.0–33.0 mm (29.5 ± 1.5 mm; n = 29)

205

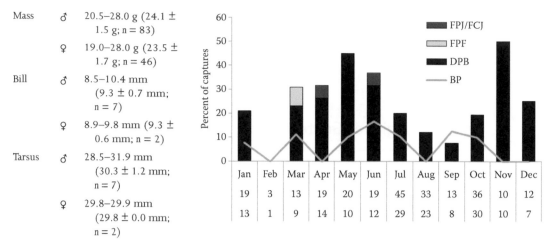

| | | | Jan | Feb | Mar | Apr | May | Jun | Jul | Aug | Sep | Oct | Nov | Dec |
|---|---|---|---|---|---|---|---|---|---|---|---|---|---|---|
| Mass | ♂ | 20.5–28.0 g (24.1 ± 1.5 g; n = 83) | 19 | 3 | 13 | 19 | 20 | 19 | 45 | 33 | 13 | 36 | 10 | 12 |
| | ♀ | 19.0–28.0 g (23.5 ± 1.7 g; n = 46) | 13 | 1 | 9 | 14 | 10 | 12 | 29 | 23 | 8 | 30 | 10 | 7 |
| Bill | ♂ | 8.5–10.4 mm (9.3 ± 0.7 mm; n = 7) | | | | | | | | | | | | |
| | ♀ | 8.9–9.8 mm (9.3 ± 0.6 mm; n = 2) | | | | | | | | | | | | |
| Tarsus | ♂ | 28.5–31.9 mm (30.3 ± 1.2 mm; n = 7) | | | | | | | | | | | | |
| | ♀ | 29.8–29.9 mm (29.8 ± 0.0 mm; n = 2) | | | | | | | | | | | | |

Similar species: None.

Skull: Completes probably during FCF (n = 37).

Brood patch: Both sexes incubate and develop BPs, which are probably more extensive in ♀♀. May have a prolonged breeding season, but data are unclear; additional area breeding records have been reported from Jul and Dec, as well as wet-season records from French Guiana (Whitney 2003, Leite et al. 2012).

Sex: ♂ has a black face and throat with a bold silvery-white supercilium. ♀ has a chestnut face and throat with a bold silvery-white supercilium.

Molt: Group 3/Complex Basic Strategy; FPF partial, DPBs complete. The FPF includes less and med covs, but rarely gr covs (or sometimes 1 or 2)? Molt limits are cryptic and further study is needed to confirm this pattern and variations. The double-peak in timing may be an artifact of low sample sizes; more study is needed.

FCJ     Mantle mottled, dusky underparts, lacks eyeline. Skull is unossified.

FPF     Replacing juvenile body feathers, less, med, and/or inner gr covs with adult-like feathers. Skull is unossified.

FCF     Look for molt limits within the med covs or between the med and gr covs, or sometimes within the gr covs. The crown may also average more dusky and supercilium above the eye may average less silvery-white. The skull is probably either ossified or unossified.

Less and med covs new, all gr covs, alulas, and flight feathers retained juvenile (left); note subtle differences in facial pattern between FCF (right) and DCB.

SPB     Unknown how to distinguish from DPB (use UPB), but look for differences in rect shape between retained and replaced feathers. Skull is ossified.

DCB        Without molt limits among the gr covs. Skull is ossified.

DPB        Unknown how to distinguish from SPB (use UPB), but look for similar rect shape between retained and replaced feathers. Skull is ossified.

## LITERATURE CITED

Alves, M. A., C. F. D. Rocha, M. V. Sluys, and M. Vecchi. 2013. Nest, eggs, and effort partitioning in incubation and rearing by a pair of the Black-cheeked Gnateater, *Conopophaga melanops* (Passeriformes, Conopophagidae), in an Atlantic Rainforest area of Rio de Janeiro, Brazil. Ararajuba 10:67–71.

Leite, G. A., F. B. R. Gomes, and D. B. MacDonald. 2012. Description of the nest, nestling and broken-wing behavior of *Conopophaga aurita* (Passeriformes: Conopophagidae). Revista Brasileira de Ornitologia 20:128–131.

Ryder, T. B., and J. D. Wolfe. 2009. The current state of knowledge on molt and plumage sequences in selected Neotropical bird families: a review. Ornitología Neotropical 20:1–18.

Whitney, B. M. 2003. Family Conopophagidae (Gnateaters). Pp. 732–745 in J. del Hoyo, A. Elliott, and J. Sargatal (editors), Handbook of the birds of the world, Broadbills to Tapaculos. Lynx Edicions, Barcelona, Spain.

Willis, E. O., Y. Oniki, and W. R. Silva. 1983. On the behaviour of Rufous Gnateaters (*Conopophaga lineata*, Formicariidae). Naturalia 8:67–83.

# CHAPTER TWENTY-FOUR

# Grallariidae (Antpittas)

# Species in South America: 53

# Species recorded at BDFFP: 3

# Species captured at BDFFP: 3

Antpittas are small- to medium-sized ground-dwelling insectivores characterized by long legs and small rounded tails. Antpittas are often monochromatic (like the three species found at our study site), although some species exhibit sex-related plumage differences where males have brighter and more colorful crowns. Antpittas have 10 pp, 9 ss, and 12 rects.

However, Ryder and Wolfe (2009) and Dickey and Van Rossem (1938) suggest that antpittas adhere to a Complex Basic Strategy where preformative molts are often partial, resulting in molt limits among the median and greater coverts. This appears consistent with at least two of the species covered here. Definitive prebasic molts consistently appear complete.

Skull ossification appears to be useful for aging where complete ossification probably occurs during the formative plumage. For the few species where information is available, it appears that males and females exhibit brood patches because both sexes typically incubate eggs, similar to the closely related Thamnophilidae (Greeney et al. 2008).

## SPECIES ACCOUNTS

*Grallaria varia varia*
Variegated Antpitta • Tovacuçu

Band Size: H
# Individuals Captured: 36

| | |
|---|---|
| Wing | 103.0–123.0 mm (114.5 ± 4.0 mm; n = 25) |
| Tail | 38.0–47.0 mm (42.4 ± 2.2 mm; n = 28) |
| Mass | 103.0–145.0 g (120.1 ± 10.1 g; n = 25) |

Similar species: Closest to *Hylopezus macularius*, but larger and note plumage differences.

Skull: Completes possibly during FCF (n = 2).

Brood patch: Both sexes likely incubate. Has only been observed in Jun. May typically breed during the late wet to early dry season. This is supported by juveniles captured Jun–Jul and a fresh FCF captured in Oct.

Sex: ♂ = ♀.

Molt: Group 3/Complex Basic Strategy; FPF partial, DPBs complete. FPF includes all body feathers, less covs, most (or sometimes all?) med covs, and some (±5) gr covs, but not remiges (or occasionally 1–3 tertiaries?). Perhaps molts at least Apr–Sep based on five captures observed with wing molt.

| FCJ | Gr covs have triangular buffy tips. Skull is unossified. |
|---|---|
| FPF | Replacing body feathers and ss covs with feathers that lack bold triangular buffy tips and instead are more olive-brown with a buffy streak along the rachis (in body feathers especially). Skull is unossified. |
| FCF | Molt limits among the gr covs with the replaced longer inner gr covs lacking buffy tips contrasting with the shorter retained outer gr covs with buffy tips. The skull is either ossified or unossified. |

Inner 5 gr covs replaced, s9 replaced, outer gr covs, outer 3 med covs, all alulas, and all other flight feathers retained juvenile.

| SPB | Replacing dull brown juvenile flight feathers with brighter olive-brown adult flight feathers. Skull is ossified. |
|---|---|
| DCB | Without molt limits among the gr covs. All ss covs lack buffy triangular spots at the tip. Skull is ossified. |
| DPB | Replacing olive-brown flight feathers with brighter olive-brown adult flight feathers. Skull is ossified. |

## Hylopezus macularius macularius

Spotted Antpitta • Torom-carijó

Band Size: F, E

# Individuals Captured: 52

Wing 76.0–89.0 mm (83.0 ± 2.7 mm; n = 35)

Tail 28.0–38.0 mm (34.0 ± 2.0 mm; n = 34)

Mass 36.3–50.0 g (42.9 ± 3.1 g; n = 45)

Bill 11.3–12.2 mm (11.8 ± 0.6 mm; n = 2)

Tarsus 39.9–40.3 mm (40.1 ± 0.3 mm; n = 2)

Similar species: Closest to *Grallaria varia*, but smaller and note many differences in plumage.

Skull: Completes, possibly during FCF (n = 8).

Brood patch: Both sexes may incubate. May be mainly a wet-season breeder as BPs have been observed from Dec–Apr, but also once in Sep.

Sex: ♂ = ♀.

Molt: Group 3(?)/Complex Basic Strategy; FPF partial(?), DPBs complete. FPF assumed to be partial as in many other antpittas and antthrushes, including some to all med and some inner gr covs. Based on 10 captures with birds in molt, wing molt may occur mainly Mar–Oct, but perhaps also earlier or later.

(Courtesy of Aida Rodrigues.)

| FCJ | Rusty brown crown, nape gray, prominent buff eyering, yellowish flanks, two dark stripes on throat, and small necklace of black (fide M. Cohn-Haft). Skull is unossified. |
|---|---|
| FPF | Look for evidence of juvenile-like body plumage being replaced with an adult-like plumage. Skull is unossified. |

**FCF**   Unknown, but look for molt limits among the ss covs. Beware of pseudolimits among the gr covs. Skull can be ossified or unossified.

**SPB**   Unknown how to distinguish from DPB, but at a minimum look for differences in rect shape between more pointed retained feathers and more rounded replaced feathers (use UPB when uncertain). Skull is ossified.

**DCB**   Gray crown and nape. Would lack molt limits in the ss covs. Beware of pseudolimits among the gr covs. Skull is ossified.

**DPB**   Unknown how to distinguish from SPB, but at a minimum look for similar rect shape between rounded retained and replaced feathers (use UPB when uncertain). Skull is ossified.

## *Myrmothera campanisona campanisona*

Band Size: G

Thrush-like Antpitta • Tovaca-patinho

# Individuals Captured: 8

Wing   78.0–82.0 mm (79.8 ± 1.5 mm; n = 5)

Tail   31.0–37.0 mm (35.4 ± 2.5 mm; n = 5)

Mass   46.0–54.0 g (48.5 ± 2.7 g; n = 7)

Similar species: Superficially plumaged like a *Catharus* thrush, but with longer legs and a much shorter tail.

Skull: Completes possibly during FCF (n = 5).

Brood patch: Both sexes may incubate. Has only been observed in Nov.

Sex: ♂ = ♀.

Molt: Group 3/Complex Basic Strategy; FPF partial, DPBs complete. FPF includes body feathers, some (or all?) less covs, some med covs, and some (±3) inner gr covs, but not remiges (or sometimes terts?). One bird was captured about halfway through wing molt in Jun.

**FCJ**   Ss covs are brown without olive tones. Skull is unossified.

**FPF**   Replacing juvenile body feathers, ss covs, and sometimes some terts or rects with adult-like feathers. Skull is unossified.

**FCF**   Molt limits are found within the med and gr covs, with the replaced inner covs relatively long and olive-brown contrasting against duller brown outer covs. The skull is either ossified or unossified.

Inner 3 gr covs, med covs, and less covs replaced, outer gr covs, all alulas, and all flight feathers retained juvenile.

SPB       Replacing juvenile flight feathers with adult flight feathers. May be difficult to distinguish from DPB (use UPB). Skull is ossified.

DCB       Without molt limits; all ss covs are brown with an olive wash, but beware of heavily worn birds. Skull is ossified.

DPB       Replacing adult-like body and flight feathers with adult-like feathers. May be difficult to distinguish from SPB (use UPB). Skull is ossified.

Rect shape is similar between retained reddish-brown feathers and replaced olive-brown feathers (left); p7–p10 retained definitive basic, p6 missing, p5 molting, p1–p4 new, s1–s2 new, s3–s5 retained definitive basic, s6 missing, s7–s9 new (right).

## LITERATURE CITED

Dickey, D. R., and A. J. Van Rossem. 1938. The birds of El Salvador. Field Museum of Natural History, Zoological Series, Chicago, IL.

Greeney, H. F., R. C. Dobbs, P. R. Martin, and R. A. Gelis. 2008. The breeding biology of *Grallaria* and *Grallaricula* antpittas. Journal of Field Ornithology 79:113–129.

Ryder, T. B., and J. D. Wolfe. 2009. The current state of knowledge on molt and plumage sequences in selected Neotropical bird families: A review. Ornitología Neotropical 20:1–18.

# CHAPTER TWENTY-FIVE

# Formicariidae (Antthrushes)

# Species in South America: 11

# Species recorded at BDFFP: 2

# Species captured at BDFFP: 2

Antthrushes are a small group of "galinha-do-mato" (forest chickens) closely related to Thamnophilidae and Conopophagidae, and are represented by 11 species in two genera (*Formicarius* and *Chamaeza*) exclusively in the Neotropics. Antthruses are small- to medium-sized birds well adapted to foraging near and on the ground with long legs and a short, rounded tail. Antthrushes have 10 pp, 9 ss, and 12 rects.

All antthrushes follow a Complex Basic Strategy (Ryder and Wolfe 2009), although they vary in sexual dichromatism and preformative molt extent. At our study site *Formicarius colma* undergoes a partial to incomplete preformative molt and is sexually dichromatic, where males exhibit more black on the throat and face. Conversely, *Formicarius analis* apparently undergoes a complete preformative molt and sexes are similar in all plumages. Does this correspondence between the extent of preformative molt and sexual dichromatism adhere to other *Formicarius* or Formicariidae?

At least at our sites and in a few other well-documented species (Krabbe and Schulenberg 2013), both male and female antthrushes exhibit brood patches incubating two to three egg clutches, which is consistent with their close taxonomic relationship with Thamnophilidae in which both sexes share in incubation. Skulls appear to ossify during the preformative molt or formative plumage.

---

## SPECIES ACCOUNTS

*Formicarius colma colma*                                                                 Band Size: G

Rufous-capped Antthrush • Galinha-do-mato                                    # Individuals Captured: 571

| | | |
|---|---|---|
| Wing | ♂ | 75.0–89.0 mm (83.3 ± 2.4 mm; n = 153) |
| | ♀ | 75.0–89.0 mm (82.8 ± 2.5 mm; n = 118) |
| Tail | ♂ | 42.0–57.0 mm (50.0 ± 2.6 mm; n = 166) |
| | ♀ | 42.0–58.0 mm (49.8 ± 2.7 mm; n = 114) |

| Mass | ♂ | 38.0–53.0 g (45.7 ± 2.5 g; n = 253) |
| | ♀ | 39.0–53.5 g (46.9 ± 2.8 g; n = 165) |
| Bill | ♂ | 12.0–14.2 mm (12.8 ± 0.5 mm; n = 36) |
| | ♀ | 11.6–14.0 mm (12.7 ± 0.6 mm; n = 21) |
| Tarsus | ♂ | 32.0–37.4 mm (34.9 ± 1.3 mm; n = 36) |
| | ♀ | 32.2–37.3 mm (34.6 ± 1.2 mm; n = 21) |

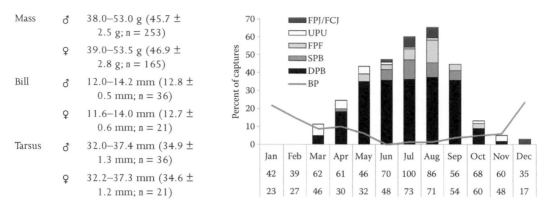

Legend: FPJ/FCJ, UPU, FPF, SPB, DPB, BP

| | Jan | Feb | Mar | Apr | May | Jun | Jul | Aug | Sep | Oct | Nov | Dec |
|---|---|---|---|---|---|---|---|---|---|---|---|---|
| | 42 | 39 | 62 | 61 | 46 | 70 | 100 | 86 | 56 | 68 | 60 | 35 |
| | 23 | 27 | 46 | 30 | 32 | 48 | 73 | 71 | 54 | 60 | 48 | 17 |

Similar species: *F. analis* has bright orange (not dusky olive-brown) undertail coverts, a white loral spot (unlike white loral bar in ♀ *F. colma*), and crown not contrasting with back.

Skull: Completes during FCF (n = 107).

Brood patch: Develops in both sexes most frequently from Dec–Feb, but perhaps occasionally year-round.

Sex: ♂ has black lores and throat, which are white in ♀, although some ♀♀ can be quite variable with some black mixed in the lores and throat.

Molt: Group 3/Complex Basic Strategy; FPF partial-(incomplete), DPBs complete. FPF includes body feathers, less and med covs, inner 3–7 gr covs, 0–all rects, and often (always?) 1–3 terts. Rarely up to 2 inner pp are replaced. Most frequent from May–Sep and least frequent from Oct–Mar.

FCJ    Throat and lores are orange. Forehead is orange. Less, med, and gr covs and remiges are brown without olive tinge. Underwing covs are relatively narrow and short with blurry borders between the pale yellow bases and dusky black only at the tips. Skull is unossified.

FPF    Has a mix of juvenile orange and incoming black forehead feathers. Replaced olive ss covs and/or terts contrast against retained brownish juvenile ss covs and ss. Skull is unossified.

Body feathers, but not ss covs or flight feathers replaced.

FCF   Has molt limits with the replaced inner gr covs and terts (sometimes) brighter olive-brown than the dull brown retained outer gr covs, terts, and ss. Often, but not always, has some juvenile-like orange feathering in the forehead, lores, or around eyes. The skull is either ossified or unossified.

Outer 5 gr covs retained, inner gr covs replaced or missing, s8 replaced, all other flight feathers retained (left); right four images show variation in the amount of white, orange, and black on the faces of presumed females (top row) and males (bottom row).

SPB   Replacing dull brownish flight feathers with olive-brown flight feathers. Sometimes still with some orange juvenile feathers in the forehead, lores, or around eyes. Retained juvenile ss are typically slightly shorter than replaced definitive basic ss. May be difficult to distinguish from SPB especially late in molt (use UPB). Skull is unossified.

p9–p10 retained juvenile, p8 missing, p7 molting, p1–p6 new, s1–s2 new, s3 and s5 missing, s4 retained juvenile, s6–s9 new.

DCB   Less, med, gr covs, and ss are uniformly olive-brown. Underwing covs are relatively long and broad with distinct boundaries between the pale yellow bases and narrow black tips. The forehead is solid black. Rects relatively rounded. Skull is ossified.

DPB   Replacing dull olive-brown ss covs and flight feathers with olive-brown covs and flight feathers. Be cautious of very worn retained definitive feathers appearing like more brown juvenile feathers; thus, may often be difficult to distinguish from SPB (use UPB). Skull is ossified.

## Formicarius analis crissalis

Black-faced Antthrush • Pinto-do-mato-de-cara-preta

| | |
|---|---|
| Wing | 85.0–97.0 mm (90.4 ± 2.6 mm; n = 41) |
| Tail | 44.0–56.0 mm (51.3 ± 2.7 mm; n = 40) |
| Mass | 50.5–74.5 g (62.1 ± 5.2 g; n = 52) |
| Bill | 12.8–14.2 mm (13.4 ± 0.6 mm; n = 6) |
| Tarsus | 35.8–37.9 mm (36.7 ± 0.8 mm; n = 6) |

Similar species: *F. colma* has dusky olive-brown (not orange) undertail coverts and has strong crown/back contrast, unlike *F. analis*. Furthermore, compared with the white loral spot on *F. analis*, *F. colma* exhibits either a white loral bar (in ♀♀) or lacks white in the lores altogether (in ♂♂).

Skull: Apparently completes, probably toward the end of FPF or FCF (n = 6).

Brood patch: Both sexes incubate (Krabbe and Schulenberg 2013). Likely breeds during the late dry and early wet seasons.

Sex: ♂ = ♀.

Molt: Group 4/Complex Basic Strategy; FPF complete, DPBs complete. Most frequent from May–Aug and has not been observed from Nov–Feb.

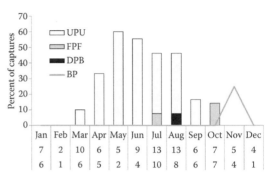

| | Jan | Feb | Mar | Apr | May | Jun | Jul | Aug | Sep | Oct | Nov | Dec |
|---|---|---|---|---|---|---|---|---|---|---|---|---|
| | 7 | 2 | 10 | 6 | 5 | 9 | 13 | 13 | 6 | 7 | 5 | 4 |
| | 6 | 1 | 6 | 5 | 2 | 4 | 10 | 8 | 6 | 7 | 4 | 1 |

FCJ    The throat is white (always?) with black mottling. Ss covs, ss, and pp brown. Rects relatively pointed. Undertail covs relatively soft. Skull is unossified.

FPF    Replaces all brown juvenile body and flight feathers with slightly more olive adult-like body and flight feathers. Skull may become ossified by the end of the molt, when it may be very challenging to distinguish from DPB, thus UPU should be used.

Body feathers, but not ss or flight feathers replaced (upper left); p9–p10 retained juvenile, p8 missing, p1–p7 new, s1–s2 new, s3 molting, s4–s6 retained juvenile, s7–s8 new, s9 retained juvenile (upper right); underwing early in the FPF showing juvenile underwing covs being replaced by adult-like underwing covs (lower left); right side of tail with r4 more rounded and molting compared to retained juvenile r5–r6 and r1–r2, r3 missing (lower right).

FAJ    Throat is all black. Ss covs, ss, and pp brownish, but with olive tones. Rects relatively rounded. Skull is ossified.

UPB    Replaces worn olive-brown adult body and flight feathers with more olive adult body and flight feathers. Skull is ossified. May be very challenging to distinguish from FPF, thus UPU should be used in these cases.

## LITERATURE CITED

Krabbe, N. K., and T. S. Schulenberg. 2013. Family Formicariidae (Antthrushes). In J. del Hoyo, A. Elliott, J. Sargatal, D. A. Christie, and E. de Juana (editors), Handbook of the birds of the world alive. Lynx Edicions, Barcelona, Spain. <http://www.hbw.com>.

Ryder, T. B., and J. D. Wolfe. 2009. The current state of knowledge on molt and plumage sequences in selected Neotropical bird families: a review. Ornitología Neotropical 20:1–18.

# CHAPTER TWENTY-SIX

# Furnariidae (Ovenbirds, Leaftossers, and Woodcreepers)

# Species in South America: 292

# Species recorded at BDFFP: 24

# Species captured at BDFFP: 23

An extremely variable family in terms of plumage and morphology, especially in the bill and tail, the furnariids have adapted to nearly all environments in South and Central America, and display a wide array of foraging techniques. Three major groups are recognized: Sclerurinae (leaftossers), Dendrocolaptinae (woodcreepers), and Furnariinae (everything else, collectively referred to as ovenbirds). For their size, furnariids often have relatively large and strong legs and feet. Most species are variably brown with streaks, throat patches, or wing patches, and many species (especially woodcreepers and leaftossers) have strengthened tail feather shafts. Most species are sexually monochromatic, including all at the BDFFP, but males often average slightly or notably larger than females (Vaurie 1980, Winker et al. 1994). Furnariids have 10 pp, 9 ss, and 12 rects.

Most, if not all, species follow a Complex Basic Strategy and most tropical species appear to have complete preformative molts, but at least some Central American and austral temperate species or populations appear to have partial or incomplete preformative molts (Guallar et al. 2009, Ryder and Wolfe 2009, Pyle et al. 2015). Dickey and Van Rossem (1938) report the retention of 1–3 central rectrices during otherwise complete molts in some woodcreepers, although we are skeptical of this unprecedented replacement pattern (P. Pyle, pers. comm.) because central rectrices can wear faster than outer rectrices, giving the false appearance of molt limits. In at least some individuals of some woodcreeper species (e.g., *Glyphorynchus spirurus*), rectrix replacement follows a centripetal replacement pattern with r1 replaced last (Howell 1957), similar to some woodpeckers, and perhaps delayed replacement of r1 would allow outer rectrices to fully grow and prop tree-clinging species against the sides of trunks and branches. Thus, such a suspension limit may mimic a molt limit. Rarely, flight feathers are also retained in otherwise complete molts, but this should be used for aging with caution because we have documented these in known-age *Glyphorynchus* both in formative and third basic plumages. In addition, prealternate molts have been suggested as incomplete (including rectrices, but not primaries or secondaries) in *Synallaxis erythrothorax* in El Salvador (Dickey and Van Rossem 1938), and absent to limited in two species of woodcreepers in Mexico (Guallar et al. 2009), but are not otherwise known to occur in this diverse family (Marantz et al. 2003, Remsen 2003, Ryder and Wolfe 2009, Pyle et al. 2015). Prealternate molts may be anticipated in other near-equatorial dry forest or open country species. A single molt can last 4–6 months, but limited data exist on only a few species and some evidence suggests that some species or populations molt in as fast as 2 months. Distinct juvenile plumages are often found in some species, although in many cases the juvenile plumage is only subtly different. Typically the juvenile plumage lasts for just a few months or less, and may be identified by either an "ochraceous wash on the underparts, an increase in dusky feather margins that produces a scaly appearance, usually on the throat and breast, and an absence of, or reduction in, the contrasting throat or crown patches that characterizes adult plumage" (Remsen 2003).

In some species of woodcreepers, the undertail coverts are distinctly softer and less patterned in the juvenile plumage, and should be evaluated (especially) more broadly in the family.

Although it has been reported that the skull does not ossify completely in many genera (Remsen 2003), it appears that most regularly captured species at the BDFFP have completely ossified skulls as an adult (see also Winker et al. 1994). For these species, it may therefore be possible to identify FCF birds by the degree of ossification, but an ossified skull could be a FCF or DCB bird and are best left as "after juvenile" (FAJ). Many species in Central America breed near the onset or during the wet season (Remsen 2003), but timing appears to be more variable among taxa at the BDFFP (Johnson et al. 2012). Both sexes incubate in many species, but in some ant-following species, incubation is done exclusively by the female.

## SPECIES ACCOUNTS

### *Sclerurus mexicanus peruvianus*

Band Size: E

Tawny-throated Leaftosser • Vira-folha-de-peito-vermelho

# Individuals Captured: 86

| | |
|---|---|
| Wing | 72.0–84.0 mm (79.0 ± 2.7 mm; n = 60) |
| Tail | 51.0–66.0 mm (58.8 ± 2.7 mm; n = 62) |
| Mass | 20.0–30.0 g (24.7 ± 1.8 g; n = 79) |
| Bill | 17.1–18.1 mm (17.9 ± 0.4 mm; n = 5) |
| Tarsus | 26.0–27.2 mm (26.7 ± 0.6 mm; n = 3) |

Similar species: Nearly identical in plumage to other *Sclerurus*, but with richer red, less streaked throat than *S. rufigularis* and considerably smaller than *S. caudacutus*. Best separated from *S. rufigularis* by bill length. "Small" *Sclerurus* (i.e., not *S. caudacutus*) with bill lengths above 15.0 mm are *S. mexicanus*. Tarsi above 25.0 mm are also probably *S. mexicanus*, but more study is needed; if true, this may be useful in identifying immatures with incompletely grown bills. Bill shape has been used as a reliable characteristic for separating the three *Sclerurus*, but this is subjective and probably not as reliable as measurements.

Skull: Completes (n = 28) probably during FPF or soon thereafter.

Brood patch: All three BPs have been observed in Apr, but breeding may occur throughout the wet season as in *S. rufigula*.

Sex: ♂ = ♀.

Molt: Group 4/Complex Basic Strategy; FPF complete, DPBs complete. The rects may be the last major feather group to complete. Molt peaks from Aug–Oct and is least frequent from Jan–May.

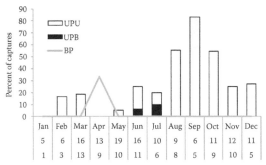

| | Jan | Feb | Mar | Apr | May | Jun | Jul | Aug | Sep | Oct | Nov | Dec |
|---|---|---|---|---|---|---|---|---|---|---|---|---|
| | 5 | 6 | 16 | 13 | 19 | 16 | 10 | 9 | 6 | 11 | 12 | 11 |
| | 1 | 3 | 13 | 9 | 10 | 11 | 6 | 8 | 5 | 9 | 10 | 5 |

FCJ    Compared to FAJ, it is duller, with paler shaft streaks on the breast, dusky scaling on the throat and breast (Remsen 2003), but these characteristics are subtle at best. Skull is unossified.

FPF    Replacing adult-like juvenile body and flight feathers with adult-like formative feathers. Skull is unossified, but usually ossifies before completion of the FPF, which then makes this difficult to distinguish from UPB.

FAJ     Averages brighter, with few or no shaft streaks on the breast, and lacking dusky scaling on chestnut-reddish throat and breast (Remsen 2003). Skull is ossified.

UPB     Replacing adult-like body and flight feathers with adult-like feathers. May not be possible to distinguish from FPF during replacement of the last flight feathers (use UPU). Skull is ossified.

## *Sclerurus rufigularis furfurosus*

Short-billed Leaftosser • Vira-folha-de-bico-curto

Band Size: E

\# Individuals Captured: 309

| | |
|---|---|
| Wing | 69.5–83.0 mm (76.0 ± 2.8 mm; n = 210) |
| Tail | 51.0–65.0 mm (58.1 ± 2.7 mm; n = 183) |
| Mass | 17.5–25.0 g (21.2 ± 1.4 g; n = 268) |
| Bill | 10.7–14.2 mm (12.9 ± 0.8 mm; n = 34) |
| Tarsus | 21.6–25.7 mm (23.5 ± 0.9 mm; n = 23) |

Similar species: From *S. caudacutus* by small size. See *S. mexicanus*.

Skull: Completes (n = 67) during FPF or soon thereafter.

Brood patch: Both sexes probably develop BPs. Unlike most species at the BDFFP, BPs peak during the wet season, from Dec–Jun.

Sex: ♂ = ♀.

Molt: Group 4/Complex Basic Strategy; FPF complete, DPBs complete. Rects may be the last major feather group to be replaced. Molt peaks from May–Aug and is least frequent from Oct–Feb.

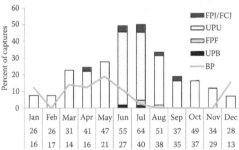

| | Jan | Feb | Mar | Apr | May | Jun | Jul | Aug | Sep | Oct | Nov | Dec |
|---|---|---|---|---|---|---|---|---|---|---|---|---|
| | 26 | 26 | 31 | 41 | 47 | 55 | 64 | 51 | 37 | 49 | 34 | 28 |
| | 16 | 17 | 14 | 16 | 21 | 27 | 40 | 38 | 35 | 37 | 29 | 13 |

FCJ     Compared to FAJ, it is duller overall, has a browner rump and uppertail-covs (Remsen 2003), and probably averages blacker on belly. Some may also have dusky scaling on the throat (e.g., LSUMZ specimen of *S. r. fulvigularis* from Guyana).

FPF      Replacing adult-like juvenile body and flight feathers with adult-like formative feathers. Skull is unossified, but should usually ossify before completion of the FPF, which then makes this difficult to distinguish from UPB (use UPU).

p9–p10 retained juvenile, p7–p8 molting, p1–p6 new, s1–s2 new, s3 missing, s4–s7 retained juvenile, s8–s9 new (left); right r6 retained juvenile, r5 missing, r4 molting, r1–r3 new (right).

FAJ      Probably brighter overall and less black in the underparts. Lacks dusky scaling on the throat and instead is brighter tawny-orange with faint pale streaking. Skull is ossified.

UPB      Replacing adult-like body and flight feathers with adult-like feathers. Skull is ossified. May not be possible to distinguish from FPF during replacement of the last flight feathers (use UPU).

*Sclerurus caudacutus insignis*

Black-tailed Leaftosser • Vira-folha-pardo

| | |
|---|---|
| Wing | 80.0–99.0 mm (91.8 ± 3.6 mm; n = 84) |
| Tail | 57.0–76.0 mm (66.8 ± 4 mm; n = 80) |
| Mass | 31.0–47.5 g (39.4 ± 3.1 g; n = 122) |
| Bill | 15.7–17.6 mm (16.7 ± 0.9 mm; n = 5) |
| Tarsus | 23.8–27.3 mm (26.1 ± 1.4 mm; n = 5) |

Similar species: Larger than other *Sclerurus* spp. Bill tends to appear slightly upturned because of a straight maxilla and upward-curving mandible, but measurements are probably more reliable.

Skull: Completes (n = 24) probably during FPF or soon thereafter.

Brood patch: Both sexes probably incubate and develop BPs. It may primarily breed during the wet season as in other *Sclerurus*.

Sex: ♂ = ♀.

Molt: Group 4/Complex Basic Strategy; FPF probably complete like other *Sclerurus*, DPBs complete. A well-defined molt schedule, peaking from Apr–Jul.

FCJ May be slightly darker overall with a duller throat compared to FAJ (Remsen 2003), but these features have not been confirmed at the BDFFP. Skull is unossified.

FPF Replacing adult-like juvenile body and flight feathers with adult-like formative feathers. Skull is unossified, but should usually ossify before completion of the FPF, which then makes this difficult to distinguish from UPB (use UPU).

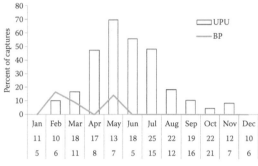

| | Jan | Feb | Mar | Apr | May | Jun | Jul | Aug | Sep | Oct | Nov | Dec |
|---|---|---|---|---|---|---|---|---|---|---|---|---|
| | 11 | 10 | 18 | 17 | 13 | 18 | 25 | 22 | 19 | 22 | 12 | 10 |
| | 5 | 6 | 11 | 8 | 7 | 5 | 15 | 12 | 16 | 21 | 7 | 6 |

FAJ May be slightly lighter overall with a brighter throat compared to the juvenile plumage (Remsen 2003), but these features have not been confirmed at the BDFFP. Skull is ossified.

UPB Replacing adult-like body and flight feathers with adult-like feathers. Skull is ossified. May not be possible to distinguish from FPF during replacement of the last flight feathers (use UPU).

*Certhiasomus stictolaemus clarior*

Spot-throated Woodcreeper • Arapaçu-de-garganta-pintada

Band Size: D, E

# Individuals Captured: 513

| | |
|---|---|
| Wing | 66.0–89.0 mm (79.3 ± 5.6 mm; n = 341) |
| Tail | 66.0–96.0 mm (80.9 ± 6.6 mm; n = 292) |
| Mass | 12.0–23.0 g (16.8 ± 2.4 g; n = 454) |
| Bill | 9.5–14.2 mm (12.1 ± 1.0 mm; n = 51) |
| Tarsus | 18.0–23.8 mm (21.5 ± 1.3 mm; n = 45) |

Similar species: See *D. longicauda*. Small *C. stictolaema* can resemble *Glyphorynchus spirurus*, but note the difference in bill shape and the lack of a tawny-orange wing stripe in *C. stictolaema*.

Skull: Completes (n = 163) probably during FPF or possibly FCF, but little information is available.

Brood patch: Observed from Jul–Dec. The 33 individuals with brood patches and wing measurements all fall within the "♀ group" (wing: 69–78 mm), suggesting that only ♀♀ incubate.

Sex: ♂ > ♀, but unknown amount of overlap (Figure 26.1).

Molt: Group 4/Complex Basic Strategy; FPF complete, DPBs complete. Peaks from Sep–Dec and least frequent from May–Jul.

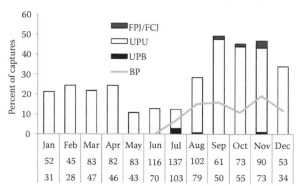

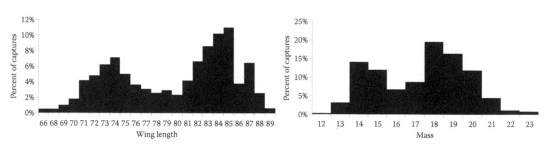

Figure 26.1 Frequency distributions of wing length (left) and mass (right) for *C. stictolaemus* captured at the BDFFP, suggesting average differences in body size between ♂♂ and ♀♀, although the degree of overlap is unknown.

FCJ    Not previously described; nearly identical to FAJ. Appears to lack chestnut-red in the less covs and has blacker bill than adult, and has reduced scaly markings in the throat. Skull is unossified.

FPF    Replacing adult-like juvenile body and flight feathers with adult-like formative feathers. Skull is unossified, but should usually ossify before completion of the FPF, which then makes this difficult to distinguish from UPB (use UPU).

FAJ    Has chestnut-red in the less covs, more scaling in the throat, and a lighter, browner bill than FCJ. ♂♂ and ♀♀ are monochromatic, but ♂♂ average larger than ♀♀ in wing length and mass measurements, although the degree of overlap is unknown. Skull is ossified.

UPB    Replacing adult-like body and flight feathers with adult-like feathers. Skull is ossified. May not be possible to distinguish from FPF during replacement of the last flight feathers (use UPU).

p5–p10 retained, p4 missing, p2–p3 molting, p1 new, all ss retained.

*Sittasomus griseicapillus axillaris*

Olivaceous Woodcreeper • Arapaçu-verde

| | |
|---|---|
| Wing | 68.0–85.0 mm (77.2 ± 5.1 mm; n = 36) |
| Tail | 66.0–80.0 mm (72.4 ± 4.0 mm; n = 35) |
| Mass | 11.3–17.0 g (14.2 ± 1.5 g; n = 48) |
| Bill | 10.4–11.3 mm (10.8 ± 0.4 mm; n = 4) |
| Tarsus | 18.4–20.5 mm (19.3 ± 0.8 mm; n = 5) |

Similar species: No other woodcreeper has the combination of unmarked grayish-olive head, throat, and back and stripe down center of spread wing.

Skull: Completes (n = 6) probably during the FPF or perhaps FCF.

Brood patch: Unclear if both sexes incubate, but may be only ♀♀, as in *S. g. jaliscensis* (Guallar et al. 2009). BPs have been observed in Jun and Sep (n = 2), but should be expected at least throughout the dry season.

Sex: ♂ (>) ♀.

Molt: Group 4/Complex Basic Strategy; FPF complete(?), DPBs complete. One bird at the BDFFP was seen molting flight feathers with an open skull, suggesting a complete FPF molt, although in Mexican *S. g. jaliscensis*, the FPF was suggested to be limited (Guallar et al. 2009). Probably most frequent from Sep–Mar and least frequent from Apr–Aug.

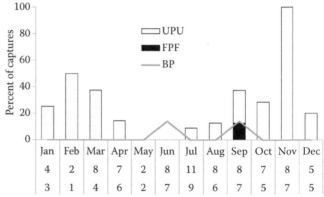

| | Jan | Feb | Mar | Apr | May | Jun | Jul | Aug | Sep | Oct | Nov | Dec |
|---|---|---|---|---|---|---|---|---|---|---|---|---|
| | 4 | 2 | 8 | 7 | 2 | 8 | 11 | 8 | 8 | 7 | 8 | 5 |
| | 3 | 1 | 4 | 6 | 2 | 7 | 9 | 6 | 7 | 5 | 7 | 5 |

FCJ — Similar to FAJ, but perhaps slightly paler below and brighter above, especially on rump; ss covs may be edged rufous (Marantz et al. 2003). Skull is unossified.

FPF — Replacing adult-like juvenile body and flight feathers with adult-like formative feathers. Skull is unossified, but should usually ossify before completion of the FPF, which then makes this difficult to distinguish from UPB (use UPU).

FAJ — Perhaps slightly brighter below and paler above than in FCJ, especially on rump; ss covs not edged rufous (Marantz et al. 2003). Skull is ossified.

UPB — Replacing adult-like body and flight feathers with adult-like feathers. Skull is ossified. May not be possible to distinguish from FPF during replacement of the last flight feathers.

*Deconychura longicauda longicauda*

Long-tailed Woodcreeper • Arapaçu-rabudo

Wing  91.0–116.0 mm (103.0 ± 6.3 mm; n = 96)

Tail  89.0–116.0 mm (102.1 ± 6.5 mm; n = 83)

Mass  22.0–36.3 g (29.1 ± 3.8 g; n = 127)

Bill  14.1–19.3 mm (16.4 ± 1.6 mm; n = 14)

Tarsus  23.7–28.1 mm (25.7 ± 1.5 mm; n = 13)

Similar species: A generic-looking woodcreeper with an intermediate amount of head and throat streaking. Most similar in size to *Xiphorhynchus pardalotus*, but less well marked, especially on the back. Most similar in plumage to *Certhiasomus stictolaema*, but has longer wing measurements, bill, and tarsus with minimal overlap. *C. stictolaema* also has more brown and less chestnut in the wings and a chestnut (not brown) lower back, concolor with the rump and rects (not with the upper back).

Skull: Completes (n = 31) probably during FPF or FCF, but little information available.

Brood patch: Have been observed from Jun–Sep, perhaps peaking before *C. stictolaema*. Five individuals with brood patches and wing measurements all fall within the "female group" (wing: 94–98 mm), suggesting that only ♀♀ may incubate, but more study is needed.

Sex: ♂ > ♀, probably with unknown amount of overlap.

Molt: Group 4/Complex Basic Strategy; FPF complete, DPBs complete. Has a poorly defined molt schedule, but possibly most frequent from Jul–Jan and least frequent from Feb–Apr, consistent with having earlier breeding and molting periods than *C. stictolaema* (see "Brood Patch" section).

Band Size: E, D

# Individuals Captured: 147

*D. longicauda*

*C. stictolaema*

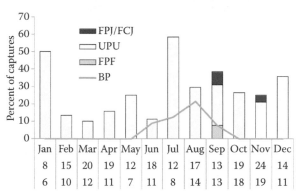

| | Jan | Feb | Mar | Apr | May | Jun | Jul | Aug | Sep | Oct | Nov | Dec |
|---|---|---|---|---|---|---|---|---|---|---|---|---|
| | 8 | 15 | 20 | 19 | 12 | 18 | 12 | 17 | 13 | 19 | 24 | 14 |
| | 6 | 10 | 12 | 11 | 7 | 11 | 8 | 14 | 13 | 18 | 19 | 11 |

Legend: FPJ/FCJ, UPU, FPF, BP

FCJ    Nearly identical to FAJ. It may be worth examining the amount of red in p10, which may be less extensive in FCJ. The juvenile plumage may also have brownish scaling in the throat and larger and less distinct breast spots (Marantz et al. 2003), but these features appear to be quite subtle.

FPF    Replacing adult-like juvenile body and flight feathers with adult-like formative feathers. Skull is unossified, but should usually ossify before completion of the FPF, which then makes this difficult to distinguish from UPB.

FAJ    Essentially identical to the FCJ. ♂♂ and ♀♀ are monochromatic, but ♂♂ are typically larger than ♀♀ in wing length and mass measurement, but the degree of overlap is unknown.

UPB    Replacing adult-like body and flight feathers with adult-like feathers. Skull is ossified. May not be possible to distinguish from FPF during replacement of the last flight feathers.

p5–p10 retained, p4 missing, p3 molting (not visible here), p1–p2 new, s1–s7 and s9 retained, s8 missing.

## *Dendrocincla fuliginosa fuliginosa*

Band Size: F, G

Plain-brown Woodcreeper • Arapaçu-pardo

# Individuals Captured: 340

| | |
|---|---|
| Wing | 95.0–115.0 mm (106.3 ± 4.3 mm; n = 244) |
| Tail | 80.0–107.0 mm (93.7 ± 5.5 mm; n = 233) |
| Mass | 31.0–50.0 g (40.2 ± 3.5 g; n = 293) |
| Bill | 18.9–24.0 mm (21.3 ± 1.2 mm; n = 47) |
| Tarsus | 26.3–31.0 mm (28.5 ± 1.0 mm; n = 46) |

Similar species: *Dendrocincla* spp. do not have curved rachises that protrude well beyond the feather barbules, thus their tail appears fairly flat from the side. *D. merula* is more dark chocolate brown with a more contrastingly white throat.

Skull: Can be difficult to determine in some individuals because of thick cranial skin. The skull completely ossifies (n = 82) and, although the timing is not well documented, it probably completes during FPF or FCF.

Brood patch: Only ♀♀ incubate and develop BPs. Mainly breeds from Aug–Nov, but may breed occasionally throughout the year.

Sex: ♂ > ♀, but with substantial overlap (Marantz et al. 2003).

Molt: Group 4/Complex Basic Strategy; FPF complete, DPBs complete. Has a well-defined molt schedule peaking from Sep–Dec and least frequent from May–Jul.

FCJ    Perhaps slightly brighter above and paler below than FAJ with the face and throat more conspicuously tinged buff, bill shorter and more solidly black, and gray irises (Marantz et al. 2003). Skull is unossified.

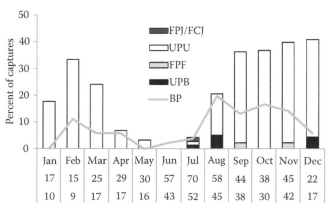

| | Jan | Feb | Mar | Apr | May | Jun | Jul | Aug | Sep | Oct | Nov | Dec |
|---|---|---|---|---|---|---|---|---|---|---|---|---|
| | 17 | 15 | 25 | 29 | 30 | 57 | 70 | 58 | 44 | 38 | 45 | 22 |
| | 10 | 9 | 17 | 17 | 16 | 43 | 52 | 45 | 38 | 30 | 42 | 17 |

Legend: FPJ/FCJ, UPU, FPF, UPB, BP

FPF    Replacing adult-like juvenile body and flight feathers with adult-like formative feathers. Skull is unossified, but should usually ossify before completion of the FPF, which then makes this difficult to distinguish from UPB (use UPU).

p9–p10 retained juvenile, p8 missing, p7 molting, p1–p6 new, s1–s3 new, s4–s7 retained juvenile, s8 molting, s9 new.

FAJ    The bill is black, but perhaps paler at the base and tip than FCJ (Marantz et al. 2003). The iris is grayish-brown, but it is unclear if iris color is useful for aging at the BDFFP. Skull is ossified.

UPB    Replacing adult-like body and flight feathers with adult-like feathers. Skull is ossified. May not be possible to distinguish from FPF during replacement of the last flight feathers (use UPU).

## *Dendrocincla merula obidensis*

White-chinned Woodcreeper • Arapaçu-da-taoca

Band Size: F, G

\# Individuals Captured: 562

Wing    94.0–117.0 mm (105.3 ± 3.9 mm; n = 294)

Tail    64.0–91.0 mm (76.5 ± 4.9 mm; n = 275)

Mass    40.0–64.0 g (53.0 ± 4.3 g; n = 462)

Bill    16.6–20.4 mm (18.6 ± 1.2 mm; n = 24)

Tarsus    29.0–33.5 mm (31.0 ± 1.3 mm; n = 24)

Similar species: No other woodcreeper at the BDFFP is this dark chocolate brown without barring or streaking. See also *D. fuliginosa*. Note that the iris color is brown at the BDFFP, whereas it is blue for other subspecies, as is often illustrated in field guides.

Skull: Very difficult to see due to dark skin, but appears to complete (n = 120), likely during the FPF.

Brood patch: Only ♀♀ incubate and develop BPs (Willis and Oniki 1978). May occur year-round, but mainly from Sep–Jan.

Sex: ♂ = ♀.

Molt: Group 4/Complex Basic Strategy; FPF complete, DPBs complete. Peaks from Sep–Feb and least frequent from Jun–Aug.

FCJ May have dirty white or dingy buff throat, slightly darker underparts than FAJ, and an entirely dark bill (Marantz et al. 2003), but these features need to be confirmed at the BDFFP. Skull is unossified.

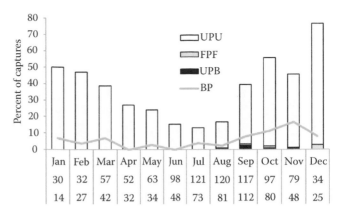

| | Jan | Feb | Mar | Apr | May | Jun | Jul | Aug | Sep | Oct | Nov | Dec |
|---|---|---|---|---|---|---|---|---|---|---|---|---|
| | 30 | 32 | 57 | 52 | 63 | 98 | 121 | 120 | 117 | 97 | 79 | 34 |
| | 14 | 27 | 42 | 32 | 34 | 48 | 73 | 81 | 112 | 80 | 48 | 25 |

FPF Replacing adult-like juvenile body and flight feathers with adult-like formative feathers. Skull is unossified, but should usually ossify before completion of the FPF, which then makes this difficult to distinguish from UPB (use UPU).

FAJ The white throat highly contrasts with chest and face. The dark bill may have a pale base (Marantz et al. 2003), but differences between FCJ and FAJ are subtle and more study is needed at the BDFFP. Skull is ossified.

UPB Replacing adult-like body and flight feathers with adult-like feathers. May not be possible to distinguish from FPF during replacement of the last flight feathers (use UPU). Skull is ossified.

## Glyphorynchus spirurus

Wedge-billed Woodcreeper • Arapaçu-de-bico-de-cunha

Band Size: D

# Individuals Captured: 2236

Wing 59.0–76.0 mm (67.8 ± 2.8 mm; n = 1386)

Tail 52.0–77.0 mm (65.0 ± 4.2 mm; n = 1291)

Mass 10.0–18.0 g (13.7 ± 1.1 g; n = 1876)

Bill 8.1–11.2 mm (9.5 ± 0.6 mm; n = 285)

Tarsus 16.5–20.1 mm (18.5 ± 0.7 mm; n = 278)

Similar species: The smallest woodcreeper; the wedge-shaped bill is unique. Most similar to *Xenops minutus*, which differs in having a bicolored tail, semiconcealed silvery-white malar stripe, and lacks stiffened rachises that extend beyond barbules.

Skull: Completes (n = 756), usually during FPF or sometimes FCF.

Brood patch: Both sexes apparently incubate mainly from Oct–Feb, but can breed at lower frequencies the rest of the year.

Sex: ♂ = ♀.

Molt: Group 4/Complex Basic Strategy; FPF (incomplete)-complete, DPB (incomplete)-complete. A fairly well-defined molt schedule peaking from Mar–Jul and least frequent from Oct–Dec. From specimens collected in Nicaragua, Howell (1957) noted that r1 is the last rect to be replaced. Very rarely, one or more pp covs and/or the outer 3–5 (usually 4) pp can be retained, but this should not be used to determine age because it can occur in FPF as well as DPB molts, which is based on known-age banded individuals.

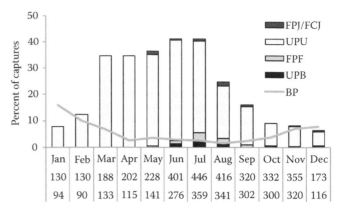

| | Jan | Feb | Mar | Apr | May | Jun | Jul | Aug | Sep | Oct | Nov | Dec |
|---|---|---|---|---|---|---|---|---|---|---|---|---|
| | 130 | 130 | 188 | 202 | 228 | 401 | 446 | 416 | 320 | 332 | 355 | 173 |
| | 94 | 90 | 133 | 115 | 141 | 276 | 359 | 341 | 302 | 300 | 320 | 116 |

FCJ    Similar to FAJ, but breast streaking is less pronounced, throat scaling is darker and heavier, and crown feathers have dusky edges (Marantz et al. 2003), but these features are very difficult to discern and may vary with wear. The underwing covs may average whiter than in FAJ. Ss covs and remiges nearly identical to subsequent plumages. Undertail covs are duskier and weaker textured than in subsequent plumages. Skull is unossified.

  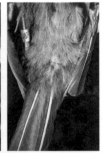

FPF    Replacing adult-like juvenile body and flight feathers with adult-like formative feathers. Especially early in molt, look at the throat for a mix of replaced rounder, shorter, brighter, and less broadly edged feathers contrasting against darker and more pointed juvenile feathers. Skull is unossified, but should usually ossify before completion of the FPF, which then makes this difficult to distinguish from UPB (use UPU).

Mix of dull buff juvenile and ochraceous formative throat feathers.

FAJ    Similar to FCJ, but breast streaking averages more pronounced, the throat is brighter with narrower scaling, and crown feathers lack dusky edges (Marantz et al. 2003); these features are very difficult to discern and may vary with wear. The undertail covs are bolder, more golden in color, and better structured than in FCJ. Skull is ossified.

UPB    Replacing adult-like body and flight feathers with adult-like feathers. May not be possible to distinguish from FPF during replacement of the last flight feathers (use UPU). Skull is ossified.

p8–p10 retained, p7 missing, p6 molting, p1–p5 new, s1–s2 new, s3 molting, s4–s6 retained, s7 missing, s8 retained, s9 new.

## Dendrocolaptes certhia certhia

Amazonian Barred-Woodcreeper • Arapaçu-barrado

Wing    119.0–135.0 mm (125.4 ± 3.9 mm; n = 85)

Tail     106.0–127.0 mm (118 ± 4.9 mm; n = 67)

Mass    54.5–81.0 g (66.6 ± 5.1 g; n = 119)

Bill     26.1–31.5 mm (29.4± 2.0 mm; n = 8)

Tarsus  29.1–32.0 mm (30.5 ± 1.0 mm; n = 8)

Similar species: Averages smaller than *Dendrocolaptes picumnus*, but measurements overlap. The plumage is similar to *D. picumnus*, but without bold streaking in the throat. The bill in *D. picumus* is black (not red), but beware of juvenile *D. certhia* with black bills. From *Xiphohrynchus pardalotus* by lacking buff streaking on crown and chest.

Skull: Completes (n = 28) probably during FPF or soon thereafter.

Brood patch: Both sexes probably incubate and develop BPs (Marantz et al. 2003). Observed from Jun–Jan, but probably more sporadic after Sep; probably at best rarely nests during the wet season like other large woodcreepers.

Sex: ♂ = ♀.

Molt: Group 4/Complex Basic Strategy; FPF complete, DPBs complete. Molt is probably most frequent from Aug–Apr.

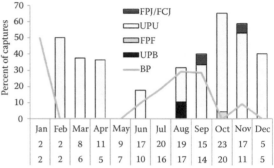

| | Jan | Feb | Mar | Apr | May | Jun | Jul | Aug | Sep | Oct | Nov | Dec |
|---|---|---|---|---|---|---|---|---|---|---|---|---|
| | 2 | 2 | 8 | 11 | 9 | 17 | 20 | 19 | 15 | 23 | 17 | 5 |
| | 2 | 2 | 6 | 5 | 7 | 10 | 16 | 17 | 14 | 20 | 11 | 5 |

FCJ     Similar to FAJ, but with black (not red) bill. Skull is unossified.

FPF     Replacing adult-like juvenile body and flight feathers with adult-like formative feathers. Skull is unossified, but should usually ossify before completion of the FPF, and bill is black, but transitions to red probably before the completion of the FPF, which then makes this difficult to distinguish from UPB (use UPU).

Note blackish-red bill (left); p8–p10 retained juvenile, p6–p7 molting, p1–p5 new, s1 molting, s2–s6 retained juvenile, s7–s9 new (right).

FAJ    The bill is variably reddish-brown, but never mostly black. Skull is ossified. With experience and caution, some birds with blackish bills and ossified skulls may be reliably aged as FCF.

UPB    Replacing adult-like body and flight feathers with adult-like feathers. May not be possible to distinguish from FPF during replacement of the last flight feathers (use UPU). Skull is ossified.

## *Dendrocolaptes picumnus picumnus*

Band Size: H, G

Black-banded Woodcreeper • Arapaçu-meio-barrado

# Individuals Captured: 21

Wing    130.0–139.5 mm (136.2 ± 3.0 mm; n = 9)

Tail    110.0–126.0 mm (120.0 ± 5.4 mm; n = 9)

Mass    67.0–90.0 g (78.5 ± 5.7 g; n = 20)

Similar species: See *Dendrocolaptes certhia*.

Skull: Probably completes during FPF or soon thereafter.

Brood patch: Both sexes develop BPs, which have been observed in Aug and Sep, but little data available during other months.

Sex: ♂ = ♀.

Molt: Group 4/Complex Basic Strategy; FPF probably complete, DPBs complete. Molt has been observed from Sep–Jan, but there are few data available.

FCJ    Similar to FAJ, but throat is described to be more scaly, streaking on back and underparts is bolder, pattern of spotting underlying breast streaks is bolder, underparts are more weakly barred, crown is darker overall, and more spotted tan streaked often with tips of feathers darker (Marantz et al. 2003), but these features have not been confirmed at the BDFFP. Skull is unossified.

FPF    Replacing adult-like juvenile body and flight feathers with adult-like formative feathers, but may be difficult to distinguish from UPB (use UPU). Skull is unossified, but should usually ossify before completion of the FPF.

FAJ     See FCJ. Skull is ossified.

UPB     Replacing adult-like body and flight feathers with adult-like feathers. May not be possible to distinguish
        from FPF during replacement of the last flight feathers. Skull is ossified.

## *Hylexetastes perrotii perrotii*

Red-billed Woodcreeper • Arapaçu-de-bico-vermelho

| | |
|---|---|
| Wing | 120.0–140.0 mm (128.3 ± 4.7 mm; n = 60) |
| Tail | 105.0–126.0 mm (113.2 ± 5.9 mm; n = 40) |
| Mass | 100.0–130.0 g (115.4 ± 7.2 g; n = 89) |
| Bill | 29.5–30.6 mm (30.1 ± 0.5 mm; n = 4) |
| Tarsus | 36.3–37.6 mm (36.9 ± 0.5 mm; n = 5) |

Similar species: Large size, unstreaked body, red bill, brown malar dividing white subauricular streak and white throat, and bluish orbital feathers are unique.

Skull: Completes (n = 8) probably during FPF or soon thereafter.

Brood patch: Unknown if both sexes incubate. BPs have only been seen from Aug–Dec, but sample size is limited.

Sex: ♂ = ♀.

Molt: Group 4/Complex Basic Strategy; FPF probably complete, DPBs complete. Molt appears to be most frequent from Feb–May and least frequent from Jun–Sep.

Band Size: H, L

# Individuals Captured: 111

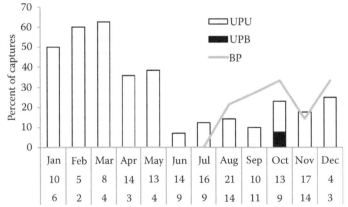

FCJ    Similar to FAJ, but crown is weakly streaked, bluish tones around eye are reduced or absent, possibly has more rufescent tones below, iris is brownish, and bill averages shorter and is more dusky with a paler lower mandible (Marantz et al. 2003). Skull is unossified.

(Courtesy of Aída Rodrigues.)

FPF    Replacing adult-like juvenile body and flight feathers with adult-like formative feathers. Evaluate iris and bill color; these should be FCJ-like, but we expect their transition to adult-like by the completion of the FPF. Skull is unossified, but should usually ossify before completion of the FPF, which may make them difficult to distinguish from UPB (use UPU).

FAJ    Similar to FCJ, but crown is not streaked, averages less heavily barred and possibly less rufescent below, iris is dark red, and bill is longer and reddish-black (Marantz et al. 2003). Skull is ossified.

UPB    Replacing adult-like body and flight feathers with adult-like feathers. Skull is ossified. May not be possible to distinguish from FPF during replacement of the last flight feathers (use UPU).

## *Xiphorhynchus pardalotus pardalotus*

Chestnut-rumped Woodcreeper • Arapaçu-assobiador

Band Size: F

# Individuals Captured: 942

| | |
|---|---|
| Wing | 89.0–116.0 mm (101.3 ± 4.7 mm; n = 561) |
| Tail | 71.0–104.0 mm (86.7 ± 5.6 mm; n = 512) |
| Mass | 30.0–46.5 g (37.7 ± 2.6 g; n = 821) |
| Bill | 23.2–29.1 mm (26.5 ± 1.4 mm; n = 94) |
| Tarsus | 22.8–26.5 mm (24.7 ± 0.8 mm; n = 101) |

Similar species: No other woodcreeper is so boldly marked with pale teardrop-shaped streaks, especially on the back. It is also the only woodcreeper with its bill regularly longer than its tarsus, although in 16% (n = 38) of individuals aged as FAJ or UPB, the tarsus was just slightly longer than the bill; how many of these were due to measurement error, aging error, or real differences is unknown. By measuring bill and tarsus lengths, this can help confirm the ID and/or age of this species; beware of FCJ or FCF with incompletely grown bills.

Skull: Completes (n = 251) during FPF or soon thereafter.

Brood patch: Both sexes may incubate and develop BPs, which are most frequent from Jul–Nov, but may rarely occur the rest of the year.

Sex: ♂ = ♀.

Molt: Group 4/Complex Basic Strategy; FPF complete, DPBs complete. Molt peaks from Sep–Jan and is least frequent from May–Aug.

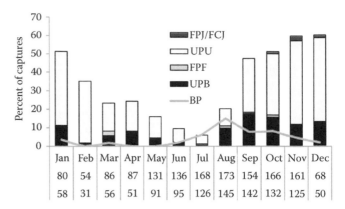

| | Jan | Feb | Mar | Apr | May | Jun | Jul | Aug | Sep | Oct | Nov | Dec |
|---|---|---|---|---|---|---|---|---|---|---|---|---|
| | 80 | 54 | 86 | 87 | 131 | 136 | 168 | 173 | 154 | 166 | 161 | 68 |
| | 58 | 31 | 56 | 51 | 91 | 95 | 126 | 145 | 142 | 132 | 125 | 50 |

FCJ    Bill is always shorter than tarsi, and the back appears to be less marked with spots than FAJ. The juvenile plumage may also have streaks with weaker dark borders (Marantz et al. 2003), but this may be subtle and has not been confirmed at the BDFFP. The less covs may average less chestnut-red than FAJ.

FPF    Replacing adult-like juvenile body and flight feathers with adult-like formative feathers. Skull is unossified, but should usually ossify before completion of the FPF, and bill is black, but transitions to red probably occur before the completion of the FPF, which then makes this difficult to distinguish from UPB (use UPU).

FAJ    Bills are usually longer than tarsi (84%, n = 38) once the skull ossifies and perhaps by the completion of FPF. This plumage may also have streaks with stronger dark borders than in juvenile plumage (Marantz et al. 2003), but this may be subtle and has not been confirmed at the BDFFP. The less covs may average more chestnut-red than in FCJ. There is variation in the presence and distinctness of teardrop-shaped markings on gr covs, but the significance of this is unknown.

| UPB | Replacing adult-like body and flight feathers with adult-like feathers. Bills are usually longer than tarsi (84%, n = 38). Skull is ossified. May not be possible to distinguish from FPF during replacement of the last flight feathers (use UPU). |
|---|---|

p6–p10 retained, p5 missing, p1–p4 new, s1 molting, s2–s6 retained, s7–s9 new.

## *Campylorhamphus procurvoides procurvoides*

Curve-billed Scythebill • Arapaçu-de-bico-curvo

Band Size: F

\# Individuals Captured: 88

| Wing | 86.0–102.0 mm (94.9 ± 3.7 mm; n = 59) |
|---|---|
| Tail | 75.0–99.0 mm (84.5 ± 4.8 mm; n = 48) |
| Mass | 29.0–41.5 g (34.4 ± 2.5 g; n = 75) |
| Bill | 51.7–57.9 mm (54.9 ± 2.7 mm; n = 6) |
| Tarsus | 21.6–23.6 mm (22.8 ± 0.8 mm; n = 8) |

Similar species: No other woodcreeper at the BDFFP has such an extremely curved bill. Most similar to *C. trochilirostris*, which has not been found at the BDFFP but is found nearby in várzea/igapó forest and may be a potential vagrant to the BDFFP. Voice is the best distinguishing field mark, but in hand there may be differences in bill length (and degree of curvature, which may be difficult to quantify) and tarsus, although direct comparisons from the region are not available.

Skull: Completes (n = 21) probably during FPF or soon thereafter.

Brood patch: Not known if both sexes incubate. BPs observed Oct, Nov, Jan, Feb, and May, suggesting a late dry/early wet season breeding season, but perhaps occasionally nests at other times of the year; more study is needed.

Sex: ♂ = ♀.

Molt: Group 4/Complex Basic Strategy; FPF complete(?), DPBs complete. May molt at any time of year, perhaps most frequent from Dec–Aug, but the pattern is difficult to discern because of low monthly sample sizes.

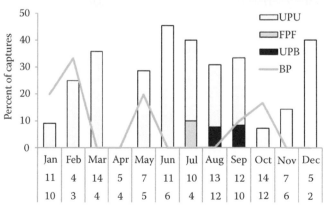

| | Jan | Feb | Mar | Apr | May | Jun | Jul | Aug | Sep | Oct | Nov | Dec |
|---|---|---|---|---|---|---|---|---|---|---|---|---|
| | 11 | 4 | 14 | 5 | 7 | 11 | 10 | 13 | 12 | 14 | 7 | 5 |
| | 10 | 3 | 4 | 4 | 5 | 6 | 4 | 12 | 10 | 12 | 6 | 2 |

Notes: *C. procurvoides* appears to be polyphyletic based on recent genetic evidence and relationships of some subspecies with *C. trochilirostris* are not well resolved. *C. p. procurvoides* may be best grouped with *C. p. sanus* and *C. p. gyldenstolpei* (Zimmer 1934, Marantz et al. 2003, Aleixo et al. 2013, Remsen et al. 2016).

**FCJ**    Previously undescribed, but apparently very similar to FAJ with less distinctly streaked and more grayish throat. Skull is unossified.

**FPF**    Replacing adult-like juvenile body and flight feathers with adult-like formative feathers. Skull is unossified, but should usually ossify before completion of the FPF, and bill is black, but transitions to red probably before the completion of the FPF, which then makes this difficult to distinguish from UPB (use UPU).

**FAJ**    From FCJ by more distinctly streaked throat. Bill may average redder. Skull is ossified.

**UPB**    Replacing adult-like body and flight feathers with adult-like feathers. Skull is ossified. May not be possible to distinguish from FPF during replacement of the last flight feathers (use UPU).

*Lepidocolaptes albolineatus* (monotypic)

Band Size: D

Guianan Woodcreeper • Arapaçu-de-listras-brancas

# Individuals Captured: 3

| | |
|---|---|
| Wing | 82.0–85.5 mm (83.8 ± 2.5 mm; n = 2) |
| Tail | 71.5 mm (n = 1) |
| Mass | 16.9–20.0 g (18.6 ± 1.6 g; n = 3) |
| Bill | 18.2 mm (n = 1) |
| Tarsus | 18.4 mm (n = 1) |

Similar species: Long, thin, slightly curved bill, bold white streaks on underparts, and small size distinctive.

Skull: Probably completes during FPF or soon thereafter as in other woodcreepers, but no data available.

Brood patch: Unknown if both sexes develop BPs. Timing unknown.

Sex: ♂ = ♀.

Molt: Group 4/Complex Basic Strategy; FPF complete (?), DPBs complete. Timing unknown.

Notes: Recent genetic and vocal analyses were sufficient to support that the former *L. albolineatus*, which included several subspecies across the Amazon (*sensu* Zimmer 1934), is actually a species complex representing, instead, five good species (Marantz et al. 2003, Rodrigues et al. 2013, Remsen et al. 2016).

FCJ      Like FAJ, but upperparts darker, crown grayer with bolder spotting that extends weakly as streaks to upper back, streaking on underparts whiter, somewhat reduced, with weaker borders (Marantz et al. 2003), but this needs to be confirmed at the BDFFP. Skull is unossified.

FPF      Replacing adult-like juvenile body and flight feathers with adult-like formative feathers. Skull is unossified, but should usually ossify before completion of the FPF, which then makes this difficult to distinguish from UPB (use UPU).

FAJ      Compared to FCJ, upperparts average lighter, crown is less gray with weaker spotting, and increased bolder streaking on underparts (Marantz et al. 2003), but this needs to be confirmed at the BDFFP. Birds with retained central rects may be aged as FCF, but beware of excessive wear in these feathers and suspended molts. Also beware of pseudolimits among the gr covs. Skull is ossified.

This bird was completing a molt, but skull information was not collected (UPU); otherwise it is representative of FAJ.

UPB      Replacing adult-like body and flight feathers with adult-like feathers. May not be possible to distinguish from FPF during replacement of the last flight feathers (use UPU). Skull is ossified.

*Xenops minutus ruficaudus*

Plain Xenops • Bico-virado-miúdo

| | |
|---|---|
| Wing | 59.0–71.0 mm (64.8 ± 2.8 mm; n = 216) |
| Tail | 42.0–60.0 mm (50.4 ± 3.5 mm; n = 188) |
| Mass | 10.0–15.5 g (12.3 ± 1.0 g; n = 292) |
| Bill | 9.6–11.1 mm (10.3 ± 0.5 mm; n = 17) |
| Tarsus | 16.2–18.5 mm (17.1 ± 0.6 mm; n = 16) |

Similar species: Most similar to *Glyphorynchus spirurus* including the pale stripe through the spread wing, but *X. minutus* has a black and orange tail and a semiconcealed silvery-white malar stripe. *X. milleri* lacks black in the tail, a silvery-white malar stripe, and has heavier streaking below.

Skull: Completes (n = 97) probably during FPF or just after.

Brood patch: Both sexes share in incubation and develop BPs. Observed mainly from Jul–Nov, but may occasionally breed until Feb.

Sex: ♂ (>) ♀; in *X. m. mexicanus*, the length of p8 (♂: 45.0–50.0 mm, ♀: 42.0–45.0 mm) was reliable in distinguishing sex in 90.3% individuals, although wing chord (♂: 61.4–67.0 mm, ♀: 58.7–62.0 mm) and tail length (♂: 47.2–55.3 mm, ♀: 44.5–48.2 mm) may also be useful, especially when used in combination (Winker et al. 1994).

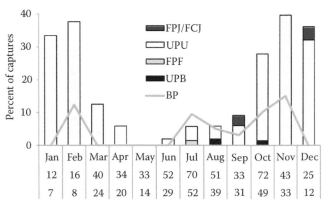

| | Jan | Feb | Mar | Apr | May | Jun | Jul | Aug | Sep | Oct | Nov | Dec |
|---|---|---|---|---|---|---|---|---|---|---|---|---|
| | 12 | 16 | 40 | 34 | 33 | 52 | 70 | 51 | 33 | 72 | 43 | 25 |
| | 7 | 8 | 24 | 20 | 14 | 29 | 52 | 39 | 31 | 49 | 33 | 12 |

Legend: FPJ/FCJ, UPU, FPF, UPB, BP

Molt: Group 4/Complex Basic Strategy; FPF complete, DPBs complete. Peaks from Oct–Feb and is least frequent from Apr–Jul. Drops rects frequently during handling and probably also in the wild, thus tail molt should be carefully assigned.

FCJ     Has a slightly less distinct pattern throughout and the throat may be slightly clouded grayish than FAJ (Ridgway 1911, Remsen 2003). The less and med covs may average more rufous and less olive than FAJ. The gape is not reliable to age birds as a bright yellow gape is retained. Skull is unossified.

FPF     Replacing adult-like juvenile body and flight feathers with adult-like formative feathers. Skull is unossified, but should usually ossify before completion of the FPF, which then makes this difficult to distinguish from UPB (use UPU).

FAJ    It has a slightly more distinct pattern throughout and the throat is not clouded grayish (Remsen 2003). Skull is ossified. Note that a yellowish gape is commonly retained, probably for life.

UPB    Replacing adult-like body and flight feathers with adult-like feathers. Skull is ossified. May not be possible to distinguish from FPF during replacement of the last flight feathers (use UPU).

## *Microxenops milleri* (monotypic)

Band Size: Unknown

Rufous-tailed Xenops • Bico-virado-da-copa

\# Individuals Captured: 4

Wing    62.0–66.0 mm (64.3 ± 2.1 mm; n = 3)

Tail    36.0–37.0 mm (36.5 ± 0.7 mm; n = 2)

Mass    11.0–12.5 g (11.7 ± 0.8 g; n = 3)

Similar species: Heavy streaking on throat and chest, bicolored rufous and black wings, all rufous tail, and small size distinct.

Skull: Completes (n = 5, LSUMZ) probably during FPF or just after.

Brood patch: Both sexes may share in incubation and develop BPs. Timing is unknown.

Sex: ♂ = ♀.

Molt: Group 4(?)/Probably Complex Basic Strategy as in other Furnariids, and FPF and DPBs are expected to be complete, but no data available. Timing is unknown.

FCJ    Juvenile plumage is undescribed. Based on two LSUMZ specimens, birds with ≤10% skull ossification had relatively broad pale streaks on crown, nape, and back with a relatively pale belly compared to specimens with fully ossified skulls. Skull is unossified.

FPF    Replacing adult-like juvenile body and flight feathers with adult-like formative feathers. Skull is unossified, but might be expected to ossify before completion of the FPF like most other Furnariids at the BDFFP, which then makes this difficult to distinguish from UPB (use UPU).

FAJ    Based on a small sample of LSUMZ specimens, it has weaker streaking on the crown, nape, and back with a darker belly than FCJ. Skull is ossified.

UPB    Replacing adult-like body and flight feathers with adult-like feathers. Skull is ossified. May not be possible to distinguish from FPF during replacement of the last flight feathers (use UPU).

## *Philydor erythrocercum erythrocercum*

Band Size: F, G

Rufous-rumped Foliage-gleaner • Limpa-folha-de-sobre-ruivo

\# Individuals Captured: 188

Wing    76.0–92.0 mm (83.9 ± 4.6 mm; n = 100)

Tail    58.0–74.0 mm (64.9 ± 3.8 mm; n = 97)

Mass    19.6–31.0 g (23.9 ± 2.2 g; n = 158)

Bill    11.2–13.0 mm (12.3 ± 0.5 mm; n = 7)

Tarsus    21.7–23.8 mm (23.0 ± 0.8 mm; n = 7)

Similar species: Most similar to *Automolus ochrolaemus* and *A. infuscatus*, but has stronger contrast in supercilium and between upperparts and underparts.

Skull: Completes (n = 27) probably during FPF or soon thereafter.

Brood patch: Unknown if both sexes develop BPs. Observed from Jun–Nov, peaking in Jul–Sep.

Sex: ♂ = ♀.

Molt: Group 4/Complex Basic Strategy; FPF complete, DPBs complete. Peaks from Aug–Feb and is least frequent from Mar–Jun.

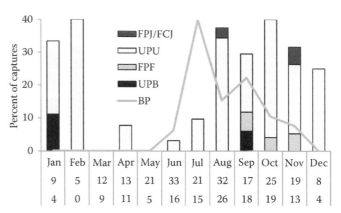

| | Jan | Feb | Mar | Apr | May | Jun | Jul | Aug | Sep | Oct | Nov | Dec |
|---|---|---|---|---|---|---|---|---|---|---|---|---|
| | 9 | 5 | 12 | 13 | 21 | 33 | 21 | 32 | 17 | 25 | 19 | 8 |
| | 4 | 0 | 9 | 11 | 5 | 16 | 15 | 26 | 18 | 19 | 13 | 4 |

FCJ    Like FAJ, but with slightly more ochraceous-orange in the supercilium. Rects, on average, are more pointed than the FAJ, but differences may be subtle. This plumage is probably retained only briefly. The ss covs are weakly structured and duller brown than in subsequent plumage. The bill may also average shorter. Skull is unossified.

FPF    Replacing adult-like juvenile body and flight feathers with adult-like formative feathers. Early in the FPF look for mixed ochraceous and yellowish feathers in the supercilium. Skull is unossified, but probably usually ossifies before the completion of the FPF, which then makes this difficult to distinguish from UPB (use UPU).

FAJ    Has pale yellowish supercilium concolor with the throat and underparts. The ss covs are richer olive-brown than in FCJ. Rects average more truncate than in FCJ, but differences may be subtle. Skull is ossified.

UPB    Replacing adult-like body and flight
       feathers with adult-like feathers. Skull is
       ossified. May not be possible to
       distinguish from FPF during
       replacement of the last flight feathers
       (use UPU).

p6–p10, p5 missing, p1–p4 molting, ss retained.

## *Philydor pyrrhodes* (monotypic)

Cinnamon-rumped Foliage-gleaner • Limpa-folha-vermelho

Band Size: F

\# Individuals Captured: 56

| | |
|---|---|
| Wing | 77.0–93.0 mm (84.3 ± 5.2 mm; n = 27) |
| Tail | 55.0–72.0 mm (63.3 ± 4.6 mm; n = 27) |
| Mass | 24.0–38.0 g (30.7 ± 3.8 g; n = 47) |
| Bill | 13.1–13.6 mm (13.3 ± 0.3 mm; n = 3) |
| Tarsus | 22.9–24.4 mm (23.8 ± 0.8 mm; n = 3) |

Similar species: None.

Skull: Completes (n = 10) probably during FPF or soon
thereafter, but few data available.

Brood patch: Unknown if both sexes incubate.
Probably breeds during the dry season, but more
study is needed.

Sex: ♂ = ♀.

Molt: Group 3(?) and/or Group 4(?)/Complex Basic Strategy; FPF incomplete(?)-complete(?), DPBs complete.
One bird was seen with retained s4–s6, presumably after an incomplete FPF molt, although we cannot rule
out this was not a molt suspension; however, like other Furnariidae, a complete FPF molt might be expected,
or we might expect to occasionally see similar retention patterns in other species. Little data available to
determine timing of molt, but expected to follow breeding, and possibly peaking during the late dry and
early wet seasons.

FCJ    Subtly different from the FAJ and some individuals may not be safely aged by plumage alone. May have a slightly paler crown, back, and throat on average. The dusky postocular eye stripe may be less well defined than in subsequent plumages and blend into the auriculars. Pp and ss covs minimally contrast and are brownish-olive. Skull is unossified.

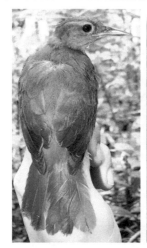

FPF    Replacing adult-like juvenile body and flight feathers with adult-like formative feathers. Skull is unossified, but might be expected to ossify before completion of the FPF like most other Furnariids at the BDFFP, which then makes this difficult to distinguish from UPB (use UPU).

FCF    Based on one bird, molt limits may occur among the inner ss, although more study is needed to rule out this being in suspension of a complete molt. Skull would likely be ossified.

All pp replaced, s1–s3 and s7–s9 replaced, s4–s6 retained and presumably juvenile given the contrast in wear.

FAJ    Without molt limits among the ss. Pp covs are blackish and slightly contrast against the outer ss covs, which are edged rufous-olive. The dark postocular is well defined, especially against the auriculars. Skull is ossified.

SPB        May be possible to occasionally see two generations of ss (juvenile and adult-like formative) being replaced by adult-like feathers. Skull is ossified.

UPB        Replacing adult-like body and flight feathers with adult-like feathers. Skull is ossified. May not be possible to distinguish from FPF during replacement of the last flight feathers (use UPU).

## *Automolus ochrolaemus turdinus*

Band Size: G

Buff-throated Foliage-gleaner • Barranqueiro-camurça

\# Individuals Captured: 148

Wing     79.0–92.0 mm (86.3 ± 3.0 mm; n = 98)

Tail       67.0–81.0 mm (74.7 ± 3.1 mm; n = 88)

Mass     28.9–39.0 g (34.1 ± 2.4 g; n = 130)

Bill        13.7–16.5 mm (14.8 ± 0.9 mm; n = 13)

Tarsus   23.8–27.9 mm (26.0 ± 1.2 mm; n = 15)

Similar species: *A. infuscatus* is very similar in plumage and no morphometric features are known to distinguish it from *A. ochrolaemus*. Popular field guides should be used with caution as they poorly cover the *Automolus* subspecies found at the BDFFP. *A. ochrolaemus* has more orange tones, especially in the face and auriculars while *A. infuscatus* has more chestnut-brown tones in the face and auriculars. The throat in *A. infuscatus* tends to be paler yellowish-white rather than pale buffy-orange. *A. infuscatus* also tends to have a less distinct supercilium than *A. ochrolaemus*. *A. ochrolaemus* is typically found in disturbed areas and second growth, whereas *A. infuscatus* is typically found in forest interior and usually with mixed-flocks, but there is apparently some overlap in habitat tolerance such that a strict habitat basis for identification should not be assumed.

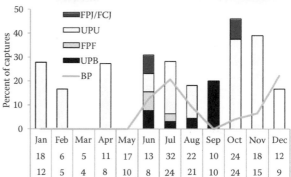

Skull: Completes (n = 61) during FPF or soon thereafter.

Brood patch: Both sexes incubate and develop BPs. Have been observed from Jun–Dec (except Sep) and probably should be expected mainly during the dry season.

Sex: ♂ = ♀.

Molt: Group 4/Complex Basic Strategy; FPF complete, DPBs complete. Molt probably peaks from Oct–Jan, but data are limited.

FCJ Slightly duller than FAJ with a less distinct eyering, crown tinged chestnut, face tinged rufous, throat and breast slightly mottled or streaked (Winker et al. 1994, Remsen 2003). The bill may average shorter than the definitive plumage. Rects are more pointed than in FAJ. Skull is unossified.

FPF Replacing adult-like juvenile body and flight feathers with adult-like formative feathers. Skull is unossified but might be expected to ossify before completion of the FPF like most other Furnariids at the BDFFP, which then makes this difficult to distinguish from UPB (use UPU).

FAJ Possibly averages brighter than FCJ with a bolder eyering and with an unmottled throat and breast (Remsen 2003), but these features may be very difficult to judge without a direct comparison. The bill may average longer than in juveniles. Rects are more truncate than in FCJ. Skull is ossified.

UPB Replacing adult-like body and flight feathers with adult-like feathers. Skull is ossified. May not be possible to distinguish from FPF during replacement of the last flight feathers.

p5–p10 retained, p4 missing, p1–p3 new, s1 molting, s2–s7 retained, s8–s9 new.

*Automolus infuscatus cervicalis*

Olive-backed Foliage-gleaner • Barranqueiro-pardo

| | |
|---|---|
| Wing | 78.0–95.0 mm (85.6 ± 2.9 mm; n = 312) |
| Tail | 62.0–80.0 mm (70.6 ± 3.5 mm; n = 282) |
| Mass | 25.5–37.5 g (31.6 ± 2.1 g; n = 470) |
| Bill | 12.3–14.9 mm (13.8 ± 0.7 mm; n = 50) |
| Tarsus | 23.1–25.2 mm (25.2 ± 1.0 mm; n = 50) |

Similar species: See *A. ochrolaemus*.

Skull: Completes (n = 157) probably during FPF or soon thereafter.

Brood patch: Both sexes incubate and develop BPs (Remsen 2003). May breed year-round, but probably mainly from Jul–Oct, as most juveniles are seen from Sep–Nov.

Sex: ♂ = ♀.

Molt: Group 4/Complex Basic Strategy; FPF complete, DPBs complete. Molt peaks from Aug–Nov and is least frequent in Jun and Jul.

FCJ     Slightly darker and duller with shorter crown feathers than FAJ (Remsen 2003). Rects are relatively pointed. Skull is unossified.

Band Size: G

# Individuals Captured: 560

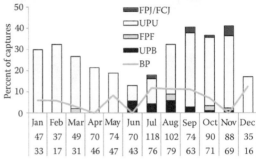

| | Jan | Feb | Mar | Apr | May | Jun | Jul | Aug | Sep | Oct | Nov | Dec |
|---|---|---|---|---|---|---|---|---|---|---|---|---|
| | 47 | 37 | 49 | 70 | 74 | 70 | 118 | 102 | 74 | 90 | 88 | 35 |
| | 33 | 17 | 31 | 46 | 47 | 43 | 76 | 79 | 63 | 71 | 69 | 16 |

FPF     Replacing adult-like juvenile body and flight feathers with adult-like formative feathers. Skull is unossified, but might be expected to ossify before completion of the FPF like most other Furnariids at the BDFFP, which then makes this difficult to distinguish from UPB (use UPU).

p6–p10 retained juvenile (note relative lack of wear), p5 molting, p1–p4 new, all ss retained juvenile (left); outer four pairs of rects molting, inner two pairs of rects retained juvenile, suggesting a centripetal replacement pattern (right).

FAJ    Slightly brighter with longer crown feathers than FCJ (Remsen 2003). Rects are more truncate than FCJ. Skull is ossified.

UPB    Replacing adult-like body and flight feathers with adult-like feathers. Skull is ossified. May not be possible to distinguish from FPF during replacement of the last flight feathers (use UPU).

p5–p10 retained (note relative amount of wear compared to FPF), p4 molting, p1–p3 new, all ss retained (left); r1 and r6 molting, r2–r5 retained (right).

*Clibanornis rubiginosus obscurus*                                   Band Size: G, F

Ruddy Foliage-gleaner • Barranqueiro-ferrugem              # Individuals Captured: 191

| | |
|---|---|
| Wing | 74.0–91.0 mm (80.9 ± 3.4 mm; n = 128) |
| Tail | 64.5–79.0 mm (70.2 ± 3.2 mm; n = 111) |
| Mass | 29.0–44.0 g (36.4 ± 2.6 g; n = 170) |
| Bill | 12.5–16.8 mm (14.3 ± 1.1 mm; n = 12) |
| Tarsus | 26.3–29.3 mm (27.5 ± 1.0 mm; n = 11) |

Similar species: Relatively distinct among foliage-gleaners with a deep chestnut red throat. Most like *Sclerurus* in plumage, but with a chestnut-red (not black) tail, and a heavier, stouter bill.

Skull: Completes (n = 42) probably during FPF or soon thereafter.

Brood patch: Both sexes may incubate and develop BPs as in other *Automolus*. Breeding may mainly occur from Aug–Dec either continuing into Apr or with a secondary season in Mar–Apr.

Sex: ♂ = ♀.

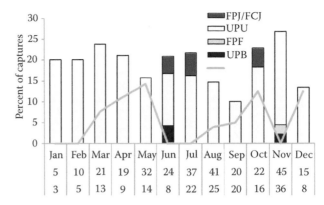

| | Jan | Feb | Mar | Apr | May | Jun | Jul | Aug | Sep | Oct | Nov | Dec |
|---|---|---|---|---|---|---|---|---|---|---|---|---|
| | 5 | 10 | 21 | 19 | 32 | 24 | 37 | 41 | 20 | 22 | 45 | 15 |
| | 3 | 5 | 13 | 9 | 14 | 8 | 22 | 25 | 20 | 16 | 36 | 8 |

Molt: Group 4/Complex Basic Strategy; FPF complete, DPBs complete. One recaptured bird retained juvenile feathers in throat for about 4 months and appeared to be commencing FPF at this age. Molt can occur at any time of year, but may peak from Nov–Feb and be least frequent in Sep.

FCJ    Duller brown above, paler dusky-brown below, and with a pale ochraceous (not deep reddish-brown) throat. Rects are more pointed than in FAJ. Skull is unossified.

FPF    Replacing adult-like juvenile body and flight feathers with adult-like formative feathers. Skull is unossified but might be expected to ossify before completion of the FPF like most other Furnariids at the BDFFP, which then makes this difficult to distinguish from UPB (use UPU).

FAJ    Has deep reddish-brown throat. Rects are more truncate than FCJ. Skull is ossified.

UPB    Replacing adult-like body and flight feathers with adult-like feathers. Skull is ossified. May not be possible to distinguish from FPF during replacement of the last flight feathers.

p6–p10 retained, p4–p5 molting, p3 retained (why?), p1–p2 new, s1 molting, s2–s7 retained, s8 new, s9 molting.

## *Synallaxis rutilans dissors*

Band Size: E

Ruddy Spinetail • João-teneném-castanho

\# Individuals Captured: 138

| | |
|---|---|
| Wing | 51.0–60.0 mm (55.4 ± 1.9 mm; n = 95) |
| Tail | 55.0–72.0 mm (63.4 ± 3.6 mm; n = 54) |
| Mass | 14.0–20.8 g (16.9 ± 1.3 g; n = 117) |
| Bill | 8.1–10.0 mm (9.0 ± 0.5 mm; n = 25) |
| Tarsus | 19.8–22.8 mm (21.3 ± 0.6 mm; n = 25) |

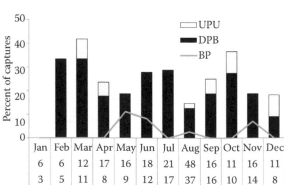

Similar species: No other bird at the BDFFP is plain rich rufous overall with black chin.

Skull: Completes (n = 26), but no information on timing.

Brood patch: Unknown if BPs develop in both sexes. Breeding observed from May–Nov. May have two breeding seasons per year, but data are limited.

Sex: ♂ = ♀.

Molt: Group 3/Complex Basic Strategy; FPF probably partial or incomplete as in several other *Synallaxis* spp. (Ryder and Wolfe 2009), DPBs complete. In the FCF, look for retained lesser and greater alulas, pp covs, and flight feathers, with most or all gr covs and perhaps also some to all rects replaced. Can molt at any time of year, but perhaps least frequent in Dec–Jan.

FCJ    Dull brown overall with a gray throat and vaguely streaked below (Remsen 2003, LSUMZ specimens). Skull is unossified.

FPF    Probably replacing brown contour feathers and ss covs with more olive and reddish contour feathers and ss covs. Skull is unossified.

FCF    Possibly with molt limits among gr covs or between gr covs and pp covs as in other *Synallaxis* spp. (Dickey and Van Rossem 1938, Ryder and Wolfe 2009). Skull may be unossified or ossified.

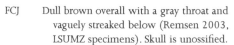

| | Jan | Feb | Mar | Apr | May | Jun | Jul | Aug | Sep | Oct | Nov | Dec |
|---|---|---|---|---|---|---|---|---|---|---|---|---|
| | 6 | 6 | 12 | 17 | 16 | 18 | 21 | 48 | 16 | 11 | 16 | 11 |
| | 3 | 5 | 11 | 8 | 9 | 12 | 17 | 37 | 16 | 10 | 14 | 8 |

SPB     Unknown how to distinguish from DPB (use UPB). Retained pp and ss may be relatively worn compared to the same retained feathers in DPB. Skull is ossified.

DCB     Rufous head and underparts, washed olive on the back, and black throat. Ss covs uniform in wear, but note how inner less, med, and gr are more olive than the outer rufous-red covs. ♀♀ may be slightly paler than ♂♂ and more washed with olive (Remsen 2003). Skull is ossified.

DPB     Unknown how to distinguish from SPB (use UPB). Retained pp and ss may be relatively less worn compared to the same retained feathers in SPB. Skull is ossified.

p8–p10 retained, p6–p7 molting, p1–p5 new, all ss retained (left); four retained rects, outer left rects molting, other rects missing.

## LITERATURE CITED

Aleixo, A., C. E. B. Portes, A. Whittaker, J. D. Weckstein, L. P. Gonzaga, K. J. Zimmer, C. C. Ribas, and J. M. Bates. 2013. Molecular systematics and taxonomic revision of the Curve-billed Scythebill complex (*Campylorhamphus procurvoides*: Dendrocolaptidae), with description of a new species from western Amazonian Brazil. Pp. 253–257 in J. del Hoyo, A. Elliott, and D. Christie (editors), Handbook of the birds of the world, Special volume: new species and global index. Lynx Edicions, Barcelona, Spain.

Dickey, D. R., and A. J. Van Rossem. 1938. The birds of El Salvador. Field Museum of Natural History, Zoological Series, Chicago, IL.

Guallar, S., E. Santana, S. Contreras, H. Verdugo, and A. Gallés. 2009. Paseriformes del Occidente de México: Morfometría, datación y sexado. Monografies del Museu de Ciències Naturals 5 (in Portuguese).

Howell, T. R. 1957. Birds of a second-growth rain forest area of Nicaragua. Condor 59:73–111.

Johnson, E. I., P. C. Stouffer, and R. O. Bierregaard. 2012. The phenology of molting, breeding and their overlap in central Amazonian birds. Journal of Avian Biology 43:141–154.

Marantz, C. A., A. Aleixo, L. R. Bevier, and M. A. Patten. 2003. Family Dendrocolaptidae (Woodcreepers). Pp. 358–447 in J. del Hoyo, A. Elliott, and D. Christie (editors), Handbook to the birds of the world, Vol. 8: Broadbills to Tapaculos. Lynx Edicions, Barcelona, Spain.

Pyle, P., A. Engilis, and D. A. Kelt. 2015. Manual for ageing and sexing birds of Bosque Fray Jorge National Park and Northcentral Chile, with notes on range and breeding seasonality. Special Publication of the Occasional Papers of the Museum of Natural Science, Louisiana State University, Baton Rouge, LA.

Remsen, J. V. 2003. Family Furnariidae (Ovenbirds). Pp. 162–357 in J. del Hoyo, A. Elliot, and D. Christie (editors), Handbook of the birds of the world, Vol. 8: Broadbills to Tapaculos. Lynx Edicions, Barcelona, Spain.

Remsen, J. V., C. D. Cadena, A. Jaramillo, M. Nores, J. F. Pacheco, M. B. Robbins, T. S. Schulenberg, F. G. Stiles, D. F. Stotz, and K. J. Zimmer. [online]. 2016. A classification of the bird species of South America. American Ornithologists' Union. Version 30 November 2016. <http://www.museum.lsu.edu/~Remsen/SACCBaseline.htm>.

Ridgway, R. 1911. The birds of North and Middle America, Part III. Bulletin of the U.S. Natural History Museum 50:1–859.

Rodrigues, E. B., A. Aleixo, A. Whittaker, and L. N. Naka. 2013. Molecular systematics and taxonomic revision of the Lineated Woodcreeper complex (*Lepidocolaptes albolineatus*: Dendrocolaptidae), with description of a new species from southwestern Amazonia. *In* J. del Hoyo, and D. Christie (editors), Handbook of the birds of the world, Special volume: New species and global index. Lynx Edicions, Barcelona, Spain.

Ryder, T. B., and J. D. Wolfe. 2009. The current state of knowledge on molt and plumage sequences in selected Neotropical bird families: a review. Ornitología Neotropical 20:1–18.

Vaurie, C. 1980. Taxonomy and geographical distribution of the Furnariidae (Aves, Passeriformes). Bulletin of the American Museum of Natural History 166.

Willis, E. O., and Y. Oniki. 1978. Birds and army ants. Annual Review of Ecology and Systematics 9:243–263.

Winker, K., G. A. Voelker, and J. T. Klicka. 1994. A morphometric examination of sexual dimorphism in the *Hylophilus*, *Xenops*, and an *Automolus* from southern Veracruz, Mexico. Journal of Field Ornithology 65:307–323.

Zimmer, J. T. 1934. Studies of Peruvian birds, No. 14. Notes on the genera *Dendrocolaptes*, *Hylexetastes*, *Xiphocolaptes*, *Dendroplex*, and *Lepidocolaptes*. American Museum Novitates 753:1–26.

# CHAPTER TWENTY-SEVEN

# Tyrannidae (Tyrant Flycatchers and Allies)

\# Species in South America: 366

\# Species recorded at BDFFP: 53

\# Species captured at BDFFP: 27

Flycatchers are an extremely diverse family of birds that have adapted to a variety of lifestyles: migratory or resident, ground- to canopy-dwelling, frugivorous or insectivorous, shy or gregarious, and so forth. Flycatchers are small-sized (*Myiornis ecaudatus* being the smallest known passerine on Earth) to medium-sized birds. The general form has an upright posture with a somewhat wide bill, highly accentuated in some species, and having rictal bristles. Many are drab with olives, browns, or dull yellows, but others are brightly patterned and even ornamented, like the long rectrices of *Tyrannus savana* or the jeweled crest of *Onycorhynchus*. Flycatchers have 10 pp, 9 ss, and 12 rects.

Depending on the species, flycatchers can be sexually dichromatic (usually varying in crown color), dimorphic (e.g., in shape of outer primaries), or monochromatic/monomorphic. All adhering to either the Complex Basic or Complex Alternate Strategy (Pyle 1997, Howell 2010). The prealternate molt is most commonly exhibited in migratory species, although much is to be learned about the presence or absence of prealternate molts in austral migrants. Furthermore, the timing and location of where preformative and prealternate molts occur in austral migrants is not well known, and may occur either on the breeding grounds, nonbreeding grounds, or on both, and suspending (or sometimes not?) for migration; this level of variation exists in boreal migrants. The extent of the preformative molt varies from partial, to incomplete and regular, to incomplete and eccentric, to complete (Pyle 1997, 1998; Guallar et al. 2009; Pyle et al. 2015). In some genera (e.g., *Contopus*, *Myiarchus*, and *Elaenia*), pp covs are partially or entirely retained when most or all primaries and secondaries are replaced (Pyle 1998, Burton 2002, Pyle et al. 2004, Guallar et al. 2009, Wolfe and Frey 2011, Pyle et al. 2015).

Skull ossification seems to vary among species; it appears to ossify during the formative plumage for some species, and rarely or never completely ossifies in others. Females may exclusively exhibit brood patches and incubate eggs except, maybe, in ground-dwelling species such as *Corythopis torquatus*.

## SPECIES ACCOUNTS

*Elaenia parvirostris* (monotypic)

Small-billed Elaenia • Guaracava-de-bico-curto

Band Size: C, D (CEMAVE 2013)

\# Individuals Captured: 1

| | |
|---|---|
| Wing | 68.0 mm (n = 1) |
| | 72–75 mm (n = 2; Junge and Mees 1961) |
| Tail | 61.0 mm (n = 1) |
| | 61–64 mm (n = 2; Junge and Mees 1961) |
| Mass | 13.0 g (n = 1) |
| | 15.5 g (Hilty 2003) |
| | 12.5–16.0 g (n = 2; Junge and Mees 1961) |

Similar species: Other elaenia are more crested (not round-headed) and do not have as strong of an eyering. From other flycatchers, especially small, greenish canopy flycatchers, by concealed white crown patch and lack of rictal bristles.

Skull: Completely ossifies (n = 92, LSUMZ specimens), apparently toward the end of the FPF or during the FCF.

Brood patch: Unknown, but absent at BDFFP because it is a nonbreeding austral migrant visitor.

Sex: ♂ = ♀.

Molt: Group 3 (or Group 8?)/Complex Basic Strategy(?); FPF incomplete, DPBs complete, PAs unknown, but should be considered. The FPF includes the replacement of some or all pp (if retained, among p1–p4), pp covs, and/or ss as in *E. mesoleuca* (E. Johnson and J. D. Wolfe, unpublished data) and some other *Elaenia* (P. Pyle, unpublished data). Timing is unknown, but two specimens from late Jul and early Aug in Trinidad and Tobago were fresh and finishing wing molt, suggesting that molt occurs on the nonbreeding grounds (Junge and Mees 1961). This may also suggest that a protracted "winter" molt may negate the need to have inserted a prealternate molt. As far as records from the BDFFP, one early May record at the BDFFP did not find molt, which may have been a migrant or recently arrived nonbreeding resident.

Notes: An austral migrant at the BDFFP, to be expected between Apr and Sep.

FCJ    Probably not possible to observe at BDFFP given it is a nonbreeding visitor only, unless they molt entirely on the wintering grounds, in which case they should arrive relatively fresh compared to DCB. Without molt limits and much like adult, but skull is unossified.

FPF    Look for differences in rect shape between new and replaced feathers. Relative wear between new and replaced flight feathers should be less dramatic than DPB. Skull normally unossified, but may become ossified toward the end of the FPF. Skull is variably ossified.

FCF    Typically at least with one or more pp covs retained, but also sometimes (always?) with one or more ss and/or inner pp retained. Skull is ossified or nearly so.

Outer 3 and 5 pp covs retained; this individual also retained p1 and s1–s2. (LSUMZ 175432.)

SPB    Look for two generations of retained pp covs being replaced by a typically sequential complete molt. Relative wear between new and replaced flight feathers should be more dramatic than FPF (use UPB when uncertain). Skull is ossified.

DCB    Without molt limits among the pp covs, pp, and ss. Skull is ossified.

DPB    Look for one generation of retained pp covs being replaced by a typically sequential complete molt. Relative wear between new and replaced flight feathers should be more dramatic than FPF (use UPB when uncertain). Skull is ossified.

## *Corythopis torquatus anthoides*                                    Band Size: D, E

Ringed Antpipit • Estalador-do-norte                    # Individuals Captured: 405

Wing    55.0–71.0 mm (63.8 ± 3.0 mm; n = 222)

Tail    42.5–58.0 mm (51.2 ± 3.6 mm; n = 212)

Mass    10.8–19.5 g (14.9 ± 1.2 g; n = 318)

Bill    8.2–9.5 mm (8.9 ± 0.4 mm; n = 21)

Tarsus    27.1–30.0 mm (28.4 ± 0.9 mm; n = 22)

Similar species: *Hylophylax naevia* also has white underparts with black necklace, but it is much more patterned in the upperparts than *C. torquatus*.

Skull: Does not fully ossify, terminating around 70%–90% complete (n = 93). Retained windows are typically symmetrical on the top of the crown. With caution, the degree of ossification may be useful to separate FCF from DCB, especially Oct–Apr.

Brood patch: Unknown if both sexes
develop BPs. Breeds mainly from
Jun–Jan, but juveniles have also been
seen Mar–May, and it is not clear
whether breeding is regular during
the wet season, or only occurs in
some years.

Sex: ♂ = ♀.

Molt: Group 3/Complex Basic Strategy; FPF
partial(?), DPBs complete. The FPF may
include 0–2 inner gr covs, resulting in
cryptic molt limits (F. Newell, pers. comm.).
Beware of pseudolimits among ss covs and
ss, because the plumage typically blends
from gray to brownish-gray or olive-gray.

| | Jan | Feb | Mar | Apr | May | Jun | Jul | Aug | Sep | Oct | Nov | Dec |
|---|---|---|---|---|---|---|---|---|---|---|---|---|
| | 41 | 30 | 40 | 46 | 45 | 59 | 91 | 75 | 61 | 74 | 52 | 12 |
| | 23 | 16 | 26 | 18 | 21 | 23 | 52 | 52 | 48 | 54 | 37 | 5 |

The frequency of molt peaks from Nov–Mar and is least frequent from Jun–Sep.

FCJ     Like subsequent plumages, but with a brown necklace. Rects are more pointed than DCB. Skull is unossified.

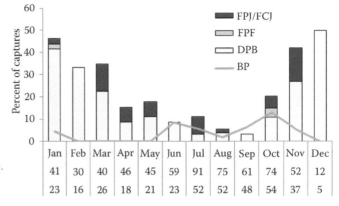

FPF     Replacing juvenile body feathers with adult-like feathers. Chest
markings show a mix of brown and black. Skull is unossified.

Note mix of brown and black breast band.

FCF     Look for cryptic molt limits among the inner gr covs or
between the gr and med covs. Rects relatively pointed.
Skull is unossified.

SPB     Replacing body and flight feathers and like DPB, but replacing pointed rects with adult-like truncate rects.
Skull is unossified.

TYRANNIDAE (TYRANT FLYCATCHERS AND ALLIES)     257

DCB    Chest markings are black. Ss covs and flight feathers are olive-brown, blending to gray distally and without molt limits. Rects are more truncate than FCJ. Retains gape-like color along edge of bill. Skull is nearly ossified (>75%).

DPB    Replacing adult-like body and flight feathers with adult-like feathers. Skull is nearly ossified. May not be possible to distinguish from SPB during replacement of the last flight feathers (use UPB).

## *Mionectes oleagineus oleagineus*

Ochre-bellied Flycatcher • Abre-asa

Band Size: D

# Individuals Captured: 27

Wing    53.0–67.0 mm (58.1 ± 3.5 mm; n = 24)

Tail    39.0–51.0 mm (44.9 ± 3.0 mm; n = 24)

Mass    9.4–13.0 g (10.8 ± 0.9 g; n = 23)

Bill    7.2–7.5 mm (7.4 ± 0.2 mm; n = 2)

Tarsus    16.6–17.1 mm (16.9 ± 0.3 mm; n = 2)

Similar species: Best distinguished from *M. macconnelli* by pale yellowish (not dark gray) lining inside bill. Usually also has pale yellowish edging to tertials, which *M. macconnelli* does not. *M. oleagineus* probably strictly occupies second growth (or perhaps very large gaps) at the BDFFP, but it occasionally wanders into small forest fragments, whereas *M. macconnelli* is a forest species that regularly wanders into second growth.

Skull: Unknown, but probably does not complete as in *M. macconnelli* (n = 1), thus aging using skull criteria should be done with caution.

Brood patch: Only ♀♀ develop BPs (Skutch 1960 in Hilty 2003). May breed in the early wet season as in *M. macconnelli*.

Sex: ♂♂ have narrower tips on outer pp (p8–p10) than ♀♀, but are otherwise similar in plumage aspect.

*M. oleagineus*       *M. macconnelli*

Molt: Group 3/Complex Basic Strategy; FPF partial, DPBs complete. FPF probably includes 0–4 inner gr covs, resulting in cryptic molt limits as possibly in other *Mionectes* (F. Newell, pers. comm.). The timing is unknown as only 1 of 24 birds examined was in molt (it was finishing wing molt in Apr), thus timing may be similar to *M. macconnelli*.

FCJ    ♂ = ♀. Unknown, but probably duller version of adult. Unknown if shape of outer pp (p8–p10) varies by sex in this plumage, but may average narrower in ♂♂. Rects relatively pointed. Skull is unossified.

FPF    ♂ = ♀. Replacing adult-like juvenile body feathers with adult-like feathers. Skull is relatively unossified, but beware of differentiating from DPB with unossified skulls.

FCF    ♂ = ♀. Look for cryptic molt limits among the inner gr covs or between the med covs and gr covs. Rect shape may be the best clue for aging, being relatively pointed. Skull unossified as in DCB.

SPB    ♂ ≈ ♀. Replacing body and flight feathers and like DPB, but replacing pointed rects with adult-like truncate rects. Skull is ossified.

DCB    ♂ ≈ ♀. Without molt limits among the ss covs. Retains gape-like yellow color along edge of bill. Rects relatively truncate. Skull is nearly ossified.

DPB    ♂ ≈ ♀. Replacing adult-like body and flight feathers with adult-like feathers. Skull is nearly ossified. May not regularly be possible to distinguish from SPB during replacement of the last flight feathers (use UPB).

## Mionectes macconnelli macconnelli

McConnell's Flycatcher • Abre-asa-da-mata

Band Size: D

# Individuals Captured: 1566

Wing    54.0–71.0 mm (62.5 ± 2.8 mm; n = 1126)

Tail    38.0–56.0 mm (47.1 ± 2.9 mm; n = 1075)

Mass    9.0–15.5 g (12.3 ± 1.1 g; n = 1320)

Bill    7.3–9.4 mm (8.1 ± 0.4 mm; n = 180)

Tarsus    16.0–20.2 mm (18.1 ± 0.8 mm; n = 182)

Similar species: Best separated from *M. oleagineus* by dark gray (not pale yellow) lining inside bill and lacking pale edging to terts.

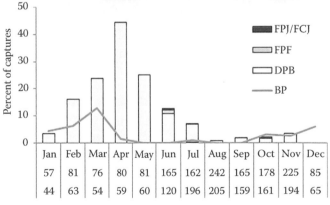

Skull: Does not fully ossify (n = 336). Retained windows are typically symmetrical on the top of the crown. May be useful to distinguish FPF from DPB.

Brood patch: Only ♀♀ develop BPs (Willis et al. 1978). Willis et al. (1978) found that breeding mainly occurs from Jan–Mar, which is consistent with our BP data, but breeding can also occur less often as early as Oct.

Sex: ♂♂ have narrower p8–p10 than ♀♀, but this difference may be more pronounced in DCB than FCF (more study is needed). Sexes are otherwise similar in plumage aspect.

| | Jan | Feb | Mar | Apr | May | Jun | Jul | Aug | Sep | Oct | Nov | Dec |
|---|---|---|---|---|---|---|---|---|---|---|---|---|
| | 57 | 81 | 76 | 80 | 81 | 165 | 162 | 242 | 165 | 178 | 225 | 85 |
| | 44 | 63 | 54 | 59 | 60 | 120 | 196 | 205 | 159 | 161 | 194 | 65 |

Molt: Group 3/Complex Basic Strategy; FPF partial, DPBs complete. FPF probably includes 0–4 inner gr covs, resulting in cryptic molt limits as in other *Mionectes* (F. Newell, pers. comm.). Has a well-defined molting season from Feb–Jun with nearly no birds molting from Aug–Dec.

FCJ   ♂ = ♀. Adult-like, but colors are duller and upperparts are grayer. The orange in the underparts may also extend higher into the chest/throat on average. Gr covs are more washed in orange and brown tones. Unknown if shape of outer pp (p8–p10) varies by sex in this plumage, but may average slightly narrower in ♂♂. Rects relatively pointed. Skull is mostly unossified.

FPF   ♂ = ♀. Replacing adult-like juvenile body feathers with adult-like feathers. Skull is relatively unossified, but beware of differentiating from DPB with unossified skulls.

FCF   ♂ = ♀. Look for cryptic molt limits among the inner gr covs or between the med covs and gr covs. Rect shape may be the best clue for aging, being relatively pointed. Skull unossified as in DCB.

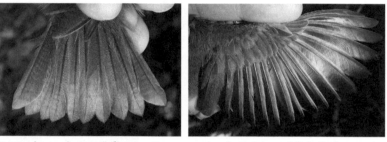

Retained juvenile rects (left); inner 3 gr covs replaced, all alulas and flight feathers retained (right).

SPB   ♂ ≈ ♀. Replacing body and flight feathers, and like DPB, but replacing pointed rects with adult-like truncate rects. Skull is relatively ossified.

DCB   ♂ ≈ ♀. Brighter colored and with more contrast in plumage with more green on upperparts and gr covs than FCJ. Retains gape-like yellow color along edge of bill. Skull is nearly ossified.

DPB   ♂ ≈ ♀. Replacing adult-like body and flight feathers with adult-like feathers. Skull is nearly ossified. May not be possible to distinguish from SPB during replacement of the last flight feathers (use UPB).

*Lophotriccus galeatus* (monotypic)

Band Size: C

Helmeted Pygmy-Tyrant • Caga-sebinho-de-penacho

# Individuals Captured: 8

Wing     42.0–50.0 mm (45.5 ± 3.7 mm; n = 4)

Tail      31.0–43.0 mm (37.2 ± 4.4 mm; n = 5)

Mass    6.0–8.0 g (6.8 ± 0.7 g; n = 7)

Bill      6.4 mm (n = 1)

Tarsus   17.1 mm (n = 1)

Similar species: Most like *Hemitriccus zosterops*, but with three-pronged crest of black feathers edged olive and pale tan (not creamy white) iris. Also note the shortened outer three primaries (♂♂ only?). *L. vitiosus*, although never captured at the BDFFP, should be considered a possible source of confusion. The centers to the crest feathers in *L. vitiosus* are not jet black and are instead more brownish.

Skull: Unknown.

Brood patch: Unknown.

Sex: ♂ like ♀ in plumage aspect, but ♂♂ have outer three pp shortened.

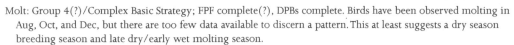

Molt: Group 4(?)/Complex Basic Strategy; FPF complete(?), DPBs complete. Birds have been observed molting in Aug, Oct, and Dec, but there are too few data available to discern a pattern. This at least suggests a dry season breeding season and late dry/early wet molting season.

FCJ      Unknown. Pp covs should be dull and relatively brownish. Skull is mostly unossified.

FPF     Look for relatively fresh, but brownish retained feathers being replaced by adult-like feathers. Skull is unossified, but may become nearly or completely ossified toward completion.

FAJ     Relatively greenish and glossy pp covs. Skull is nearly or completely ossified.

UPB     Look for relatively worn adult-like retained feathers being replaced by adult-like feathers. Skull may be nearly or completely ossified.

*Hemitriccus josephinae* (monotypic)

Band Size: D

Boat-billed Tody-Tyrant • Maria-bicudinha

# Individuals Captured: 2

Wing    47.5–53.5 mm (50.5 ± 4.2 mm; n = 2)

Tail    38.5–45.0 mm (41.8 ± 4.6 mm; n = 2)

Mass    8.6–9.6 g (9.1 ± 0.7 g; n = 2)

Bill    7.2 mm (n = 1)

Tarsus    20.6 mm (n = 1)

Similar species: From *Lophotriccus galeatus* by chestnut-brown (not pale tan) iris and lack of crest. *H. zosterops* has a creamy white iris, no nodules on tarsi, and a narrower bill. *Tyranneutes virescens* is also similar, but lacks yellow edging to tertials, has a shorter tail, and a gray iris.

Skull: Unknown.

Brood patch: Unknown.

Sex: ♂ = ♀.

Molt: Group 4(?)/Complex Basic Strategy; FPF complete(?), DPBs complete. The only captures were of a pair (probably ♂ and ♀) caught in Sep; one was molting and the other was not.

FCJ    Unknown. Pp covs should be relatively dull and brownish. Skull is mostly unossified.

FPF    Look for relatively fresh, but brownish retained feathers being replaced by adult-like feathers. Skull is unossified, but may become nearly or completely ossified toward completion.

FAJ    Relatively greenish and glossy pp covs. Skull may be nearly or completely ossified.

UPB    Look for relatively worn adult-like retained feathers being replaced by adult-like feathers. Skull may be nearly or completely ossified.

p3–p10 retained, p1–p2 molting, all ss retained.

*Hemitriccus zosterops zosterops*

White-eyed Tody-tyrant • Maria-de-olho-branco

| | |
|---|---|
| Wing | 43.0–56.0 mm (48.1 ± 3.3 mm; n = 25) |
| Tail | 34.0–48.0 mm (40.6 ± 4.7 mm; n = 23) |
| Mass | 6.5–10.5 g (8.4 ± 1.1 g; n = 31) |
| Bill | 7.0–7.2 mm (7.1 ± 0.1 mm; n = 2) |
| Tarsus | 16.8–17.4 mm (17.1 ± 0.5 mm; n = 2) |

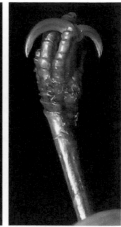

Similar species: From both *Lophotriccus* by lacking a crest. *H. josephinae* has a chestnut-brown iris, nodules on tarsi, and a wider bill. *Tyranneutes virescens* is also similar, but lacks yellow edging to tertials, has shorter tail, and a grayer iris.

Skull: Probably does not entirely complete (n = 2 BDFFP, n = 3 *H. z. zosterops* LSUMZ, and n = 11 *H. z. flaviviridis* LSUMZ).

Brood patch: Unknown if both sexes incubate. BPs have been seen from Jul–Oct and Jan, probably breeding throughout the dry season.

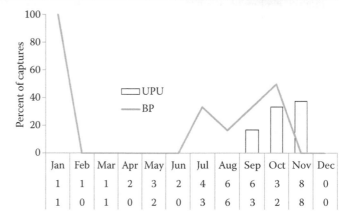

| | Jan | Feb | Mar | Apr | May | Jun | Jul | Aug | Sep | Oct | Nov | Dec |
|---|---|---|---|---|---|---|---|---|---|---|---|---|
| | 1 | 1 | 1 | 2 | 3 | 2 | 4 | 6 | 6 | 3 | 8 | 0 |
| | 1 | 0 | 1 | 0 | 2 | 0 | 3 | 6 | 3 | 2 | 8 | 0 |

Sex: ♂ = ♀.

Molt: Group 4(?)/Complex Basic Strategy; FPF complete(?), DPBs complete. Wing molt has been observed Sep–Nov; but there are few data to discern a pattern. Probably molts mainly during the late dry/early wet seasons following breeding.

FCJ     Unknown. Pp covs should be dull and relatively brownish. Skull is mostly unossified.

FPF     Look for relatively fresh, but brownish retained feathers being replaced by adult-like feathers. Skull is unossified, but may become nearly or completely ossified toward completion.

FAJ     Look for relatively greenish and glossy pp covs. Skull is relatively ossified.

UPB     Look for relatively worn adult-like retained feathers being replaced by adult-like feathers. Skull is relatively ossified.

*Rhynchocyclus olivaceus guianensis*

Olivaceous Flatbill • Bico-chato-grande

Band Size: E

# Individuals Captured: 141

| | |
|---|---|
| Wing | 64.0–76.0 mm (69.7 ± 2.5 mm; n = 90) |
| Tail | 49.0–64.0 mm (56.3 ± 3.0 mm; n = 91) |
| Mass | 17.0–24.0 g (20.0 ± 1.2 g; n = 114) |
| Bill | 8.3–10.3 mm (9.2 ± 0.7 mm; n = 10) |
| Tarsus | 18.0–20.7 mm (19.4 ± 0.8 mm; n = 11) |

Similar species: Most similar to *Tolmomyias* spp., but has an even broader bill, is larger, and is less patterned with less contrasting edging to coverts and remiges and no contrast between olive head and back. *Ramphotrigon ruficauda* is similar in size and form, but has extensive rufous on remiges and rectrices and a narrower bill. From *Hemitriccus* spp. by large size, less pattern in wings, and dark iris.

Skull: May typically retain small windows or completely ossify during FCF.

Brood patch: Unknown if both sexes incubate. BPs have been seen from Jun–Nov, perhaps peaking early in the dry season given the timing of molt.

Sex: ♂ = ♀.

Molt: Group 4/Complex Basic Strategy; FPF complete, DPB complete. Molt is most evident during the dry season, especially from Aug–Dec.

FCJ Similar to adult, but duller with more ochraceous rump (Bates 2004). Pp covs relatively dull and brownish. Iris is brownish. Skull is unossified.

| | Jan | Feb | Mar | Apr | May | Jun | Jul | Aug | Sep | Oct | Nov | Dec |
|---|---|---|---|---|---|---|---|---|---|---|---|---|
| | 10 | 8 | 9 | 12 | 17 | 18 | 32 | 26 | 17 | 13 | 21 | 14 |
| | 6 | 2 | 4 | 4 | 7 | 8 | 7 | 17 | 10 | 9 | 6 | 2 |

FPF Retained pp covs, ss, and pp relatively fresh, but with minimal luster and brownish. Iris is brownish to grayish. Skull is unossified, but may occasionally become ossified or nearly ossified before FPF completes, making it difficult to distinguish from UPB (use UPU).

Body and less covs replaced, med and gr covs retained juvenile, p3–p10 retained juvenile, p2 missing, p1 molting, all ss retained juvenile (left); p8–p10 molting, p1–p7 new, s1–s2 new, s3 molting, s4–s6 retained, s7 missing, s8–s9 new (right).

FAJ More bright olive overall and greener rump than FCJ. Iris is gray. Skull is ossified or with small windows.

UPB Retained pp covs, ss, and pp relatively worn, glossy, and greenish. Iris is gray. Skull is ossified or nearly so, and birds late in molt may be difficult to distinguish from FPF (use UPU).

*Tolmomyias assimilis examinatus*

Band Size: E, D

Yellow-margined Flycatcher • Bico-chato-da-copa

# Individuals Captured: 21

| | |
|---|---|
| Wing | 58.0–64.0 mm (61.1 ± 2.0 mm; n = 18) |
| Tail | 46.0–56.0 mm (50.8 ± 3.0 mm; n = 19) |
| Mass | 12.3–15.1 g (13.8 ± 0.7 g; n = 19) |
| Bill | 6.7–8.6 mm (7.4 ± 0.6 mm; n = 7) |
| Tarsus | 18.1–20.4 mm (19.4 ± 0.9 mm; n = 7) |

Similar species: *Tolmomyias* spp. from other small olive or yellowish flycatchers by wide, flat bill and no crown stripe. It is smaller and has stronger yellow edging to coverts than *Rhynchocyclus olivaceus*.

From *T. poliocephalus* by more olive (with limited gray) crown, larger size, and usually (always?) by combination of wing chord (>57 mm) and weight (>13 g). Some authors (e.g., Hilty 2003) suggest that *T. assimilis* has a pale spot or "speculum" in the folded wing caused by pale edging at the base of the outer 3 or 4 pp, which *T. poliocephalus* does not, but we have found this to be difficult to use. Based on *Tolmomyias* spp. captured with identifications verified in 2007–2013 (n = 14), we treat any potential *Tolmomyias* misidentifications in the BDFFP database as *T. assimilis* if the wing is greater than 57 mm.

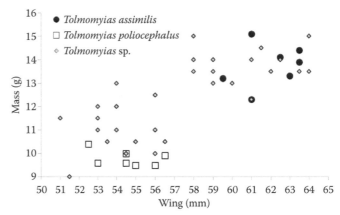

Skull: Probably completes (or sometimes with small windows?) likely during FCF, but perhaps during FPF in some cases.

Brood patch: Unknown if both sexes incubate. BPs have been seen from Aug–Dec.

Sex: ♂ = ♀.

Molt: Group 4/Complex Basic Strategy; FPF complete as in other *Tolmomyias* (Dickey and Van Rossem 1938), DPBs complete. The only bird observed in molt was in Jan, and probably mainly molts during the late dry and early wet seasons.

FCJ Like adult, but perhaps with more ochraceous ss covs (Caballero 2004). Pp covs should be relatively dull and brownish. Skull is unossified.

FPF Retained pp covs, ss, and pp relatively fresh, but with minimal luster and brownish. Skull is unossified, but may occasionally become ossified or nearly ossified before FPF completes, making it difficult to distinguish from UPB (use UPU).

FAJ Retains gape into adult plumage and should not be used to age birds. Ss covs more yellowish than FCJ. Skull is ossified or perhaps sometimes with small windows.

UPB Retained pp covs, ss, and pp relatively worn, glossy, and greenish. Skull is ossified or nearly so, and birds late in molt may be difficult to distinguish from FPF (use UPU).

*Tolmomyias poliocephalus poliocephalus*                                 Band Size: D

Gray-crowned Flycatcher • Bico-chato-de-cabeça-cinza          # Individuals Captured: 27

Wing     50.5–56.5 mm (54.0 ± 1.8 mm; n = 23)

Tail      43.0–50.0 mm (45.8 ± 1.9 mm; n = 22)

Mass    9.0–13.0 g (10.7 ± 1.1 g; n = 24)

Bill       6.3–7.5 mm (7.0 ± 0.6 mm; n = 5)

Tarsus   18.0–19.0 mm (18.5 ± 0.5 mm; n = 3)

Similar species: *Tolmomyias* from other small olive or yellowish flycatchers by wide, flat bill and no crown stripe. From *T. assimilis* by gray crown contrasting against olive-green back and smaller size, specifically combination of wing (<57 mm) and weight (<12 g). Wing length may not overlap (Hilty 2003) or may rarely overlap between *T. assimilis* and *T. poliocephalus*, but more study is needed for the populations at the BDFFP (see also *T. assimilis*).

Skull: Probably completes (or sometimes with small windows?) likely during FCF, but perhaps during FPF in some cases.

Brood patch: Unknown if both sexes incubate. The only BPs were seen from Aug–Oct, but probably breeds through the dry season.

Sex: ♂ = ♀.

Molt: Group 4/Complex Basic Strategy; FPF complete as in other *Tolmomyias* (Dickey and Van Rossem 1938), DPBs complete. The only four birds observed in molt were in Sep, Oct, and Jan, and probably mainly molts during the late dry and early wet seasons. May have slightly earlier nesting and molting seasons than *T. assimilis*, but sample sizes are limited.

FCJ       Undescribed. Pp covs should be relatively dull and brownish. Skull is unossified.

FPF       Retained pp covs, ss, and pp relatively fresh, but with minimal luster and brownish. Skull is unossified, but may occasionally become ossified or nearly ossified before FPF completes, making it difficult to distinguish from UPB (use UPU).

FAJ       Retains gape into adult plumage and should not be used to age birds. Pp covs should be relatively glossy and greenish. Skull is ossified.

UPB       Retained pp covs, ss, and pp relatively worn, glossy, and greenish. Skull is ossified or nearly so, and birds late in molt may be difficult to distinguish from FPF (use UPU).

*Platyrinchus saturatus saturatus*

Band Size: D

Cinnamon-crested Spadebill • Patinho-escuro

# Individuals Captured: 413

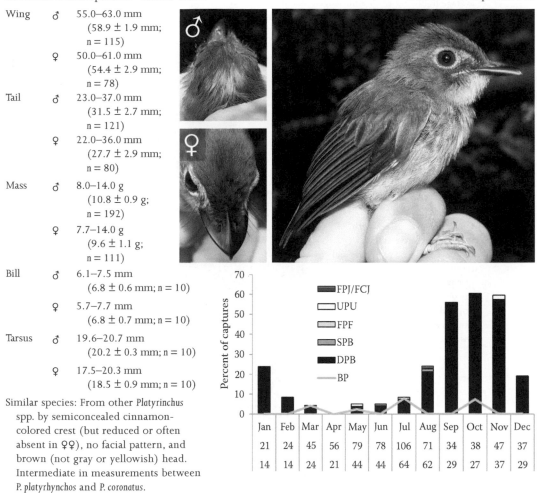

| Wing | ♂ | 55.0–63.0 mm (58.9 ± 1.9 mm; n = 115) |
| | ♀ | 50.0–61.0 mm (54.4 ± 2.9 mm; n = 78) |
| Tail | ♂ | 23.0–37.0 mm (31.5 ± 2.7 mm; n = 121) |
| | ♀ | 22.0–36.0 mm (27.7 ± 2.9 mm; n = 80) |
| Mass | ♂ | 8.0–14.0 g (10.8 ± 0.9 g; n = 192) |
| | ♀ | 7.7–14.0 g (9.6 ± 1.1 g; n = 111) |
| Bill | ♂ | 6.1–7.5 mm (6.8 ± 0.6 mm; n = 10) |
| | ♀ | 5.7–7.7 mm (6.8 ± 0.7 mm; n = 10) |
| Tarsus | ♂ | 19.6–20.7 mm (20.2 ± 0.3 mm; n = 10) |
| | ♀ | 17.5–20.3 mm (18.5 ± 0.9 mm; n = 10) |

Similar species: From other *Platyrinchus* spp. by semiconcealed cinnamon-colored crest (but reduced or often absent in ♀♀), no facial pattern, and brown (not gray or yellowish) head. Intermediate in measurements between *P. platyrhynchos* and *P. coronatus*.

Chart legend: FPJ/FCJ, UPU, FPF, SPB, DPB, BP. Y-axis: Percent of captures.

| | Jan | Feb | Mar | Apr | May | Jun | Jul | Aug | Sep | Oct | Nov | Dec |
|---|---|---|---|---|---|---|---|---|---|---|---|---|
| | 21 | 24 | 45 | 56 | 79 | 78 | 106 | 71 | 34 | 38 | 47 | 37 |
| | 14 | 14 | 24 | 21 | 44 | 44 | 64 | 62 | 29 | 27 | 37 | 29 |

Skull: Completes during FCF (or perhaps sometimes retains small windows into DCB; n = 98).

Brood patch: Has only been observed in ♀♀, but not known if it may also occur in ♂♂ to a lesser extent. Only seen sporadically from Mar–Oct.

Sex: ♂♂ have a metallic rufous-orange semiconcealed crown patch, whereas ♀♀ may have limited dull orange semiconcealed crown patch, but this is often lacking. Distinguishable during the FPF.

Molt: Group 3/Complex Basic Strategy; FPF partial, DPBs complete. FPF includes body feathers and less covs, med covs, 0–2 gr covs, but not rects(?), or remiges. DPB rect molt may occasionally or even regularly be simultaneous. Has a well-defined molting schedule being most frequent from Sep–Nov and least frequent from Mar–Jun.

FCJ   ♂ = ♀. Probably without rufous-orange crown patch. The body feathers, and especially ss covs, average lightly more chestnut-edged than in subsequent plumages. This plumage is probably only briefly retained and most birds in early stages of FPF may be sexed according to newly replaced crown feathers as in FCF and DCB. Rects are pointed. Skull is unossified.

FPF   ♂ ≠ ♀. Replacing slightly chestnut-edged, dull brownish body feathers, less, med, and sometimes inner gr covs with richer brown (slightly olive-toned) feathers. Rects are pointed. Skull is unossified.

FCF   ♂ ≠ ♀. Adult-like, except with molt limits between the browner replaced med covs and more chestnut-edged retained gr covs, or sometimes among the gr covs. Rects relatively pointed, retained from FCJ. Skull is either unossified or ossified.

Less and med covs replaced, inner 2 gr covs replaced, all alulas and flight feathers retained.

SPB   ♂ ≠ ♀. Replaced rects relatively rounded relative to retained juvenile rects, retained pp and ss relatively worn compared to DPB, and retained pp covs relatively dull and brown. Note that in some birds all rects may be replaced simultaneously. Skull is ossified (or occasionally with small windows retained?).

p8–p10 retained juvenile (note increased wear relative to DPB), p7 molting, p1–p6 new, s1–s2 new, s3 molting, s4–s7 retained juvenile, s8–s9 new.

DCB   ♂ ≠ ♀. Lacks molt limits between the med and gr covs. Rects are relatively rounded. Retains gape into adult plumage and should not be used to age birds. Skull is ossified (or occasionally with small windows retained?).

DPB    ♂ ≠ ♀. Replaced and retained rects both relatively rounded, retained pp and ss relatively fresh compared to SPB, and retained pp covs relatively glossy and olive. Note that in some birds all rects may be replaced simultaneously. Skull is ossified (or occasionally with small windows retained?).

All pp and all ss replaced including s6 molting; this would more appropriately be called a UPB.

## Platyrinchus coronatus coronatus

Band Size: C

Golden-crowned Spadebill • Patinho-de-coroa-dourada

# Individuals Captured: 462

| | | |
|---|---|---|
| Wing | ♂ | 48.0–58.0 mm (53.8 ± 2.0 mm; n = 168) |
| | ♀ | 48.0–56.0 mm (51.7 ± 1.6 mm; n = 61) |
| Tail | ♂ | 20.0–28.0 mm (24.1 ± 1.8 mm; n = 183) |
| | ♀ | 19.0–27.0 mm (22.0 ± 1.8 mm; n = 60) |
| Mass | ♂ | 7.0–10.0 g (8.6 ± 0.6 g; n = 238) |
| | ♀ | 7.0–10.0 g (8.4 ± 0.7 g; n = 91) |
| Bill | ♂ | 5.3–7.6 mm (6.5 ± 0.6 mm; n = 20) |
| | ♀ | 5.7–7.7 mm (6.6 ± 0.5 mm; n = 10) |
| Tarsus | ♂ | 13.3–15.9 mm (14.9 ± 0.8 mm; n = 18) |
| | ♀ | 13.2–15.7 mm (14.8 ± 0.8 mm; n = 8) |

Similar species: From other Platyrinchus spp. by having bold facial pattern. Smallest Platyrinchus with minimal overlap in measurements with P. platyrhynchos; larger birds overlap smaller P. saturatus.

Skull: Completes (n = 80) during FCF (or sometimes with small windows retained into DCB?).

Brood patch: May occur in both sexes from Jul–Dec.

Sex: ♂♂ have bright metallic orange crown with metallic yellow feathers partially concealed. ♀♀ have bright metallic orange crown patch without or with limited yellow. Distinguishable during the FPF.

Molt: Group 3/Complex Basic Strategy; FPF partial, DPBs complete. FPF includes body feathers and less covs, med covs, but not rects(?), gr covs, or remiges. Usually during DPBs, all or inner 1–5 rects are replaced simultaneously. Has a well-defined molting schedule being most frequent from Sep–Dec and least frequent from Mar–Jul.

Chart legend: FPJ/FCJ, UPU, FPF, DPB, BP

Percent of captures

| | Jan | Feb | Mar | Apr | May | Jun | Jul | Aug | Sep | Oct | Nov | Dec |
|---|---|---|---|---|---|---|---|---|---|---|---|---|
| | 16 | 24 | 36 | 46 | 41 | 64 | 101 | 72 | 47 | 64 | 55 | 31 |
| | 11 | 17 | 23 | 27 | 18 | 42 | 67 | 56 | 37 | 55 | 44 | 22 |

TYRANNIDAE (TYRANT FLYCATCHERS AND ALLIES)

FCJ    ♂ = ♀. Crown feathers are mainly dusky-olive. The underparts are white. This plumage is probably only briefly retained. Remiges are edged dull olive-brown. Pp covs are pale olive. Med and gr covs are dull olive with moderately narrow yellowish edging. Rects are pointed. Skull is unossified.

FPF    ♂ ≠ ♀. Replacing brownish crown feathers with orange or bright yellow-orange feathers, white body feathers in the underparts with yellow body feathers, and less, med, and sometimes inner gr covs. Rects are pointed. Skull is unossified. Some birds in early stages of FPF molt can be sexed by the color of newly replaced crown feathers.

Body, and less and med covs replaced or in molt, gr covs and all flight feathers retained juvenile (left and right).

FCF    ♂ ≠ ♀. Adult-like, except with molt limits between the bright golden-brown replaced med covs and duller retained gr covs. Pp covs are dull brownish-olive. Rects are probably relatively pointed and retained from the FCJ. Skull is either unossified or ossified.

Less and med covs replaced, all gr covs and flight feathers retained.

SPB      ♂ ≠ ♀. Retained pp and ss relatively worn compared to DPB, retained pp covs relatively dull and brown. Note that in some birds all rects may be replaced simultaneously. Skull is ossified (or occasionally with small windows retained?).

DCB      ♂ ≠ ♀. Without molt limits between the med and gr covs. Gr covs are olive with very narrow yellowish edging; these are slightly paler than the med covs, but should have less contrast in color and no contrast in wear against the gr covs compared to FCF. Remiges are edged olive. Pp covs are yellowish-olive. Rects are rounded. Retains gape into adult plumage and should not be used for aging. Skull is ossified (or occasionally with small windows retained?).

DPB      ♂ ≠ ♀. Replaced and retained rects both relatively rounded, retained pp and ss relatively fresh and greenish compared to SPB, and retained pp covs relatively glossy and olive. Note that in some birds all rects may be replaced simultaneously. Skull is ossified (or occasionally with small windows retained?).

## *Platyrinchus platyrhynchos platyrhynchos*

White-crested Spadebill • Patinho-de-coroa-branca

Band Size: D

# Individuals Captured: 47

| | |
|---|---|
| Wing | 59.0–66.0 mm (62.6 ± 2.1 mm; n = 31) |
| Tail | 27.0–37.0 mm (33.1 ± 2.2 mm; n = 33) |
| Mass | 10.5–14.0 g (11.9 ± 0.8 g; n = 41) |
| Bill | 6.3–7.7 mm (7.0 ± 0.7 mm; n = 3) |
| Tarsus | 14.6–15.2 mm (14.8 ± 0.3 mm; n = 3) |

Similar species: From other *Platyrinchus* spp. by gray head contrast with yellowish-olive body and white central crown stripe. Largest *Platyrinchus* with minimal overlap with smaller *P. coronatus*, but smaller birds overlap larger *P. saturatus*.

Skull: Probably completes (n = 5) during FCF.

Brood patch: Unknown if both sexes incubate. The only BP observed in Nov, but probably most frequent in dry season.

Sex: Both sexes have semiconcealed white crown patch, which is smaller in ♀♀ (Hilty 2003), but quantitative data are lacking. Only 2 of 47 captures have been sexed as ♀ in the BDFFP database, so differences between sexes may be subtle.

Molt: Group 3/Complex Basic Strategy; FPF partial, DPBs complete. FPF includes body feathers, less and med covs, 0 to 2 (or sometimes more?) gr

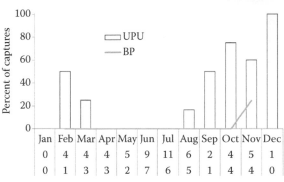

| | Jan | Feb | Mar | Apr | May | Jun | Jul | Aug | Sep | Oct | Nov | Dec |
|---|---|---|---|---|---|---|---|---|---|---|---|---|
| | 0 | 4 | 4 | 4 | 5 | 9 | 11 | 6 | 2 | 4 | 5 | 1 |
| | 0 | 1 | 3 | 3 | 2 | 7 | 6 | 5 | 1 | 4 | 4 | 0 |

covs, and sometimes the alula covert (a1), but not rectrices(?), outer gr covs, or remiges. There appears to be a well-defined molting schedule with molt being most frequent from Sep–Feb and least frequent from Apr–Jul.

FCJ      ♂ = ♀. Undescribed. Skull is unossified.

FPF      ♂ = ♀. Unknown how to distinguish from DPB by plumage. Skull is probably unossified.

FCF     ♂ ≈ ♀. As in other *Platyrinchus*, look
        for molt limits among the gr covs
        or between the gr covs and med
        covs, and sometimes among the
        alulas. Pp covs relatively brownish
        and dull. Skull may either be
        unossified or ossified.

Inner 2 gr covs and all med and less covs replaced, outer gr covs
retained juvenile, a1–a2 molt limit, all flight feathers retained
juvenile.

SPB     ♂ ≈ ♀. Retained pp and ss relatively worn compared to DPB, retained pp covs relatively dull and brownish.
        Note that in some birds all rects may be replaced simultaneously as in other *Platyrinchus*. Skull is ossified
        (or sometimes with small windows retained?).

DCB     ♂ ≈ ♀. Without molt limits among the ss covs. Pp covs relatively gray and glossy. Skull is ossified (or
        sometimes with small windows retained?).

DPB     ♂ ≈ ♀. Retained pp and ss relatively fresh and
        grayish compared to SPB, and retained pp
        covs relatively glossy and grayish. Note that
        in some birds all rects may be replaced
        simultaneously. Skull is ossified (or
        sometimes with small windows retained?).

p6–p10 retained, p5 molting, p1–p4 new, s1 new, s2–s7,
s8 molting, s9 missing.

*Onychorhynchus coronatus coronatus*                          Band Size: D, E

Royal Flycatcher • Maria-leque                               # Individuals Captured: 101

Wing    ♂   69.0–84.0 mm
            (76.6 ± 3.2 mm;
            n = 35)

        ♀   69.5–75.0 mm
            (72.3 ± 1.6 mm;
            n = 18)

Tail    ♂   56.0–66.0 mm
            (61.1 ± 2.6 mm;
            n = 31)

        ♀   55.0–63.0 mm
            (59.1 ± 2.3 mm;
            n = 20)

Mass   ♂   11.0–17.0 g (15.1 ± 1.1 g; n = 59)

    ♀   10.5–14.0 g (12.8 ± 0.9 g; n = 28)

Bill   ♂   14.1 mm (n = 1)

    ♀   12.8–14.3 mm (13.6 ± 1.1 mm; n = 2)

Tarsus   ♂   16.7 mm (n = 1)

    ♀   17.0 mm (n = 2)

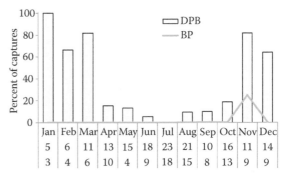

| | Jan | Feb | Mar | Apr | May | Jun | Jul | Aug | Sep | Oct | Nov | Dec |
|---|---|---|---|---|---|---|---|---|---|---|---|---|
| | 5 | 6 | 11 | 13 | 15 | 18 | 23 | 21 | 10 | 16 | 11 | 14 |
| | 3 | 4 | 6 | 10 | 4 | 9 | 18 | 15 | 8 | 13 | 9 | 9 |

Similar species: Unmistakable.

Skull: Complete (n = 22) probably during the FCF, or perhaps sometimes at the end of FPF.

Brood patch: Only the ♀ incubates and develops BPs (Farnsworth and Lebbin 2012). The only BP was observed in Nov, but probably breeds during the dry season based on the molt schedule.

Sex: ♂♂ have a bright metallic red crown with limited iridescent blue tips on central crown feathers. ♀♀ have a bright metallic orange crown patch with extensive iridescent blue tips on central crown feathers. Probably distinguishable early in the FPF.

Molt: Group 3/Complex Basic Strategy; FPF partial, DPBs complete. FPF appears to include body feathers, less covs, and some to all inner med covs, and 0–1 (or occasionally more?) inner gr covs, but not rects, ss or pp. Has a well-defined molting schedule being most frequent from Nov to Mar and least frequent from Apr to Sep.

FCJ    ♂?♀. Ss covs with broad buffy tips, not small spots. Crown is apparently not well developed and ♀-like (Hilty 2003, Farnsworth and Lebbin 2012). Skull is unossified.

FPF    ♂ ≠ ♀. Replacing broad-tipped ss covs with spotted-tipped ss covs. Skull may ossify near completion of FPF, but probably is unossified.

FCF    ♂ ≠ ♀. Crown color like DCB of same sex. Look for molt limits among the med covs or perhaps between the med and gr covs, with retained covs having broad buffy tips and replaced covs with smaller buffy triangular spots. Skull is either ossified or unossified.

Less and inner med covs replaced, outer med covs and gr covs retained juvenile, all flight feathers retained juvenile.

SPB    ♂ ≠ ♀. Retained pp and ss feathers relatively worn, dull, and brownish. May be difficult to distinguish from DPB, especially late in molt (use UPB). Skull is ossified.

DCB    ♂ ≠ ♀. All ss covs with small triangle-shaped buffy spot at the tip. Skull is ossified.

TYRANNIDAE (TYRANT FLYCATCHERS AND ALLIES)

DPB    ♂ ≠ ♀. Retained pp and ss feathers relatively fresh, glossy, and olive. May be difficult to distinguish from SPB, especially late in molt (use UPB). Skull is ossified.

p6–p10 retained, p5 missing, p4 molting, p1–p3 new, s1 molting, s2–s7 retained, s8 molting, s9 retained.

## *Myiobius barbatus barbatus*

Sulphur-rumped Flycatcher • Assanhadinho

Band Size: D

# Individuals Captured: 709

Wing   ♂   57.0–68.0 mm
       (63.3 ± 1.9 mm; n = 232)

      ♀   55.0–68.0 mm
       (59.6 ± 2.0 mm; n = 114)

Tail   ♂   47.0–60.0 mm
       (54.1 ± 2.0 mm; n = 254)

      ♀   48.0–58.0 mm
       (51.8 ± 2.0 mm; n = 112)

Mass   ♂   8.0–13.0 g
       (10.8 ± 0.7 g; n = 378)

      ♀   7.5–11.5 g
       (9.5 ± 0.8 g; n = 156)

Bill   ♂   5.8–7.5 mm
       (6.8 ± 0.4 mm; n = 44)

      ♀   5.9–7.5 mm
       (6.7 ± 0.5 mm; n = 17)

Tarsus   ♂   16.3–19.7 mm
       (18.5 ± 0.8 mm; n = 39)

      ♀   15.5–18.8 mm
       (17.0 ± 1.0 mm; n = 16)

Similar species: None.

Skull: Completes (n = 203), probably during FCF or perhaps late in the FPF.

Brood patch: Present from Jul–Nov (–Jan?) and probably only develops in ♀♀, but additional confirmation is needed.

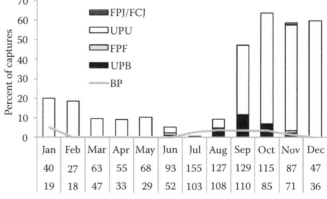

| | Jan | Feb | Mar | Apr | May | Jun | Jul | Aug | Sep | Oct | Nov | Dec |
|---|---|---|---|---|---|---|---|---|---|---|---|---|
| | 40 | 27 | 63 | 55 | 68 | 93 | 155 | 127 | 129 | 115 | 87 | 47 |
| | 19 | 18 | 47 | 33 | 29 | 52 | 103 | 108 | 110 | 85 | 71 | 36 |

Legend: FPJ/FCJ, UPU, FPF, UPB, BP

Sex: ♂♂ have an extensive yellow semiconcealed crown patch. ♀♀ have very little to no yellow in crown. Distinguishable during the FPF.

Molt: Group 4/Complex Basic Strategy; FPF complete, DPBs complete. Has a well-defined molting schedule being most frequent from Sep–Dec and least frequent from Mar–Aug.

FCJ     ♂ = ♀. Similar to subsequent plumages, but with duller yellows, more brown tones in the gr covs and remiges, and weakly structured feathers. Skull is unossified.

FPF     ♂ ≠ ♀. Replacing adult-like juvenile body and flight feathers with adult-like formative feathers. Retained pp and ss relatively fresh. Skull is unossified, but may ossify before completion of the FPF, which makes this difficult to distinguish from UPB (use UPU).

FAJ     ♂ ≠ ♀. Has brighter yellows and more greenish-olive tones in gr covs and remiges than FCJ. Skull is ossified.

UPB     ♂ ≠ ♀. Replacing adult-like body and flight feathers with adult-like feathers. Retained pp and ss relatively worn. Skull is ossified. May not be possible to distinguish from FPF during replacement of the last flight feathers (use UPU).

p8–p10 retained, p7 missing, p6 molting, p1–p5 new, s1–s3, s4–s6 retained, s7–s9 new.

*Terenotriccus erythrurus erythrurus*

Ruddy-tailed Flycatcher • Papa-moscas-uirapuru

Wing     44.0–53.0 mm (49.2 ± 1.8 mm; n = 144)

Tail      32.0–44.0 mm (38.2 ± 2.0 mm; n = 159)

Mass    5.0–8.0 g (6.6 ± 0.6 g; n = 208)

Bill      4.1–6.1 mm (5.1 ± 0.5 mm; n = 29)

Tarsus  15.9–17.8 mm (16.7 ± 0.5 mm; n = 28)

Similar species: *Neopipo* is most similar, but lacks rictal bristles, has a shorter tail, has gray (not orange) tarsi, has a concealed crown patch, and is extremely rare at the BDFFP.

Skull: Completes (n = 48, BDFFP; and n = 60 specimens, LSUMZ), but skin is dark often obscuring visibility. Probably normally completes early in the FCF.

Brood patch: Surprisingly rare to see, and unknown if both sexes incubate, but nest building and chick care done by female (Farnsworth and Lebbin 2004a). The only BPs were observed in Jul and Oct; breeding may occur mostly during the dry season based on the molt schedule.

Sex: ♀ < ♂ (Hilty 2003), but degree of overlap unknown.

Molt: Group 4/Complex Basic Strategy; FPF complete, DPBs complete. Has a well-defined molting schedule being most frequent from Sep–Dec and does not occur from Feb–Jul.

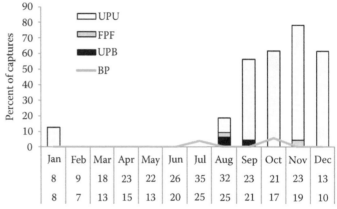

| | Jan | Feb | Mar | Apr | May | Jun | Jul | Aug | Sep | Oct | Nov | Dec |
|---|---|---|---|---|---|---|---|---|---|---|---|---|
| | 8 | 9 | 18 | 23 | 22 | 26 | 35 | 32 | 23 | 21 | 23 | 13 |
| | 8 | 7 | 13 | 15 | 13 | 20 | 25 | 25 | 21 | 17 | 19 | 10 |

FCJ      Like adult, but tail darker with a dusky tip and breast with olive wash (Farnsworth and Lebbin 2004a). Skull is unossified.

FPF      Probably difficult to distinguish from UPB (use UPU), but look for duskier retained rects compared to more orange replaced rects, and relatively fresh retained pp and ss. Skull is unossified.

FAJ      Tail more uniformly orange with oranger underparts. Skull is ossified.

Just finishing a complete molt, so technically a UPU (cannot distinguish FPF from DPB), entering FAJ.

UPB    Difficult to distinguish from FPF (use UPU), but look for similar color between retained and replaced rects, and relatively worn retained pp and ss. Skull is ossified.

Cinnamon Manakin-tyrant • Enferrujadinho

Wing    50.0 mm (n = 1)
          50–52 mm (Hellmayr 1929)

Tail     36.0 mm (n = 1)

Mass    7.0 g (n = 2)
          7 g (Hilty 2003)
          6.3–8.0 g (mean = 7.1 g; n = 5; Willard et al. 1991)

Similar species: Similar to *Terenotriccus erythrurus*, but lacks rictal bristles, legs are gray (not orange), and has concealed crown stripe.

Skull: Appears to not fully ossify (n = 3, LSUMZ).

Brood patch: Unknown.

Sex: ♂♂ have a more extensive concealed yellow crown stripe than ♀♀ (Hellmayr 1929, Farnsworth and Lebbin 2004b).

Molt: Group 4(?)/Complex Basic Strategy; FPF complete(?), DPBs complete. The FPF is assumed complete based on three specimens with relatively unossified skulls (LSUMZ). Timing unknown.

FCJ     ♂ = ♀. Like adult, with only slightly more loosely textured plumage and relatively dull ss covs, and with an orange concealed central crown stripe. Skull is largely unossified.

(LSUMZ 173097.)     (LSUMZ 173097.)

FPF     ♂ ≠ ♀. With a mix of orange and yellow in the central crown stripe. Look for relatively fresh pp being replaced by adult-like pp. Skull is relatively unossified.

FAJ     ♂ ≠ ♀. With yellow central crown stripe and relatively bright ss covs. Skull is relatively ossified.

(LSUMZ 177299.)                      (LSUMZ 173098.)

UPB    ♂ ≠ ♀. Replacing relatively worn pp with adult-like feathers.

*Contopus virens* (monotypic)

Eastern Wood-Pewee • Piui-verdadeiro

Wing    74.0–88.0 mm (80.8 ± 5.5 mm; n = 5)
      ♂ 78–90 mm (n = 100; Pyle 1997)
      ♀ 75–86 mm (n = 100; Pyle 1997)

Tail     59.0–64.0 mm (61.5 ± 2.4 mm; n = 4)
      ♂ 61–70 mm (n = 53; Pyle 1997)
      ♀ 55–69 mm (n = 50; Pyle 1997)

Similar species: Not likely to be confused with local tyrant flycatchers by size and color.

Skull: Always unossified through 15 Oct, but some FCF and DCB can retain unossified windows >2 mm wide at the rear of the skull (Pyle 1997).

Brood patch: Does not develop BPs at the BDFFP.

Sex: ♂ = ♀.

Molt: Group 3 (or sometimes Group 4?)/Complex Basic Strategy; FPF incomplete-complete(?), DPBs complete (Pyle 1997, 1998). The FPF replaces all feathers except for the pp covs, although Burton (2002) suggested that one to all pp covs can also be replaced. FPF occurs on the winter grounds, but DPB can start on the breeding or wintering grounds and last through Mar (Pyle 1997).

Notes: Boreal migrant to be expected between Sep and Apr. All captures at the BDFFP have occurred Dec–Mar.

FCJ      Adult-like, but with a brownish wash to upperparts and buff wing bars. Can still be in FCJ when arriving on wintering grounds (Pyle 1997). Skull is unossified.

FPF      Replacing adult-like juvenile body or flight feathers with adult-like formative body or flight feathers. Skull is unossified, but usually ossifies before FPF completes.

FCF      Molt limits within the pp covs (Pyle 1997). Skull probably usually ossified.

FAJ      Without molt limits among ss or pp covs. Wing bars tipped pale grayish or olive and are otherwise whitish. Skull is ossified or with small windows (Pyle 1997). It is possible that all FPF result in the retention of one or more pp covs, thus an adult-like plumage without molt limits would be DCB.

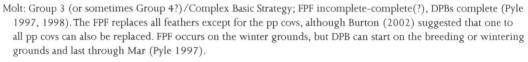

SPB      Mix of juvenile and adult-like formative pp covs being replaced by adult-like pp covs. Skull is ossified.

UPB      Difficult to distinguish from FPF (use UPU), except when skull is ossified early in the molt sequence. It is possible that all FPF result in the retention of one or more pp covs, thus an adult-like plumage replacing an adult-like plumage would be DPB.

*Tyrannus melancholicus melancholicus*

Tropical Kingbird • Suiriri

| | |
|---|---|
| Wing | 100.5–111.0 mm (107.6 ± 4.8 mm; n = 4) |
| | *T. m. melancholicus*: 116–122 mm (n = 2, Guyana; Chubb et al. 1916) |
| | *T. m. despotes*: 102–111 mm (n = 5, Surinam; Junge and Mees 1961) |
| | *T. m. chloronotus*: 107–116 mm (n = 7, Trinidad and Tobago; Junge and Mees 1961) |
| Tail | 88.0–93.0 mm (91.3 ± 2.2 mm; n = 4) |
| | *T. m. melancholicus*: 83 mm (n = 1, Guyana; Chubb et al. 1916) |
| | *T. m. despotes*: 81–89 mm (n = 5, Surinam; Junge and Mees 1961) |
| | *T. m. chloronotus*: 87–92 mm (n = 3, Trinidad and Tobago; Junge and Mees 1961) |
| Mass | 37.0–50.0 g (41.9 ± 5.7 g; n = 4) |
| | *T. m. chloronotus*: 31.5–39 g (n = 7, Trinidad and Tobago; Junge and Mees 1961) |
| Bill | 16.1–16.8 mm (16.5 ± 0.4 mm; n = 2) |
| Tarsus | 19.8–20.5 mm (20.2 ± 0.4 mm; n = 2) |

(Courtesy of Aída Rodrigues.)

Similar species: Most similar to *T. albogularis*, an austral migrant in the region that has not yet been recorded at the BDFFP, but should be expected. *T. albogularis* is paler-throated (nearly white, not gray), lacks olive on the chest, has a more conspicuous dusky mask, and is more olive on the back. Young *T. melancholicus* is most similar to this.

Skull: Probably completes (n = 2) but timing unknown (Pyle 1997).

Brood patch: ♀♀ incubate and develop BPs (Mobley 2014). Timing unknown but probably breeds at least during the dry season.

Sex: In each age class, ♂♂ have a more extensive concealed yellow or red crown stripe than ♀♀. ♂♂ also have more pointed p10, and notches on p6–p10 are ≥8 mm from the tip compared to ♀♀, at least in *T. m. occidentalis* (Pyle 1997).

Molt: Group 8(?)/Complex Alternate Strategy; FPF likely incomplete and eccentric, DPBs complete, PAs possibly limited-partial (Pyle 1997, 1998). The extent of the FPF probably varies by region and subspecies, and may be more extensive in resident subspecies than in migratory subspecies. Based on one capture at the BDFFP (fide A. Rodrigues), the FPF may involve some to all flight feathers, all rects, and none to several pp covs. In semimigratory *T. m. occidentalis*, the FPF is incomplete and eccentric, replacing 3–10 gr covs, 1–6 inner ss, 1–6 outer pp, and 1–2 central rects. The FPA includes 0–6 inner gr covs, 4–7 inner ss, outermost 2–5 pp, and 2 to all rects, but the DPA is less extensive and often partial, only rarely including 1–2 terts (Pyle 1997, 1998). In *T. m. satrapa*, the FPF and PAs are probably less extensive (Pyle 1997). See Pyle (1997) for more details about aging and sexing *T. m. occidentalis* and *T. m. satrapa*.

| | |
|---|---|
| FCJ | ♂ = ♀. Whiter throated and has less olive on chest than in subsequent plumages. Outer pp shape probably not useful for determining sex. Rects more narrow and tapered than in subsequent plumages. Skull is unossified (Pyle 1997). |
| FPF | ♂ ≈ ♀. Replacing adult-like juvenile body or flight feathers with adult-like formative body or flight feathers. Sex can be distinguished once p8 or p9 is replaced. Skull is unossified, but may ossify before FPF completes. |

♂ ≠ ♀. Look for molt limits among the flight feathers, but beware of incomplete PA (particularly FPA) and DPB molts. Concealed red or yellow crown stripe possibly minimal or absent (Pyle 1997). Shape of outer pp probably useful for sexing only when outer pp have been replaced in FPF. Skull is probably mostly or entirely complete.

p1 and all pp covs appear retained.

FPA/FCA ♂ ≠ ♀. Possibly more extensive than DPA as in *T. m. occidentalis* (Pyle 1997), but may only involve body feathers and/or ss covs. Skull is ossified.

SPB ♂ ≠ ♀. Replacing dull brownish juvenile pp covs and a mix of juvenile and adult-like flight feathers with adult-like feathers. Skull is ossified.

DCB ♂ ≠ ♀. Probably usually without molt limits among the flight feathers, but perhaps sometimes with retained adult-like inner ss. Skull is ossified.

DPB ♂ ≠ ♀. Replacing grayish adult-like pp covs and adult-like flight feathers with adult-like feathers. Skull is ossified.

DPA/DCA ♂ ≠ ♀. Perhaps less extensive than FPA as in *T. m. occidentalis*, and may only involve contour feathers and/or ss covs (Pyle 1997). Skull is ossified.

## *Rhytipterna simplex frederici*

Grayish Mourner • Vissiá

Band Size: F

# Individuals Captured: 40

Wing        91.0–101.0 mm (95.2 ± 2.6 mm; n = 29)

Tail         84.0–94.5 mm (87.8 ± 3.0 mm; n = 24)

Mass       28.5–39.0 g (33.7 ± 2.9 g; n = 30)

Bill          12.5–13.5 mm (13.0 ± 0.7 mm; n = 2)

Tarsus     24.6–24.8 mm (24.7 ± 0.1 mm; n = 2)

Similar species: A smaller version of *Lipaugus vociferans*. Lacks pale covert tips of *Laniocera hypopyrra*.

Skull: Completes (n = 8), probably during FPF or perhaps sometimes early in the FCF.

Brood patch: Unknown if both sexes incubate. The only three BPs were observed in Aug, but no other information is available.

Sex: ♂ and ♀ similar, but ♀ thought to have more fulvous edging to flight feathers and ss covs (Scholes 2004).

Molt: Group 3/Complex Basic Strategy; FPF incomplete, DPBs complete. Like closely related *Myiarchus*, replaces all less, med, and gr covs, remiges, and rects, but not pp covs during FPF. This understory forest species probably lacks PA molts unlike the *Myiarchus*. Never observed in wing molt from Jun through Nov, but surprisingly it has never been captured outside this period.

FCJ    Unknown, but ss covs probably dull brownish-gray. Iris dull brown. Skull is unossified.

FPF    Replacing adult-like juvenile body or flight feathers with adult-like formative body and flight feathers, but pp covs are not replaced with pp. Iris is dull brown, but may become brighter chestnut-brown. Skull is unossified, but may ossify before FPF completes.

FCF    Dull brownish retained pp covs contrasting against gray replaced gr covs, ss, and pp. Iris can be dull brown, but probably brightens with age. Skull is probably usually ossified, but may occasionally be unossified.

SPB    Possible to determine with substantial differences in color and wear between retained pp covs and retained pp. Iris is bright chestnut brown. Skull is ossified.

DCB    Pp covs gray, not contrasting against ss covs, pp, and ss. Iris bright chestnut-brown. Skull is ossified.

DPB    Distinguishable from SPB by a lack of difference in color and wear between retained pp covs and retained pp. Iris is bright chestnut brown. Skull is ossified.

## *Myiarchus tuberculifer tuberculifer*

Band Size: D (CEMAVE 2013)

Dusky-capped Flycatcher • Maria-cavaleira-pequena      # Individuals Captured: 5

Wing    M. t. *tuberculifer*: 75–84 mm (81.0 ± 2.5 mm; n = 11, Trinidad and Tobago; Junge and Mees 1961)
M. t. *nigriceps*: 73–81 mm (Chubb 1916)
M. t. *connectens* (♀): 75 mm (Dickey and van Rossem 1938)
M. t. *querulus*: 72.0–86.5 mm (79.8 ± 3.0 mm; n = 23; Guallar et al. 2009)

Tail    M. t. *tuberculifer*: 69–76 mm (72.9 ± 2.4 mm; n = 10, Trinidad and Tobago; Junge and Mees 1961)
M. t. *nigriceps*: 74 mm (Chubb et al. 1916)
M. t. *querulus*: 73.0–81.0 mm (77.5 ± 2.5 mm; Guallar et al. 2009)

Mass    M. t. *tuberculifer*: 15.0–21.0 g (18.5 ± 1.8 g; n = 11, Trinidad and Tobago; Junge and Mees 1961)
M. t. *tyrannulus*: 19.4 g (Hilty 2003)
M. t. *querulus*: 14.0–22.5 g (17.6 ± 1.5 g; n = 26; Guallar et al. 2009)

Tarsus    M. t. *tuberculifer*: 18–20 mm (18.9 ± 0.8 mm; n = 11, Trinidad and Tobago; Junge and Mees 1961)

Similar species: Averages smaller than *M. ferox* (wing <85 mm may be a reasonable cutoff; Chubb et al. 1916). Bill length may be best clue with *M. ferox* being larger-billed. The tail of *M. tuberculifer* may also be shorter than its wing, whereas *M. ferox* has a tail nearly equal in length to the wing. In *M. t. tuberculifer*, the wing:tail ratio ranges from 1.05 to 1.15 (1.11 ± 0.03, n = 10, Trinidad and Tobago; Junge and Mees 1961). It has less extensive (in juveniles) or no (in postjuveniles) rufous edging on rectrices, primaries, and secondaries than *T. tyrannulus*. *M. swainsoni*, an austral migrant, has also been recently captured north of Manaus near the BDFFP (fide Lindsay Wieland) and also has little to no rufous edging in the remiges and rects, but differs by having a reddish bill.

Skull: Likely completes during the FCF (Pyle 1997).

Brood patch: Can occur in both sexes (Pyle 1997). One BP observed in Aug.

Sex: ♂ (>) ♀.

Molt: Group 8(?)/Complex Alternate Strategy(?); FPF incomplete, DPBs complete, PAs limited(?)-partial(?) as in *M. t. olivascens* and *M. t. querulus* (Pyle 1997, Guallar et al. 2009). FPF may include all flight feathers and ss covs, but not pp covs. Occasionally p1–p3 and/or s1–s3 may be retained as well (Pyle 1997). It seems unlikely that nonmigratory *Myiarchus* have PA molts, but they apparently occur in near-tropical populations (Pyle 1997, Guallar et al. 2009; see also *M. ferox* species account). PAs in subtropical *M. tuberculifer* include a variable number of ss covs, and 0–2 terts (Pyle 1997), and may be similar to less extensive in Amazonian populations. Timing not known.

| | |
|---|---|
| FCJ | Has cinnamon wing-feather edging and more rufous in the rects (although probably difference less pronounced in "southern" subspecies). Outer rects more narrow than in subsequent plumages (Pyle 1997). Skull is unossified. |
| FPF | Replacing adult-like juvenile body or flight feathers with adult-like formative body or flight feathers, but not the pp covs. Skull is unossified, but may ossify before FPF completes. |
| FCF | Look for retained pp covs, which would be faded brown or tinged rufous and paler than the gr covs and bases of pp. Skull may be ossified or nearly ossified. |
| FPA/FCA | Pp covs like FCF, but with molt limits within the med covs, gr covs, and/or terts. |
| SPB | Replacing dull reddish-brown juvenile pp covs and adult-like body and flight feathers with adult-like feathers. Skull is ossified. |
| DPB | Replacing reddish adult-like pp covs and adult-like body and flight feathers with adult-like feathers. Skull is ossified. |
| DCB | Pp covs gray, not contrasting with gray less, med, gr covs, pp, and ss. Skull is ossified. |
| DPA/DCA | Pp covs like DCB, but with molt limits within the med covs, gr covs, and/or terts. |

## *Myiarchus swainsoni swainsoni/ferocior*

Band Size: D, E (CEMAVE 2013)

Swainson's Flycatcher • Irré

# Individuals Captured: 1

*M. s. swainsoni?*

(Courtesy of Lindsay Wieland.)

| | |
|---|---|
| Wing | 88.0 mm (n = 1)<br>*M. s. swainsoni*: 84–95 mm (88.4 ± 4.0 mm; n = 7; Junge and Mees 1961) |
| Tail | 74.0 mm (n = 1)<br>*M. s. swainsoni*: 75–86 mm (79.3 ± 3.8 mm; n = 7; Junge and Mees 1961) |
| Mass | 23.0 g (n = 1)<br>*M. s. swainsoni*: 18–26 g (22.8 ± 2.7 g; n = 7; Junge and Mees 1961)<br>*M. s. swainsoni*: 21.3–28.9 g (Joseph 2004)<br>*M. s. ferocior*: 34.6–37.6 g (Joseph 2004) |
| Bill | 11.7 mm (n = 1) |
| Tarsus | 24.3 mm (n = 1)<br>*M. s. swainsoni*: 19–20 mm (19.6 ± 0.5 mm; n = 7; Junge and Mees 1961) |

Similar species: The only *Myiarchus* at the BDFFP with extensive reddish at the base of the bill. Lacks rufous edging to tail, like *M. ferox* and *M. tuberculifer*, but averages darker above than these two. The tail should average shorter than in *M. ferox*. Wing:tail ratio ranges from 1.09–1.15 (1.12 ± 0.02; n = 7; Junge and Mees 1961).

Skull: Unknown, but probably completes, although perhaps occasionally retaining small (<2 mm) windows.

Brood patch: Does not occur at the BDFFP because it is an austral migrant.

Sex: ♂ (>) ♀.

Molt: Group 8(?)/Complex Alternate Strategy(?); FPF incomplete, DPBs complete, PA limited(?)-partial(?) as in other *Myiarchus*. The FPF likely includes all less, med, and gr covs, remiges, and rects, but not pp covs as in other *Myiarchus*. PA likely includes some ss covs, and perhaps 0–3 terts. Probably at least initiates, and maybe completes FPF and DPB molts on the nonbreeding grounds.

Notes: An austral migrant to the region. *M. s. swainsoni* appears genetically distinct from other subspecies (including *M. s. ferocior*) and is more closely related to *M. tuberculifer*, thus may represent a distinct species (Joseph et al. 2003, 2004; Joseph and Wilke 2004). Both *M. s. swainsoni* and *M. s. ferocior* seem like possible migrants to the central Amazon.

| | |
|---|---|
| FCJ | Does not occur at the BDFFP. |
| FPF | Replacing adult-like juvenile body or flight feathers with adult-like formative body or flight feathers, but not the pp covs. Skull is unossified, but may ossify before FPF completes. |
| FCF | Look for retained pp covs, which would be faded brown or tinged rufous and paler than the gr covs and bases of pp. Skull may be ossified or nearly ossified. |
| FPA/FCA | Pp covs like FCF, but with molt limits among the ss covs and/or the terts. |
| SPB | Replacing dull reddish-brown juvenile pp covs and adult-like body and flight feathers with adult-like feathers. Skull is ossified. |
| DPB | Replacing reddish adult-like pp covs and adult-like body and flight feathers with adult-like feathers. Skull is ossified. |
| DCB | Pp covs gray, not contrasting with gray less, med, gr covs, pp, and ss. Skull is ossified. |
| DPA/DCA | Pp covs like DCB, but with molt limits among the ss covs and/or the terts. |

## *Myiarchus ferox ferox*

Band Size: F, E

Short-crested Flycatcher • Maria-cavaleira

# Individuals Captured: 10

| | |
|---|---|
| Wing | M. f. *ferox*: 87–92 mm (Guyana; Chubb et al. 1916)<br>M. f. *ferox*: 85–90 mm (Surinam; Junge and Mees 1961) |
| Tail | M. f. *ferox*: 87 mm (♀, Guyana; Chubb et al. 1916)<br>M. f. *ferox*: 80–88 mm (Surinam; Junge and Mees 1961) |
| Mass | 21–33 g (Joseph 2004) |
| Tarsus | M. f. *ferox*: 21–23 mm (n = 12, Surinam; Junge and Mees 1961) |

Similar species: Averages larger than *M. tuberculifer*, but the degree of overlap at the BDFFP is unknown (wing >85 mm may be a reasonable cutoff; Chubb et al. 1916). Bill length may be best clue with *M. ferox* being larger-billed. The tail of *M. tuberculifer* may also be shorter than its wing, while *M. ferox* has a tail nearly equal in length to the wing (ratio about 1.00–1.06, n = 5; Chubb et al. 1916, Junge and Mees 1961). Lacks rufous edging on rects, pp, and ss in contrast to *T. tyrannulus*.

Skull: Unknown, but probably completes, although perhaps occasionally retaining small (<2 mm) windows.

Brood patch: Perhaps occurs in both sexes as in *M. tuberculifer*. Timing unknown.

Sex: ♂ (>) ♀.

Molt: Group 8/Complex Alternate Strategy; FPF incomplete, DPBs complete, PA partial. During FPF likely replaces all less, med, and gr covs, remiges, and rects, but not pp covs as in other *Myiarchus*. Interestingly, a bird captured with apparent molt limits resulting from a partial PA molt suggests that PA widespread in this genus outside of Amazonia (Pyle 1997, Guallar et al. 2009) can also occur in nonmigratory equatorial species. Timing unknown.

| | |
|---|---|
| FCJ | Unknown, but pp covs probably dull brownish-gray. Skull is unossified. |
| FPF | Replacing adult-like juvenile body or flight feathers with adult-like formative body or flight feathers, but not the pp covs. Skull is unossified, but may ossify before FPF completes. |
| FCF | Look for retained pp covs, which would be faded brown or tinged rufous and paler than the gr covs and bases of pp. Skull may be ossified or nearly ossified. |
| FPA/FCA | Probably like FCF, but occasionally or regularly with some replaced ss covs and/or terts. Skull is probably ossified, or sometimes with small windows. |
| SPB | Replacing dull reddish-brown juvenile pp covs and adult-like body and flight feathers with adult-like feathers. Skull is ossified or rarely with small windows. |
| DPB | Replacing reddish adult-like pp covs and adult-like body and flight feathers with adult-like feathers. Skull is ossified. |
| DCB | Flight feathers uniformly adult with little or no rufous edging in ss covs, ss, uppertail covs, and central rects. Skull is ossified. |
| DPA/DCA | Like DCB, but perhaps some terts and/or covs contrastingly fresh. In the bird below, it appears that less covs, most med covs, a few inner and outer gr covs, and s9 have been replaced during a DPA. Skull is ossified. |

*Myiarchus tyrannulus tyrannulus*

Band Size: E, F, G (CEMAVE 2013)

Brown-crested Flycatcher • Maria-cavaleira-de-rabo-enferrujado

# Individuals Captured: 1

Wing      95.5 mm (n = 1)
M. t. *tyrannulus*: 85–94 mm (Surinam; Junge and Mees 1961); 90 mm (Guyana; Chubb et al. 1916)
M. t. *brachyurus*: 88–101 mm (n = 77; Lanyon 1960)
M. t. *magister*: 98.5–112.5 mm (105.8 ± 4.6 mm; n = 11; Guallar et al. 2009)

Tail      Unknown at the BDFFP
M. t. *tyrannulus*: 73–79 mm (Surinam; Junge and Mees 1961); 84 mm (Guyana; Chubb et al. 1916)
M. t. *brachyurus*: 74–89 mm (n = 77; Lanyon 1960)
M. t. *magister*: 93–108 mm (101.5 ± 4.9 mm; n = 8; Guallar et al. 2009)

Mass      27.0 g (n = 1)
M. t. *tyrannulus*: 27–33 g (Surinam; Junge and Mees 1961)
M. t. *magister*: 33.7–47.2 g (41.6 ± 4.1 g; n = 11; Guallar et al. 2009)

Tarsus      Unknown at the BDFFP
M. t. *tyrannulus*: 21–22 mm (Surinam; Junge and Mees 1961)

Similar species: *T. tyrannulus* is rare at the BDFFP, but probably regular in open areas in second growth. It is the largest *Myiarchus* at the BDFFP and the only one with extensive rufous edging to the rectrices.

Skull: Unknown, but probably completes, although perhaps occasionally retaining small (<2 mm) windows.

Brood patch: Perhaps occurs in both sexes as in M. *tuberculifer*. Timing unknown.

Sex: ♂ (>) ♀.

Molt: Group 8(?)/Complex Alternate Strategy; FPF incomplete, DPBs complete, PA limited(?)-partial(?) as in M. t. *cooperi* and M. t. *magister* (Pyle 1997, Guallar et al. 2009). During FPF likely replaces all less, med, and gr covs, remiges, and rects, but not pp covs as in other *Myiarchus* (Pyle 1997, Guallar et al. 2009). PA molts in subtropical M. *tyrannulus* include less and med covs, 0–5 gr covs, and 0–3 terts, averaging more extensive in FPA than DPA (Pyle 1997), and may be similar or averaging less extensive in Amazonian populations.

Notes: M. t. *cooperi* and M. t. *magister*, boreal migrants in northern Central America and southern North America, probably do not reach the central Amazon, and measurements of those distinct subspecies are substantially larger than resident races of southern Central America and South America.

FCJ      Unknown, but pp covs probably dull brownish-gray. Skull is unossified.

FPF      Replacing adult-like juvenile body or flight feathers with adult-like formative body or flight feathers, but not the pp covs. Skull is unossified, but may ossify before FPF completes.

FCF      Look for retained pp covs, which would be faded brown or tinged rufous and paler than the gr covs and bases of pp. Skull may be ossified or nearly ossified.

FPA/FCA      Probably like FCF, but molt limits among the gr covs and/or terts. Skull is probably ossified, or sometimes with small windows.

SPB      Replacing dull reddish-brown juvenile pp covs and adult-like body and flight feathers with adult-like feathers. Skull is ossified or rarely with small windows.

DPB      Replacing reddish adult-like pp covs and adult-like body and flight feathers with adult-like feathers. Skull is ossified.

DCB      Flight feathers uniformly adult with little or no rufous edging in ss covs, ss, uppertail covs, and central rects. Skull is ossified.

DPA/DCA      Like DCB, but perhaps some terts and/or ss covs contrastingly fresh. Skull is ossified.

*Ramphotrigon ruficauda* (monotypic)

Band Size: E

Rufous-tailed Flatbill • Bico-chato-de-rabo-vermelho

# Individuals Captured: 25

Wing      68.0–79.0 mm (73.5 ± 2.7 mm; n = 20)

Tail        59.0–69.0 mm (64.5 ± 3.1 mm; n = 17)

Mass      16.5–21.2 g (18.9 ± 1.4 g; n = 22)

Bill         9.3 mm (n = 1)

Tarsus     17.7 mm (n = 1)

Similar species: *Rhynchocyclus olivaceus* lacks rufous in wings and tail.

Skull: Completely ossifies (n = 7; Parker 1984), but timing unknown.

Brood patch: Unknown if both sexes incubate. Evidence of breeding at least in dry season (Parker 1984, Walther and de Juana 2004, Gomes and Barreiros 2011).

Sex: ♂ = ♀.

Molt: Group 3(?)/Complex Basic Strategy(?); FPF incomplete(?), DPBs complete. Apparently closely related to *Myiarchus* and *Rhytipterna*, thus may replace all body and flight feathers except the pp covs during the FPF. Unknown whether it has a PA molt, but not likely as in *Rhytipterna*. Timing unknown, but four birds have been observed in molt: Sep, Oct, Nov, and Apr, which may be representative of the period of molt.

FCJ        Undescribed. Skull is unossified.

FPF        May replace body and flight feathers, but not pp covs, as in *Myiarchus* and *Rhytipterna*. Skull may be either ossified or unossified.

FCF        Perhaps with all body and flight feathers replaced except for the pp covs, which would appear relatively dull and brownish. Skull may be either ossified or unossified.

SPB        Replacing retained juvenile pp covs and adult-like pp and ss with adult-like feathers. Skull is ossified.

DCB        With replace and relatively glossy pp covs. Skull is ossified.

DPB        Replacing adult-like pp covs, pp, and ss with adult-like feathers. Skull is ossified.

*Attila spadiceus spadiceus*

Bright-rumped Attila • Capitão-de-saíra-amarelo

Wing    76.0–88.0 mm
        (80.6 ± 2.7 mm; n = 48)

Tail    57.0–68.0 mm
        (61.6 ± 2.4 mm; n = 46)

Mass    27.8–39.3 g
        (33.4 ± 2.9 g; n = 56)

Bill    12.8–15.0 mm
        (14.0 ± 0.8 mm; n = 9)

Tarsus  24.1–27.2 mm
        (25.9 ± 0.9 mm; n = 8)

Similar species: Lots of variation in plumage and iris color, but always with bright contrasting yellow rump is unique along with large size, heavy hooked bill, and large head.

Skull: Completes (n = 27) during FCF or toward end of FPF.

Brood patch: Only ♀♀ incubate and develop BPs (Walther and de Juana 2004). BPs have been observed from Jun to Oct, but we have no data from Jan to May.

Sex: ♂ = ♀.

Molt: Group 3/Complex Basic Strategy; FPF incomplete and eccentric (always?), DPBs complete. During the FPF the outer 3–8 (or sometimes fewer?) pp, inner 3–4 ss, all ss covs, all alulas (always?), and all rects are replaced, but not the pp covs (or occasionally several outer pp covs?). Molts mainly from Oct–Apr (occasionally to Jul?).

Notes: Variation in plumage and iris color is extensive and may occur primarily or only in ♀♀ (Dickey and van Rossem 1938). Three morphs evident: olive morph, gray morph, and rufous morph (Walther and de Juana 2004).

FCJ    Like adult, but with brown eyes (Walther and de Juana 2004). Skull is unossified.

FPF    Replacing adult-like juvenile body or outer pp while retaining inner browner adult-juvenile pp. Compare shape of incoming rects with older rects. Skull is unossified, but may ossify before FPF completes.

| FCF | Look for molt limits among the pp and/or pp covs in an eccentric pattern. Pp covs noticeably duller than pp. Skull is probably usually ossified. |

All ss covs and alulas replaced, p4–p10 replaced, p1–p3 and s1–s6 retained juvenile, s7–s9 replaced.

| SPB | Possible to determine with substantial differences in color and wear between retained pp covs and retained pp. Incoming rects same shape as older rects, as in DPB. Skull is ossified. |
| DCB | Without molt limits among the pp or pp covs. Pp covs glossy and similar in color to pp. Skull is ossified. |

| DPB | Possible to distinguish from SPB with lack of difference in color and wear between retained pp covs and retained pp. Incoming rects same shape as older rects, as in SPB. Skull is ossified. |

---

## LITERATURE CITED

Bates, J. [online] 2004. Olivaceous Flatbill (*Rhynchocyclus olivaceus*). In J. del Hoyo, A. Elliot, J. Sargatal, D. A. Christie, and E. de Juana, editors. Handbook of the Birds of the World Alive. Lynx Edicions, Barcelona, Spain. <http://www.hbw.com>.

Burton, K. M. 2002. Primary-covert replacement in the Eastern Wood-Pewee. North American Bird Bander 27:12–14.

Caballero, I. [online]. 2004. Yellow-margined Flycatcher (*Tolmomyias assimilis*). In J. del Hoyo, A. Elliott, J. Sargatal, D. A. Christie, and E. de Juana (editors), Handbook of the birds of the world alive. Lynx Edicions, Barcelona, Spain. <http://www.hbw.com>.

CEMAVE. 2013. Lista das espécies de aves brasileiras com tamanhos de anilha recomendados. Centro Nacional de Pesquisa e Conservação de Aves Silvestres, Cabedelo, Brasil (in Portuguese).

Chubb, C., H. Grönvold, F. V. McConnell, H. F. Milne, P. Slud, and Bale & Danielsson. 1916. The birds of British Guiana: based on the collection of Frederick Vavasour McConnell. Bernard Quaritch, London, UK.

Dickey, D. R., and A. J. Van Rossem. 1938. The birds of El Salvador. Field Museum of Natural History, Zoological Series, Chicago, IL.

Farnsworth, A., and D. Lebbin. [online]. 2004a. Ruddy-tailed Flycatcher (*Terenotriccus erythrurus*). In J. del Hoyo, A. Elliott, J. Sargatal, D. A. Christie, and E. de Juana (editors), Handbook of the birds of the world alive. Lynx Edicions, Barcelona, Spain. <http://www.hbw.com>.

Farnsworth, A., and D. Lebbin. [online]. 2004b. Cinnamon Tyrant (*Neopipo cinnamomea*). In J. del Hoyo, A. Elliott, J. Sargatal, D. A. Christie, and E. de Juana (editors), Handbook of the birds of the world alive. Lynx Edicions, Barcelona, Spain. <http://www.hbw.com>.

Farnsworth, A., and D. Lebbin. [online]. 2012. Royal Flycatcher (*Onychorynchus coronatus*). In J. del Hoyo, A. Elliott, J. Sargatal, D. A. Christie, and E. de Juana (editors), Handbook of the birds of the world alive. Lynx Edicions, Barcelona, Spain. <http://www.hbw.com>.

Gomes, F. B. R., and M. H. M. Barreiros. 2011. Observações sobre a reprodução, descrição do ninho e filote do bico-chato-de-rabo-vermelho *Ramphotrigon ruficauda* (Passeriformes: Tyrannidae) no Brasil. Atualidades Ornitológicas 160:9–10 (in Portuguese).

Guallar, S., E. Santana, S. Contreras, H. Verdugo, and A. Gallés. 2009. Paseriformes del Occidente de México: Morfometría, datación y sexado. Monografies del Museu de Ciències Naturals 5 (in Portuguese).

Hellmayr, C. E. 1929. Catalogue of Birds of the Americas and the Adjacent Islands, Part VI: Oxyruncidae, Pipridae, Cotingidae, Rupicolidae, Phytotomidae. Zoological Series XIII, Publication 266. Field Museum of Natural History, Chicago, IL.

Hilty, S. L. 2003. Birds of Venezuela (2nd ed.). Princeton University Press, Princeton, NJ.

Howell, S. N. G. 2010. Molt in North American birds. Houghton Mifflin Harcourt, Boston, MA.

Joseph, L. [online]. 2004. Swainson's Flycatcher (*Myiarchus swainsoni*). In J. del Hoyo, A. Elliott, J. Sargatal, D. A. Christie, and E. de Juana (editors), Handbook of the birds of the world alive. Lynx Edicions, Barcelona, Spain. <http://www.hbw.com>.

Joseph, L., and T. Wilke. 2004. When DNA throws a spanner in the taxonomic works: testing for monophyly in the Dusky-capped Flycatcher, *Myiarchus tuberculifer*, and its South American subspecies, M. t. *atriceps*. Emu 104:197–204.

Joseph, L., T. Wilke, and D. Alpers. 2003. Independent evolution of migration on the South American landscape in a long-distance temperate-tropical migratory bird, Swainson's Flycatcher (*Myiarchus swainsoni*). Journal of Biogeography 30:925–937.

Joseph, L., T. Wilke, E. Bermingham, D. Alpers, and R. E. Ricklefs. 2004. Towards a phylogenetic framework for the evolution of shakes, rattles, and rolls in *Myiarchus* tyrant-flycatchers (Aves: Passeriformes: Tyrannidae). Molecular Phylogenetics and Evolution 31:139–152.

Junge, G. C. A., and G. F. Mees. 1961. The avifauna of Trinidad and Tobago. E. J. Brill, Leider, Netherlands.

Lanyon, W. E. 1960. The Middle American populations of the Crested Flycatcher *Myiarchus tyrannulus*. Condor 62:341–350.

Mobley, J. [online]. 2014. Tropical Kingbird (*Tyrannus melancholicus*). In J. del Hoyo, A. Elliott, J. Sargatal, D. A. Christie, and E. de Juana (editors), Handbook of the birds of the world alive. Lynx Edicions, Barcelona, Spain. <http://www.hbw.com>.

Parker, T. A. 1984. Notes on the behavior of *Ramphotrigon* flycatchers. Auk 101:186–188.

Pyle, P. 1997. Identification guide to North American birds, Part I. Slate Creek Press, Bolinas, CA.

Pyle, P. 1998. Eccentric first-year molts in certain Tyrannid flycatchers. Western Birds 29:29–35.

Pyle, P., A. Engilis, and D. A. Kelt. 2015. Manual for ageing and sexing birds of Bosque Fray Jorge National Park and Northcentral Chile, with notes on range and breeding seasonality. Special Publication of the Occasional Papers of the Museum of Natural Science, Louisiana State University, Baton Rouge, LA.

Pyle, P., A. McAndrews, P. Veléz, R. L. Wilkerson, R. B. Sigel, and D. F. DeSante. 2004. Molt patterns and age and sex determination of selected southeastern Cuban landbirds. Journal of Field Ornithology 75:136–145.

Scholes, E. [online] 2004. Grayish Mourner (*Rhytipterna simplex*). In J. del Hoyo, A. Elliot, J. Sargatal, D. A. Christie, and E. de Juana, editors. Handbook of the Birds of the World Alive. Lynx Edicions, Barcelona, Spain. <www.hbw.com>.

Skutch, A. F. 1960. Life histories of Central American Birds, Part II: Families Vireonidae, Sylviidae, Turdidae, Troglodytidae, Paridae, Corvidae, Hirundinidae, and Tyrannidae. Pacific Coast Avifauna 34.

Walther, B., and E. de Juana. [online]. 2004. Rufous-tailed Flatbill (*Ramphotrigon ruficauda*). In J. del Hoyo, A. Elliott, J. Sargatal, D. A. Christie, and E. de Juana (editors), Handbook of the birds of the world alive. Lynx Edicions, Barcelona, Spain. <http://www.hbw.com>.

Willard, D. E., M. S. Foster, G. F. Barrowclough, R. W. Dickerman, P. F. Cannell, S. L. Coats, J. L. Cracraft, and J. P. O'Neill. The birds of Cerro de la Neblina, Territorio Federal Amazonas, Venezuela. Fieldiana, Zoology, New Series, No. 65. Field Museum of Natural History, Chicago, IL.

Willis, E. O., D. Wechsler, and Y. Oniki. 1978. On behavior and nesting of McConnell's Flycatcher (*Pipromorpha macconnelli*): does female rejection lead to male promiscuity? Auk 95:1–8.

Wolfe, J. D., and R. I. Frey. 2011. Primary covert replacement patterns in the Western Wood-Pewee (*Contopus sordidulus*). North American Bird Bander 36:113–115.

# CHAPTER TWENTY-EIGHT

# Cotingidae (Cotingas and Allies)

# Species in South America: 58

# Species recorded at BDFFP: 7

# Species captured at BDFFP: 2

Cotingas are a diverse family of mostly frugivorous, forest-dwelling birds in Central and South America. Many species within the family are quite colorful and sexually dichromatic; others, however, are quite plain where sexes look alike (Snow 2013). Among the Tyrannidae, the family limits have been controversial with species being moved in and out of the family as more morphological, behavioral, and genetic information is revealed (e.g., Ridgway 1907, Prum 1990, Ericson et al. 2006, Tello et al. 2009). Cotingas have 10 pp, 9 ss, and 12 rects.

Little is known about molt strategies among Cotingidae. They likely follow a Complex Basic Strategy in most or all cases, but the extent of the preformative molt remains undescribed for most genera and species, although it appears to be complete in at least some and partial in other species (based on few captures). Snow (1976) reviewed molt timing among many species and found that "within any local population the date of onset of molt may vary according to sex and age. In genera in which both sexes participate in nesting, males and females begin to molt at about the same time, or the males slightly in advance of the females. In genera with marked sexual dimorphism, in which only the female attends the nest, males may begin to molt well before females, at about the time that the latter begin egg-laying." Snow (1976) also noted that molt and breeding rarely overlapped in frugivorous species and, in general, the annual prebasic molt occurred when food resources were at their most abundant. Snow (1976) estimated the average time to complete molt within Cotingidae to be about 100 days. More study is needed to form a preliminary understanding of molt patterns among this neglected group of birds.

Breeding behavior is diverse across the family ranging from polygynous males forming classical to exploded leks in some species, to large family groups in others. Brood patches probably occur mainly among females, although Swallow-tailed Cotinga (*Phibalura flaviristris*) represents at least one exception (Snow 2013). Skull ossification may often complete, at least in small- to medium-sized species.

# SPECIES ACCOUNTS

*Phoenicircus carnifex* (monotypic)

Guianan Red Cotinga • Saurá

Band Size: H

# Individuals Captured: 13

| | | |
|---|---|---|
| Wing | ♂ | 92.0–108.0 mm (100.0 ± 11.3 mm; n = 2) |
| | ♀ | 98.0–107.0 mm (104.4 ± 2.8 mm; n = 8) |
| Tail | ♂ | 75.0–88.0 mm (82.7 ± 6.8 mm; n = 3) |
| | ♀ | 78.0–89.0 mm (82.9 ± 3.8 mm; n = 8) |
| Mass | ♂ | 81.0–120.0 g (98.7 ± 19.8 g; n = 3) |
| | ♀ | 75.4–130.0 g (93.3 ± 17.3 g; n = 8) |
| Bill | ♂ | 9.5 mm (n = 1) |
| | ♀ | 9.5 mm (n = 1) |
| Tarsus | ♂ | 33.2 mm (n = 1) |
| | ♀ | 31.6 mm (n = 1) |

Similar species: None.

Skull: Completes (n = 5) probably during the FCF.

Brood patch: Only develops in ♀♀ as in other lekking cotingas. Possibly a dry-season breeder with a possible BP seen in Aug and an old BP seen in Nov.

Sex: ♂♂ are unmistakable with a brilliant red cap, tail, and underparts contrasting against a dark brick red head, back, and wings. ♀♀ have a yellowish-olive back, nape, and throat with a more modest red cap, tail, and underparts. ♂♂ achieve distinctive plumage during SPB, but may be distinguishable during the FCF by being brighter red underneath. ♂♂ average smaller with shorter wings than ♀♀.

Molt: Based on an examination of the closely related *Phoenicircus nigricollis* specimens at LSUMZ, probably Group 3/Complex Basic Strategy; FPF partial, DPBs complete. FPF (n = 4, LSUMZ) includes body, less and med covs, inner 1–3 (or sometimes fewer or more?) gr covs, and 0–3 terts, but no alulas or flight feathers. Timing unknown, but probably follows breeding in late dry season and wet season.

FCJ    ♂ = ♀ Unknown; presumably ♀-like, but duller and probably with red tones reduced or absent on the back. Skull is unossified.

FPF    ♂ ≈ ♀ Replacing ♀-like body feathers with ♀-like feathers. Skull is unossified.

FCF    ♂ ≈ ♀ Both sexes ♀-like, but ♂♂ probably with brighter underparts. With molt limits among the gr covs and sometimes terts. The newer inner gr covs are notably longer than the retained outer covs. Skull is unossified or ossified.

Several inner gr covs replaced, outer gr covs and all flight feathers retained juvenile. (*C. nigricollis*, LSUMZ 92537.)

SPB    ♂ ≠ ♀ obviously replacing ♀-like body and flight feathers with black and red adult-like feathers. ♀♀ probably more difficult to distinguish from DPB, but differences in rect shape, ss length, amount of wear, and condition of pp covs may be useful. Skull is ossified.

DCB     ♂ ≠ ♀ Without molt limits among the gr covs or terts. All gr covs large and uniformly fresh. Beware of pseudolimit between gr covs and pp covs. Skull is ossified.

DPB     ♂ ≠ ♀ obviously replacing adult male-like black and red body and flight feathers with adult male-like feathers. Females probably more difficult to distinguish from SPB, but similarity in rect shape and ss length between new and old feathers, as well as amount of wear and condition of pp covs may be useful. Beware of pseudolimit between gr covs and pp covs. Skull is ossified.

## *Lipaugus vociferus* (monotypic)

Screaming Piha • Cricrió

Band Size: H, G

\# Individuals Captured: 40

Wing     107.0–125.0 mm (118.2 ± 4.7 mm; n = 24)

Tail     100.0–119.0 mm (105.9 ± 5.5 mm; n = 18)

Mass     64.4–85.0 g (73.2 ± 4.8 g; n = 33)

Bill     13.2–14.2 mm (13.8 ± 0.4 mm; n = 5)

Tarsus     24.2–26.4 mm (25.5 ± 1.0 mm; n = 5)

Similar species: From *Rhytipterna simplex* by larger size. From *Laniocera hypopyrra* by lacking orange tips to ss covs.

Skull: Completes (n = 11), probably during FPF or FCF.

Brood patch: Probably only in ♀♀ as in other lekking species. BPs observed in Oct and Nov, but probably breeds earlier into dry season.

Sex: ♂ = ♀.

Molt: Group 4/Complex Basic Strategy; FPF complete, DPBs complete. Probably regularly molts at least from Sep to Apr.

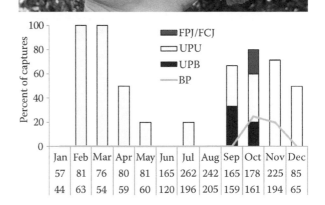

| | Jan | Feb | Mar | Apr | May | Jun | Jul | Aug | Sep | Oct | Nov | Dec |
|---|---|---|---|---|---|---|---|---|---|---|---|---|
| | 57 | 81 | 76 | 80 | 81 | 165 | 262 | 242 | 165 | 178 | 225 | 85 |
| | 44 | 63 | 54 | 59 | 60 | 120 | 196 | 205 | 159 | 161 | 194 | 65 |

Legend: FPJ/FCJ, UPU, UPB, BP

FCJ  Brownish or rusty tones to back, underparts, and wings. Skull is unossified.

FPF  Replacing brownish-gray juvenile body or flight feathers with bright gray adult-like body or flight feathers. Skull is unossified, but may ossify before FPF completes.

p4–p10 retained, p3 missing, p2 molting, p1 new, s1–s8 retained, s9 new.

FAJ  Gray overall with minimal or lacking rusty tones. Beware of pseudolimits among flight feathers with varying intensity of gray across feather tracts. Skull is ossified.

p5–p10 retained, p1–p4 new, s1–s6 and s9 retained, s7–s8 new, perhaps in a suspension and may be preferably called a DCB (left); all rects adult-like and retained from the same individual (right).

UPB  Replacing dull-gray adult-like body and flight feathers with bright gray adult-like feathers. Skull is ossified.

## LITERATURE CITED

Ericson, P. G. P., Z. Dario, J. I. Ohlson, U. S. Johansson, H. Alvarenga, and R. O. Prum. 2006. Higher-level phylogeny and morphological evolution of tyrant flycatchers, cotingas, manakins, and their allies (Aves: Tyrannida). Molecular Phylogenetics and Evolution 40:471–483.

Prum, R. O. 1990. A test of the monophyly of the manakins (Pipridae) and of the cotingas (Cotingidae) based on morphology. Occasional Papers Museum of Zoology, University of Michigan 723:1–44.

Ridgway, R. 1907. The birds of North and Middle America, part IV. Bulletin of the U.S. Natural History Museum, Washington, DC 1–973.

Snow, D. W. 1976. The relationship between climate and annual cycles in the Cotingidae. Ibis 118:366–401.

Snow, D. W. [online]. 2013. Family Cotingidae (Cotingas). In J. del Hoyo, A. Elliott, J. Sargatal, D. A. Christie, and E. de Juana (editors), Handbook of the birds of the world alive. Lynx Edicions, Barcelona, Spain. <http://www.hbw.com>.

Tello, J. G., R. G. Moylea, D. J. Marchesea, and J. Cracraft. 2009. Phylogeny and phylogenetic classification of the tyrant flycatchers, cotingas, manakins, and their allies (Aves: Tyrannides). Cladistics 25:429–467.

# CHAPTER TWENTY-NINE

# Pipridae (Manakins)

# Species in South America: 47

# Species recorded at BDFFP: 7

# Species captured at BDFFP: 7

The Pipridae, or manakins, are a group of New World suboscines closely allied with tyrant flycatchers (Tello et al. 2009). These small frugivores occupy all levels of the interior forest and tend to tolerate edges and fragmentation because of the positive effects on fruiting trees, especially Melastomataceae, which are favored by many manakins (Cramer et al. 2007, Uriatre et al. 2011). Manakins are often sexually dichromatic; males are brightly colored, whereas females are mostly green. Some species are sexually monochromatic; these tend to be mid-story to canopy-dwelling frugivores that are often variably yellow-green and flycatcher-like. Most manakins have a social structure in which males form leks, and parental care is performed solely by females. This social structure has implications for sex- and age-specific differences in bird movements and, therefore, mist-net capture rates. In general, adult males are most sedentary and are least frequently captured, whereas females are slightly more mobile as they wander between leks and raise young. Juveniles and subadults of both sexes wander widely in search of fruiting trees because they are not yet tied to a breeding location; they also probably develop and test their social skills during this time and investigate potential future leks in which to participate as breeding adults (Graves et al. 1983). Manakins have 10 pp, 9 ss, and 12 rects.

Molt in most, if not all, manakins follows a Complex Basic Strategy (Ryder and Durães 2005, Ryder and Wolfe 2009). The juvenile plumage apparently is held extremely briefly. In the most frequently captured species at the BDFFP and elsewhere, the preformative molt is typically partial resulting in molt limits among the greater coverts (Ryder and Durães 2005, Ryder and Wolfe 2009). In most species, the replaced greater coverts are a few millimeters longer and brighter green than the juvenile greater coverts. The second prebasic molt is complete, but, depending on the species, males may or may not acquire an adult plumage aspect. Not until the third prebasic molt in some species, or later in others, is it possible to separate males from females (Foster 1987, McDonald 1993, DuVal 2005, Doucet et al. 2007). Thus, a green bird without molt limits may either be a male in SCB or a female in DCB. Some green birds have a few male-like feathers among the green feathers, but these individuals could be either advanced immature males or old females (Graves 1981, Ryder and Durães 2005, Doucet et al. 2007). The extent and location of these feathers may provide clues of the sex and age, but may not be 100% reliable and are species-specific.

The skull ossifies in most (all?) species, but probably does not add much insight to aging birds because the preformative molt is often partial such that molt limit criteria are available as the skull ossifies. Only females incubate and develop brood patches, which are often less extensive compared to other passerines and usually include only the lower breast and belly, rather than proceeding up to nearly the furcular hollow (Figure 29.1). Although perhaps a little more difficult to detect, with practice, differences in skin texture between the upper and lower breast can be used to more easily determine receded brood patches compared to other families and species. Molt-breeding overlap is rare at the BDFFP probably because of highly predictable and separate molting and breeding seasons (Johnson et al. 2012). Breeding peaks during the middle to end of the dry season and molt occurs afterward and peaks during the early wet season.

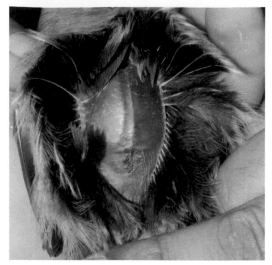

Figure 29.1 A typical brood patch seen in many Pipridae, which is often less vascularized and more restricted to the lower breast and belly compared to other passerines (see Chapter 2, Figure 2.1).

## SPECIES ACCOUNTS

*Piprites chloris chlorion*

Wing-barred Piprites • Papinho-amarelo

Wing    64.5–65.0 mm (64.8 ± 0.4 mm; n = 2)

Tail    43.0 mm (n = 2)

Mass    15.5–17.0 g (16.5 ± 0.8 g; n = 3)

Similar species: Combination of bold yellow wingbar, yellow underparts, yellow eyering, and yellow crown stripe are unique.

Skull: Probably does not ossify in adults (n = 5, LSUMZ).

Brood patch: Unknown if both sexes incubate. Timing unknown.

Sex: ♂ = ♀; both sexes have a yellow crown stripe.

Molt: Group 3/Complex Basic Strategy; FPF partial(?)-incomplete(?), DPBs complete. FPF may include all gr covs, 0–3 terts, and perhaps none to all rects. Timing unknown.

Band Size: E

# Individuals Captured: 3

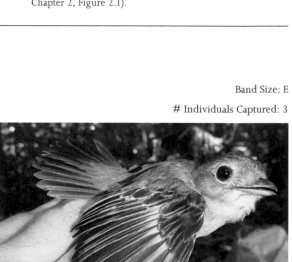

Notes: Taxonomic status is not well understood and is currently treated as *incertae sedis* in the SACC checklist (Remsen et al. 2016).

FCJ    Otherwise very similar to adult, but duller. Rects distinctly pointed (n = 1, LSUMZ). Skull is unossified.

FPF    Replacing adult-like juvenile body feathers, less, med and gr covs, and rects with adult-like ormative feathers. Look for differences in shape between retained juv rects and replaced formative rects. Skull is unossified, but might be expected to become relatively ossified before completion of the FPF.

FCF Look for molt limits between the gr covs and pp covs, among the terts, and/or among the rects. Skull is relatively ossified.

Outer 2 med covs retained, inner 1 gr cov replaced, outer gr covs retained, all alulas and flight feathers retained (left); central pair of rects retained (right).

SPB May be possible to distinguish from DPB when juv rects have been retained, but otherwise would be difficult to distinguish from DPB (use UPB). Retained pp covs may average duller and browner than retained pp covs in the DPB. Skull is relatively ossified.

DCB Without molt limits between the gr covs and pp covs and among the terts and rects. Rects rounded. Skull is relatively ossified.

DPB Replacing adult-like body and flight feathers with adult-like feathers. Retained pp covs may be brighter and more olive than in SPB. Skull is relatively ossified.

## *Neopelma chrysocephalum* (monotypic)

Saffron-crested Tyrant-Manakin • Fruxu

Band Size: Unknown

# Individuals Captured: 1

Wing    67.0 mm (n = 1)

Tail    48.5 mm (n = 1)

Mass    12.3 g (n = 1)
        14.4–16.3 g (Snow 2004)

Similar species: Drab smallish flycatcher-like bird with grayish crown, bright yellow semiconcealed crown patch, yellowish belly, and a creamy-white iris. Otherwise it is like other flycatchers with nondescript olive upperparts and dirty grayish-olive throat and chest.

Skull: Other *Neopelma* species completely ossify their skulls (n = 18, LSUMZ), and probably occurs during the FCF.

Brood patch: Only ♀♀ incubate and develop BPs (Snow 1963). Timing unknown.

Sex: Both sexes have a bright yellow concealed crown stripe, which averages more extensive in ♂♂, and sexes should be first distinguishable during the FPF.

Molt: Group 3/Complex Basic Strategy; FPF partial, DPBs complete. FPF appears to include all body feathers, less covs, med covs, and a variable number of inner gr covs, but not ss, pp, pp covs, or rects. Timing unknown.

FCJ         Undescribed. Skull is unossified.

FPF         Replacing body feathers and at least some ss covs. Look for pointed rects, perhaps newly incoming yellow concealed crown stripe, and dusky iris. Skull is unossified.

FCF         Look for molt limits among the gr covs, between the gr covs and pp covs, and/or among the alulas. Pp covs relatively brownish, dusky, and lacking much gloss. Skull is either unossified or ossified.

SPB         Retained rects may average more pointed than replaced rects, retained pp covs may average relatively dull and brown, and retained ss and pp may average more worn. Skull is ossified.

DCB         Without molt limits among the gr covs, between the gr covs and pp covs, among the alulas, or among the terts. Skull is ossified.

DPB         Retained rects similarly rounded to replaced rects, retained pp covs may average relatively glossy and greenish, and retained ss and pp may average relatively fresh. Skull is ossified.


## *Tyranneutes virescens* (monotypic)

Tiny Tyrant-Manakin • Uirapuruzinho-do-norte

Band Size: C

# Individuals Captured: 11

Wing        42.0–51.0 mm (47.5 ± 2.7 mm; n = 8)

Tail        17.0–25.0 mm (22.3 ± 3.1 mm; n = 8)

Mass        6.0–9.0 g (7.5 ± 1.1 g; n = 9)

Bill        6.4 mm (n = 1)

Tarsus      13.4 mm (n = 1)

Similar species: Flycatcher-like, but note lack of rictal bristles. Otherwise, combination of gray iris, semiconcealed yellow crown patch, very small size, and relatively short tail is unique.

Skull: Unknown if completes.

Brood patch: Only ♀♀ incubate and develop BPs (Snow 1963). Timing unknown.

Sex: ♂♂ average shorter tails and have larger yellow crown patches than ♀♀, but differences are subtle and we do not yet have reliable quantitative data to separate ♂♂ from ♀♀.

Molt: Group 3/Complex Basic Strategy; FPF partial, DPBs complete. FPF appears to include the replacement of body feathers, less covs, med covs, and inner gr covs, but not outer gr covs, pp covs, or remiges. Rect replacement during the DPB may often be simultaneous. Timing unknown, but two birds were molting in Mar suggesting a late dry to early wet season schedule.

FCJ         Unknown. Skull is unossified.

FPF         Replacing adult-like juvenile body feathers or ss covs with adult-like feathers. Skull is probably at least relatively unossified.

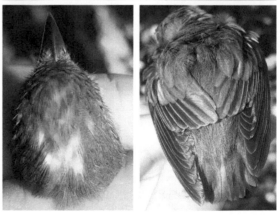

FCF    The iris is brownish-gray, but may be transitioning to being grayer. The yellow crown patch is partially concealed or absent. The inner gr covs are brighter olive-green and longer than the retained outer gr covs, which are dull olive-green and relatively short.

SPB    Retained rects may average more pointed than replaced rects, retained pp covs may average relatively dull and brown, and retained ss and pp may average more worn.

DCB    The iris is gray. Without molt limits among the gr covs. Large yellow crown patch is partially concealed.

DPB    Retained rects similarly rounded to replaced rects, retained pp covs may average relatively glossy and greenish, and retained ss and pp may average relatively fresh.

## *Corapipo gutturalis* (monotypic)

Band Size: C

White-throated Manakin • Dançarino-de-garganta-branca

\# Individuals Captured: 231

| Wing | ♂ | 51.5–57.5 mm (54.9 ± 1.1 mm; n = 79) |
| | ♀ | 50.5–56.5 mm (53.1 ± 1.3 mm; n = 78) |
| Tail | ♂ | 23.0–31.0 mm (26.5 ± 1.4 mm; n = 75) |
| | ♀ | 21.0–30.0 mm (26.3 ± 1.6 mm; n = 74) |
| Mass | ♂ | 5.8–10.0 g (7.7 ± 0.7 g; n = 91) |
| | ♀ | 6.0–11.0 g (8.5 ± 0.8 g; n = 84) |
| Bill | ♂ | 4.1–5.6 mm (4.8 ± 0.4 mm; n = 11) |
| | ♀ | 4.9–5.5 mm (5.3 ± 0.2 mm; n = 6) |
| Tarsus | ♂ | 16.4–18.0 mm (17.1 ± 0.5 mm; n = 11) |
| | ♀ | 16.1–17.5 mm (16.8 ± 0.6 mm; n = 6) |

Similar species: Adult ♂ (SAB) unmistakable. ♀♀ and subadults like other small green manakins, but note small size and pale bill.

Skull: Completes (n = 40) during FCF.

Brood patch: Only ♀♀ develop BPs, which have been observed from Jul–Nov.

Sex: Generally distinguishable only after FCF, but some FCF may be sexed with caution. See following molt and age descriptions for more details.

Molt: Group 3/Complex Basic Strategy; FPF partial, DPBs complete. FPF includes all body feathers, rects, less and med covs, and 1–5 (usually 3) gr covs. ♂♂ acquire adult-like plumage during TPB. The timing of molt remains elusive, as only 5 of 227 captures were in molt. All molting birds were seen from Nov–Jul and, despite low sample sizes, this is consistent with other manakins and BP data.

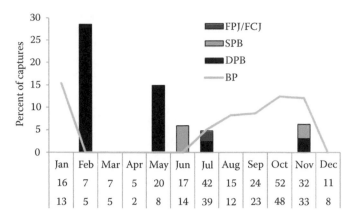

| | Jan | Feb | Mar | Apr | May | Jun | Jul | Aug | Sep | Oct | Nov | Dec |
|---|---|---|---|---|---|---|---|---|---|---|---|---|
| | 16 | 7 | 7 | 5 | 20 | 17 | 42 | 15 | 24 | 52 | 32 | 11 |
| | 13 | 5 | 5 | 2 | 8 | 14 | 39 | 12 | 23 | 48 | 33 | 8 |

FCJ     ♂ = ♀. Pale green upperparts. This plumage is probably only briefly held as some formative feathers begin to emerge while some juvenile feathers are still in pin. Skull is unossified.

FPF     ♂ = ♀. Replacing dull green body feathers and ss covs with brighter green feathers. Skull is unossified.

FCF     ♂ ≈ ♀. Molt limits are evident among the gr covs. The retained outer gr covs and pp covs are duller green and shorter than the fresher, brighter green, longer replaced inner gr covs.

    ♂: Some (all?) can be distinguished by having a whiter throat (rather than pale grayish-white) and some (most?) have darker gray or even black in the face perhaps resulting from adventitious feather replacement instead of the typical FPF. Because it is possible that not all ♂♂ are distinguishable, FCF birds without ♂-like features may be more safely sexed as "unknown."

    ♀: Throat is grayish-white with slightly grayish-olive chest, grayish-white underparts, and no gray or black in the auriculars. Skull is unossified or ossified.

Inner 3 gr covs replaced, outer gr covs retained, all flight feathers retained.

SPB     ♂ ≈ ♀. Replacing green body and flight feathers with green body and flight feathers. Some ♂♂ can be distinguished by having black in the face and a brighter white throat. Incoming p8–p10 are narrow, especially compared to juvenile p8–p10. ♀♀ from DPB by having duller green retained juv pp covs, ss, and pp. Birds early in molt without evidence of ♂ plumage should be left as unknown sex. Skull is ossified.

**SCB**  ♂ only. Green upperparts without molt limits among the gr covs. ♂♂ may always (?) have some black feathers in the face and sometimes elsewhere. Even if not, the throat is brighter white than in DCB ♀♀. The width of the outer three pp (p8–p10) may also be diagnostically narrow, even if ♂-like plumage characteristics are not obvious. Skull is ossified.

**DPB**  ♂ ≠ ♀. ♂♂ replacing black and white body and flight feathers with black and white body and flight feathers. ♀♀ replacing relatively bright green adult-like pp covs, pp, and ss with adult-like green pp covs, pp, and ss. Skull is ossified.

**DCB**  ♀ only. Green upperparts without molt limits among the gr covs. Throat is grayish-white with slightly grayish-olive chest and grayish-white underparts. More study is needed to confirm that all green birds without molt limits are ♀, because some SCB ♂♂ may lack black body feathers (see SCB). The outer three pp (p8–p10) are broader than in SCB ♂♂ and may be reliable to make sex determinations in the absence of plumage characteristics. Skull is ossified.

**TPB**  ♂ only. Replacing green body and flight feathers with glossy bluish-black body feathers and flight feathers. Skull is ossified.

**SAB**  ♂ only. Mostly glossy blue-black with a white throat and chest. Pp1–4 are marked white, but occasionally p3 and/or p4 lack white on the outer webbing, which may be related to age as often seen when adult ♂♂ have a few black body feathers that are tipped green (are these always TCB, or can some TAB also have green body feathers and lack white in p4?). Skull is ossified.

*Dixiphia pipra pipra* Band Size: D

White-crowned Manakin • Cabeça-branca # Individuals Captured: 2637

Wing ♂ 59.5–66.0 mm
(62.6 ± 1.2 mm; n = 246)

♀ 59.5–66.0 mm
(62.5 ± 1.3 mm; n = 141)

Tail ♂ 23.0–29.0 mm
(25.8 ± 1.2 mm; n = 225)

♀ 25.0–30.5 mm
(27.1 ± 1.2 mm; n = 142)

Mass ♂ 8.0–13.0 g
(10.4 ± 0.9 g; n = 273)

♀ 9.5–15.4 g
(12.4 ± 1.1 g; n = 144)

Bill ♂ 5.9–7.2 mm
(6.5 ± 0.3 mm; n = 49)

♀ 6.1–7.8 mm
(6.9 ± 0.4 mm; n = 20)

Tarsus ♂ 16.5–18.2 mm
(17.1 ± 0.5 mm; n = 43)

♀ 16.2–18.3 mm
(17.1 ± 0.5 mm; n = 14)

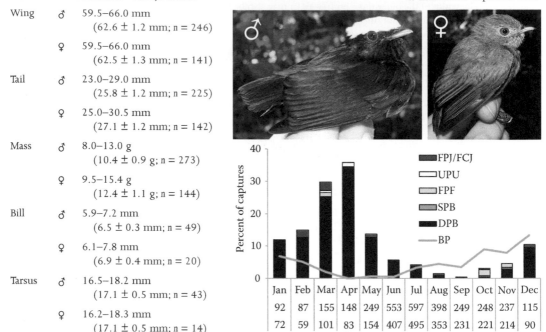

| | Jan | Feb | Mar | Apr | May | Jun | Jul | Aug | Sep | Oct | Nov | Dec |
|---|---|---|---|---|---|---|---|---|---|---|---|---|
| | 92 | 87 | 155 | 148 | 249 | 553 | 597 | 398 | 249 | 248 | 237 | 115 |
| | 72 | 59 | 101 | 83 | 154 | 407 | 495 | 353 | 231 | 221 | 214 | 90 |

Similar species: Adult ♂ (SAB) distinct, but ♀ and subadults like other green manakins. From *Lepidothrix serena* by lack of yellow on underparts. Measurements are useful to separate it from the shorter-tailed *Ceratopipra erythrocephala*. *Manacus manacus* has orange (not dull grayish-black) legs.

Skull: Completes (n = 494) during FCF, usually before Jun.

Brood patch: Only ♀♀ incubate and develop BPs. Very rarely FCF birds can have a BP (1 of 48 birds we examined with BPs was in FCF), and breeding probably does not usually occur until SCB or later. Breeding mainly occurs from Oct–Feb, but may occasionally occur as early as Jul and as late as Mar.

Sex: Distinguishable only after FCF. Even so, green birds without molt limits (SCB ♂♂ and DCB ♀♀) should be sexed with caution. See following molt and age descriptions for more details.

Molt: Group 3/Complex Basic Strategy; FPF partial, DPBs complete. FPF includes all body feathers, rects, less and med covs, and 1–6 (usually 3–4) gr covs. Rarely all gr covs are replaced. ♂♂ acquire adult-like plumage during TPB. Molt primarily occurs Jan–May, peaking in Mar–Apr.

Notes: Measurements are for DCB ♀♀ and SAB ♂♂ only. FCJ and FCF tail measurements may average slightly longer and wing measurements may average slightly shorter.

FCJ ♂ = ♀. Dull green with dull green ss covs, remiges, and rects. Begins FPF very soon after leaving the nest and may overlap with the FPJ. Iris is brown. Skull is unossified.

FPF      ♂ = ♀. Replacing dull green body feathers and ss covs with brighter green feathers. Iris is brown. Skull is unossified.

Many formative body feathers molting (left); less and med covs molting, gr covs, alulas, and all flight feathers retained juvenile (right).

FCF      ♂ = ♀. Bright green upperparts and adult ♀-like, but with molt limits often among gr covs. Retained outer gr covs are duller green and about 3 mm shorter than the brighter replaced inner covs, or occasionally all gr covs are replaced, contrasting with dull green retained pp covs. The iris is typically brown upon completion of the FPF, but becomes light orange, orange, or dark orange by Aug, before SPB begins. Skull is unossified or ossified.

Variation in FCF iris color, presumably related to age, rather than sex (upper three images); inner 4 (typical) and outer 2 (adventitious?) gr covs replaced, other gr covs and flight feathers retained juvenile (lower left); all ss covs and a1 replaced, other alulas and all flight feathers retained juvenile (lower right).

SPB ♂ = ♀. Replacing green body and flight feathers with green body and flight feathers. From DPB by having duller green retained juv pp covs, ss, and pp that are relatively worn. Skull is ossified.

p9–p10 retained juvenile, p7–p8 molting, p1–p6 new, s1 new, s2 molting, s3 missing, s4–s6 retained juvenile, s7–s9 new.

SCB ♂ only. Green upperparts without molt limits among the gr covs as in DCB ♀♀, and perhaps not always distinguishable (these should be called DCB sex unknown). The head probably averages more gray than in DCB ♀♀ and may show adult ♂-like feathers throughout the body from none to many. Beware of confusion with old DCB ♀ that may have some scattered ♂-like feathers especially on the crown; these DCB ♀♀ should have orange to dark orange irises, whereas in SCB ♂♂ the iris is red to deep red. Skull is ossified.

DPB ♂ ≠ ♀. ♂♂ replacing black and white body and black flight feathers with black and white body and black flight feathers. ♀♀ replacing relatively bright green adult-like pp covs, pp, and ss with adult-like green pp covs, pp, and ss that are relatively fresh. Skull is ossified.

DCB ♀ or unknown sex. Green upperparts without molt limits among the gr covs. The head probably averages more green or olive than in SCB ♂♂. Old ♀♀ can have some ♂-like white crown feathers or black body feathers and may not be separable from SCB ♂♂ and may be best sexed as "unknown." The iris is orange to dark orange. Presence of BPs confirms DCB ♀. Skull is ossified.

TPB ♂ only. Replacing green body and flight feathers with glossy bluish-black body feathers and flight feathers. Skull is ossified.

SAB    ♂ only. All black with a white crown and dark red iris. Some show a trace of green on the body and/or wings; these may be birds in TCB (do some TAB birds also occasionally have green body feathers?). Skull is ossified.

## *Ceratopipra erythrocephala erythrocephala*

Golden-headed Manakin • Cabeça-de-ouro

Band Size: D

# Individuals Captured: 392

| | | |
|---|---|---|
| Wing | ♂ | 51.0–58.5 mm (53.8 ± 1.6 mm; n = 48) |
| | ♀ | 52.0–59.0 mm (55.6 ± 1.2 mm; n = 64) |
| Tail | ♂ | 16.0–21.0 mm (18.2 ± 1.0 mm; n = 46) |
| | ♀ | 17.0–22.0 mm (19.9 ± 1.0 mm; n = 58) |
| Mass | ♂ | 9.0–13.2 g (11.0 ± 1.1 g; n = 50) |
| | ♀ | 10.4–13.5 g (12.0 ± 0.7 g; n = 60) |
| Bill | ♂ | 5.3–6.7 mm (6.0 ± 0.4 mm; n = 12) |
| | ♀ | 5.8–7.0 mm (6.5 ± 0.3 mm; n = 24) |
| Tarsus | ♂ | 15.5–16.9 mm (16.0 ± 0.5 mm; n = 11) |
| | ♀ | 15.0–17.7 mm (16.3 ± 0.6 mm; n = 24) |

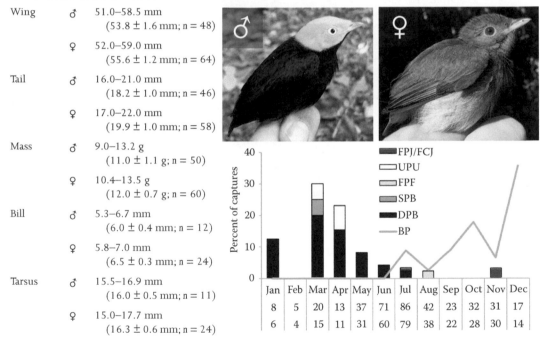

| Jan | Feb | Mar | Apr | May | Jun | Jul | Aug | Sep | Oct | Nov | Dec |
|---|---|---|---|---|---|---|---|---|---|---|---|
| 8 | 5 | 20 | 13 | 37 | 71 | 86 | 42 | 23 | 32 | 31 | 17 |
| 6 | 4 | 15 | 11 | 31 | 60 | 79 | 38 | 22 | 28 | 30 | 14 |

Similar species: ♂ distinct, but ♀♀ and juveniles very much like other dull green manakins (especially *Dixiphia pipra*). The short tail and short tarsus are the best quantitative features. Also note its pale bill and brown (not orange-red) iris, but beware of young *D. pipra*, especially from Nov–Jul, with brown eyes.

Skull: Completes (n = 92) during FCF.

Brood patch: Only ♀♀ incubate and develop BPs. Breeding occurs from Jul–Dec (and perhaps occasionally to Feb as in *D. pipra?*).

Sex: ♂♂ are black with a yellow head narrowly bordered red, have red "leggings," and a white iris. ♀♀ are dull olive-green with a brown iris, but older ♀♀ often develop minute white specs in iris (not splotches or blocks as in FCF ♂♂). Sex distinguishable during SPB, although it is sometimes possible to distinguish FCF ♂♂; see following molt and age descriptions for more details.

Molt: Group 3/Complex Basic Strategy; FPF partial, DPBs complete. FPF includes all body feathers, rects, less and med covs, and 1–5 (usually 3) gr covs. The difference between retained and replaced covs is subtler than in other manakins. Peaks in Mar and not observed from Aug–Dec.

FCJ      ♂ = ♀. Probably held for a very short time as in *D. pipra* and juvenile characteristics will be evident (e.g., weak plumage and gape). The iris is brown. Skull is unossified.

FPF      ♂ = ♀. Replacing dull green body feathers and ss covs with brighter green feathers. Iris is brown, but may become white in ♂♂. Skull is unossified.

FCF      ♂ ≈ ♀. Green with molt limits among the gr covs. The inner replaced gr covs are brighter green than the retained outer gr covs and pp covs. Skull is unossified or ossified. Some ♂♂ develop white splotches in the brown iris or occasionally the iris will be entirely white. Additionally, some ♂♂ can be identified by having traces of adult ♂-like yellow feathers in the crown or black feathers on the body. FCF lacking ♂-like features should be sexed as "unknown." Skull is either unossified or ossified.

Variation in iris color (upper left, upper center); a few mixed adult-like feathers occur in at least some FCF males (upper right); inner 6 gr covs replaced, a1 replaced, other gr covs, alula, and flight feathers retained juvenile (lower).

SPB      ♂ ≠ ♀. ♂♂ replacing green body and flight feathers with golden head and black body and flight feathers. ♀♀ like DPB ♀, but retained pp covs, pp, and ss are relatively dull brownish-green. Skull is ossified.

DCB      ♂ ≠ ♀. Without molt limits among the gr covs. Skull is ossified.

DPB      ♂ ≠ ♀. ♂♂ replacing black body and flight feathers with black body and flight feathers. ♀♀ like SPB ♀, but retained pp covs, pp, and ss are relatively bright olive-green. Skull is ossified.

*Manacus manacus manacus*

White-bearded Manakin • Rendeira

| | | |
|---|---|---|
| Wing | ♂ | 49.0–51.0 mm<br>(50.0 ± 1.4 mm; n = 2) |
| | ♀ | 50.0–54.0 mm<br>(51.6 ± 1.3 mm; n = 7) |
| Tail | ♂ | 32.0–33.0 mm<br>(32.5 ± 0.7 mm; n = 2) |
| | ♀ | 30.0–33.0 mm<br>(31.3 ± 1.2 mm; n = 6) |
| Mass | ♂ | 16.4–17.0 g<br>(16.7 ± 0.4 g; n = 2) |
| | ♀ | 12.4–17.0 g<br>(14.7 ± 1.5 g; n = 9) |

Bill    6.5–7.0 mm (6.7 ± 0.3 mm; n = 3)

Tarsus    21.9–23.6 mm (22.5 ± 0.9 mm; n = 3)

Similar species: ♂ unmistakable. ♀ and subadults like other green manakins, but with orange legs.

Skull: Completes (n = 6) during FCF.

Brood patch: Only in ♀♀, but has not been observed in the three DCB ♀♀ examined (in Jun, Sep, and Nov) so the timing remains unknown, but should be expected to be similar to other manakins.

Sex: Distinguishable only after FCF. ♂♂ have black upperparts, a white throat and cheek, and a grayish breast and belly. ♀♀ are green overall with bright orange legs. See following molt and age descriptions for more details.

Molt: Group 3/Complex Basic Strategy; FPF partial, DPBs complete. FPF includes all body feathers, rects, less and med covs, and 1–5 (usually 3) inner gr covs. The timing of molt is well defined, peaking (>30%) in Apr and is least frequent (<5%) between Jul and Nov.

| | |
|---|---|
| FCJ | ♂ = ♀. Unknown and probably only briefly held, as some FCF feathers begin to emerge while juvenile feathers are still in pin (as in *Dixiphia pipra*). Leg color is probably duller than in adults. Skull is unossified. |
| FPF | ♂ = ♀. Replacing dull green body feathers and ss covs with brighter green feathers. Skull is unossified. |
| FCF | ♂ ≈ ♀. Green overall and DCB ♀-like, but with molt limits among the gr covs. The retained outer gr covs are duller green than the brighter green replaced inner gr covs, but differences are often subtler than in other manakins. The legs are bright orange. At least some ♂♂ will have a trace of adult ♂-like plumage in the body feathers, especially in the forehead and back. Because some ♂♂ probably do not have adult ♂-like feathers, all green birds with molt limits should be sexed as "unknown." Skull is ossified or unossified. |

Some ♂♂ have adult-like feathers (left); inner 2 gr covs replaced, outer gr covs and all flight feathers retained juvenile (right).

SPB     ♂ ≠ ♀. ♂♂ replacing green body and flight feathers with black and white body and black flight feathers. ♀♀ like DPB ♀, but retained pp covs, pp, and ss are relatively dull brownish-green. Skull is ossified.

DCB     ♂ ≠ ♀. Without molt limits among the gr covs. Legs are bright orange. Skull is ossified.

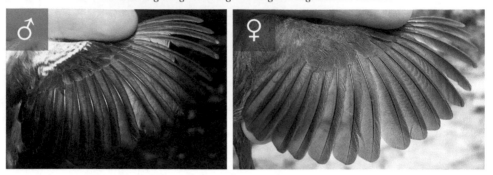

DPB     ♂ ≠ ♀. ♂♂ replacing black and white body and black flight feathers with black and white body and black flight feathers. ♀♀ like SPB ♀, but retained pp covs, pp, and ss are relatively bright olive-green. Skull is ossified.

## *Lepidothrix serena* (monotypic)

White-fronted Manakin • Uirapuru-estrela

Band Size: C

\# Individuals Captured: 735

| | | |
|---|---|---|
| Wing | ♂ | 48.0–58.0 mm (51.9 ± 1.5 mm; n = 178) |
| | ♀ | 50.0–58.0 mm (53.6 ± 1.3 mm; n = 255) |
| Tail | ♂ | 21.0–29.0 mm (25.2 ± 1.3 mm; n = 157) |
| | ♀ | 23.5–31.0 mm (27.1 ± 1.4 mm; n = 247) |
| Mass | ♂ | 8.0–12.0 g (10.1 ± 0.7 g; n = 211) |
| | ♀ | 8.0–13.0 g (10.8 ± 0.8 g; n = 308) |
| Bill | ♂ | 5.3–6.5 mm (5.9 ± 0.3 mm; n = 39) |
| | ♀ | 5.6–6.8 mm (6.3 ± 0.3 mm; n = 45) |
| Tarsus | ♂ | 17.1–19.8 mm (18.7 ± 0.7 mm; n = 37) |
| | ♀ | 16.6–19.5 mm (18.0 ± 0.6 mm; n = 43) |

Chart: Percent of captures by month with legend FPJ/FCJ, UPU, SPB, DPB, BP.

| | Jan | Feb | Mar | Apr | May | Jun | Jul | Aug | Sep | Oct | Nov | Dec |
|---|---|---|---|---|---|---|---|---|---|---|---|---|
| | 32 | 25 | 48 | 57 | 81 | 115 | 183 | 95 | 93 | 93 | 100 | 55 |
| | 23 | 17 | 24 | 37 | 47 | 78 | 155 | 82 | 84 | 76 | 85 | 46 |

Similar species: DCB ♂ unmistakable. ♀♀ and subadults like other small green manakins, but note yellow wash on belly, short wing, and dark brick red iris.

Skull: Completes (n = 193) during FCF.

Brood patch: Only ♀♀ develop BPs. Probably breeds mainly from Oct–Jan, but perhaps occasionally as early as Jul and as late as Apr.

Sex: ♂♂ are mostly black with a yellow belly, yellow spot on chest, white patch above bill, and blue rump. ♀♀ are green overall with a yellow belly. Sex distinguishable during SPB, although it is sometimes possible to distinguish FCF ♂♂. See following molt and age descriptions for more details.

Molt: Group 3/Complex Basic Strategy; FPF partial, DPBs complete. FPF includes all body feathers, rects, less and med covs, and 1–7 (usually 3) inner gr covs. The timing of molt is well defined, peaking from Jan–Apr and least frequent from Jul–Oct.

FCJ      ♂ = ♀. Generally green overall, with faint dull yellow in belly. Skull is unossified.

FPF      ♂ = ♀. Replacing dull green body feathers and ss covs with brighter green feathers. Skull is unossified.

FCF      ♂ ≈ ♀. Green upperparts and yellow underparts with molt limits among the gr covs. The retained outer gr covs are duller green and shorter than the brighter green replaced and longer inner gr covs. At least some ♂♂ will have a trace to lots of adult ♂-like plumage in the body feathers, especially in the forehead and back. Because some ♂♂ probably do not have adult ♂-like feathers, all "green and yellow" birds with molt limits should be sexed as "unknown." Skull is either unossified or ossified.

Inner 3 gr covs replaced, outer gr covs and all flight feathers retained (upper right); variation in the amount of adult ♂-like feathers (lower three images).

SPB      ♂ ≠ ♀. ♂♂ replacing green body and flight feathers with black body and flight feathers. ♀♀ like DPB ♀, but retained pp covs, pp, and ss are relatively dull green. Skull is ossified.

DCB    ♂ ≠ ♀. Without molt limits among the gr covs. In occasional ♂♂, a few black feathers will have green tips, which may indicate SCB. Some old ♀♀ may acquire scattered ♂-like feathers. Skull is ossified.

DPB    ♂ ≠ ♀. ♂♂ replacing black body and flight feathers with black body and flight feathers. ♀♀ like SPB ♀, but retained pp covs, pp, and ss are relatively bright green. Skull is ossified.

## LITERATURE CITED

Cramer, J. M., R. C. G. Mesquita, and G. B. Williamson. 2007. Forest fragmentation differentially affects seed dispersal of large and small-seeded tropical trees. Biological Conservation 137:415–423.

Doucet, S. M., D. B. McDonald, M. S. Foster, and R. P. Clay. 2007. Plumage development and molt in Long-tailed Manakins (*Chiroxiphia linearis*): variation according to sex and age. Auk 124:29–43.

DuVal, E. H. 2005. Age-based plumage changes in the Lance-tailed Manakin: a two-year delay in plumage maturation. Condor 107:915–920.

Foster, M. S. 1987. Delayed maturation, neoteny, and social system differences in two manakins of the genus *Chiroxiphia*. Evolution 41:547–558.

Graves, G. R. 1981. Brightly coloured plumage in female manakins (*Pipra*). Bulletin of the British Ornithological Club 101:270–271.

Graves, G. R., M. B. Robbins, and J. V. Remsen. 1983. Age and sexual difference in spatial distribution and mobility in manakins (Pipridae): inferences from mist-netting. Journal of Field Ornithology 54:407–412.

Johnson, E. I., P. C. Stouffer, and R. O. Bierregaard. 2012. The phenology of molting, breeding and their overlap in central Amazonian birds. Journal of Avian Biology 43:141–154.

McDonald, D. B. 1993. Delayed plumage maturation and orderly queues for status: a manakin mannequin experiment. Ethology 94:31–45.

Remsen, J. V., C. D. Cadena, A. Jaramillo, M. Nores, J. F. Pacheco, M. B. Robbins, T. S. Schulenberg, F. G. Stiles, D. F. Stotz, and K. J. Zimmer. [online]. 2016. A classification of the bird species of South America. American Ornithologists' Union. Version 30 November 2016. <http://www.museum.lsu.edu/~Remsen/SACCBaseline.htm>.

Ryder, T. B., and R. Durães. 2005. It's not easy being green: using molt and morphological criteria to age and sex green-plumage manakins (Aves: Pipridae). Ornitología Neotropical 16:481–491.

Ryder, T. B., and J. D. Wolfe. 2009. The current state of knowledge on molt and plumage sequences in selected Neotropical bird families: a review. Ornitología Neotropical 20:1–18.

Snow, D. W. 1963. The evolution of manakin displays. Pp. 553–561 in C. G. Sibley (editor), Proceedings XIII International Ornithological Congress. American Ornithologists' Union, Washington, DC.

Snow, D. W. [online]. 2004. Saffron-crested Tyrant-Manakin (*Neopelma chrysocephalum*). In J. del Hoyo, A. Elliott, J. Sargatal, D. A. Christie, and E. de Juana (editors), Handbook of the birds of the world alive. Lynx Edicions, Barcelona, Spain. <http://www.hbw.com>.

Tello, J. G., R. G. Moylea, D. J. Marchesea, and J. Cracraft. 2009. Phylogeny and phylogenetic classification of the tyrant flycatchers, cotingas, manakins, and their allies (Aves: Tyrannides). Cladistics 25:429–467.

Uriatre, M., M. Anciaes, M. D. Silva, P. Rubim, E. I. Johnson, and E. Bruna. 2011. Building a mechanistic understanding of seed dispersal in human-modified landscapes: a case study with a bird-dispersed tropical understory herb. Ecology 92:924–937.

# CHAPTER THIRTY

# Tityridae (Tityras)

# Species in South America: 30

# Species recorded at BDFFP: 8

# Species captured at BDFFP: 5

Once included within the morphologically similar Tyrannidae, but also sharing similarities with some Cotingidae, Tityridae, the tityras are in some ways a mishmash of plumages, structural characters (notably bill), and ecological guilds.

Given the recent split from other Tyranni, a pattern of molt has not yet been described in this phylogenetic grouping. Based on our observations it appears that species can either have a Complex Basic Strategy (e.g., *Schiffornis*) or a Complex Alternate Strategy (e.g., *Pachyramphus*). It may be that this distinction in the presence or absence of the prealternate molt is not necessarily related to migratory status (Howell 2010), but rather forest stratum as a response to variation in ultraviolet exposure (see also Wolfe et al. 2010). The preformative molt can be variable from partial to complete. In some species (e.g., *Schiffornis*), the juvenile plumage is highly similar to subsequent adult plumages, but in others (e.g., *Laniocera*) the juvenile plumage is unique and even elaborate relative to subsequent adult plumages. Sexual dichromatism exists in some species (e.g., *Tityra* and *Pachyramphus*) but not in others (e.g., *Schiffornis* and *Laniocera*).

Incubation appears to be performed only by females, thus the presence of a brood patch in sexually monochromatic species ought to be useful for sexing. Skulls typically ossify, but again, a full examination across all genera and species has not been summarized. Tityridae have 10 pp, 9 ss, and 12 rects.

## SPECIES ACCOUNTS

*Tityra cayana cayana*
Black-tailed Tityra • Anambé-branco-de-rabo-preto

Band Size: H
# Individuals Captured: 1

| | | |
|---|---|---|
| Wing | ♂ | Unknown at the BDFFP |
| | | 120–129 mm (n = 2; Junge and Mees 1961) |
| | ♀ | 113.0 mm (n = 1) |
| | | 120–123 mm (n = 2; Junge and Mees 1961) |
| Tail | ♂ | Unknown at the BDFFP |
| | | 70–89 mm (n = 2; Junge and Mees 1961) |
| | ♀ | 69.5 mm (n = 1) |
| | | 72 mm (n = 2; Junge and Mees 1961) |
| Mass | ♂ | Unknown at the BDFFP |
| | | 76–89 g (n = 2; Junge and Mees 1961) |
| | ♀ | 64.9 g (n = 1) |
| | | 77.5–78.5 g (n = 2; Junge and Mees 1961) |

Similar species: None.

Skull: Completes based on an examination of specimens from LSUMZ (n = 8).

Brood patch: Only ♀♀ incubates (Fitzpatrick and Mobley 2004) and develops BPs. Timing unknown.

Sex: ♂♂ lacks streaking on the head or chest, whereas ♀♀ are streaked on the head and chest.

Molt: Group 3 (or Group 8?)/Complex Basic Strategy (although Complex Alternate Strategy seems possible); FPF partial, DPBs complete, PAs unknown. FPF appears to include all gr covs and all rects, but not terts (or sometimes up to 3?) or other flight feathers (n = 1, LSUMZ). Timing unknown.

FCJ — Heavily streaked underparts, back, and crown. Flight feathers and rects brownish-black. Rect shape may not be terribly useful, as they appear relatively rounded in one specimen (LSUMZ) of *T. c. braziliensis* from Bolivia.

(LSUMZ 137480.)

FPF — Replacing brownish back and light brownish underparts with more grayish adult-like feathers. Incoming gr covs notably more blackish than retained juvenile gr covs. Skull is unossified.

FCF — Look for boldly contrasting molt limits between the gr covs and pp covs and flight feathers, or perhaps sometimes among the gr covs. Terts contrastingly brownish compared to back feathers, but beware of some birds possibly replacing one or more terts. Skull is unossified or ossified.

Male showing replaced gr covs and retained greater alula, pp covs, and flight feathers. (LSUMZ 67424.)

SPB — Replacing brownish flight feathers with black flight feathers. Skull is ossified.

DCB — Lacking molt limits between the gr covs, med covs, pp covs, and flight feathers, appearing uniformly black. Skull is ossified.

DPB — Replacing blackish flight feathers with black flight feathers. Skull is ossified.

*Schiffornis olivacea olivacea*

Olivaceous Schiffornis • Flautim-oliváceo

| Wing | 82.0–99.0 mm (90.4 ± 2.7 mm; n = 292) |
| Tail | 56.0–73.0 mm (65.4 ± 3.5 mm; n = 280) |
| Mass | 28.1–41.0 g (33.6 ± 2.5 g; n = 441) |
| Bill | 8.8–11.3 mm (10.2 ± 0.7 mm; n = 36) |
| Tarsus | 23.2–27.4 mm (25.8 ± 0.9 mm; n = 34) |

Similar species: No other bird at the BDFFP is so drab and olive-brown overall. It is larger and longer-tailed than manakins. Lack of wing bars, eyering, and bill shape separate it from flycatchers and *Piprites chloris*.

Skull: Completes (n = 140) possibly during the FPF or FCF.

Brood patch: Unknown if both sexes incubate. Breeding appears to peak during the dry season, from Jul–Oct.

Sex: ♂ = ♀.

Molt: Group 4/Complex Basic Strategy; FPF complete, DPBs complete. Rects may occasionally or regularly molt simultaneously in FPF and DPB molts. Molt peaks following breeding, from Aug–Dec.

Notes: Recently the "Thrush-like Schiffornis" (*Schiffornis turdina*) was split into five species by the SACC (Remsen et al. 2016) based on molecular and vocal evidence by Nyári (2007) and Donegan et al. (2011).

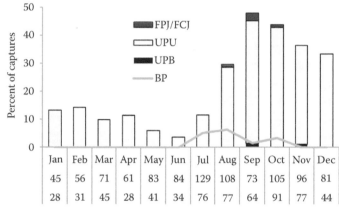

| | Jan | Feb | Mar | Apr | May | Jun | Jul | Aug | Sep | Oct | Nov | Dec |
|---|---|---|---|---|---|---|---|---|---|---|---|---|
| | 45 | 56 | 71 | 61 | 83 | 84 | 129 | 108 | 73 | 105 | 96 | 81 |
| | 28 | 31 | 45 | 28 | 41 | 34 | 76 | 77 | 64 | 91 | 77 | 44 |

FCJ    Duller and browner overall, grayer underparts, and with a grayish iris. Also with faint buffy spots on some ss covs (Wetmore 1972, Kirwan and Green 2011). Skull is unossified.

FPF    Replacing relatively brownish juvenile body and flight feathers with adult-like feathers. Retained flight feathers relatively fresh. Iris may transition from grayish-brown to rich brown. Skull is unossified, but may ossify before FPF completes.

FAJ   With dull olive body feathers and brownish-olive gr covs, pp covs, and flight feathers creating pseudolimits across feather tracts. Gape is not a good indicator of age as some yellow is retained apparently for life, as shown by a banded recapture at least 7 years old. Skull is ossified, but small windows can be retained (are these FCF?).

UPB   Replacing adult-like body and flight feathers with adult-like feathers. Retained flight feathers relatively worn. Iris is rich brown. Skull is ossified.

## *Laniocera hypopyrra* (monotypic)

Cinereous Mourner • Chorona-cinza

Band Size: G, F

# Individuals Captured: 16

| | |
|---|---|
| Wing | 96.0–109.0 mm (105.6 ± 3.8 mm; n = 10) |
| Tail | 79.0–87.0 mm (82.7 ± 2.5 mm; n = 10) |
| Mass | 40.0–55.0 g (47.6 ± 4.7 g; n = 13) |
| Bill | 12.9 mm (n = 1) |
| Tarsus | 24.6 mm (n = 1) |

Similar species: *Rhytipterna simplex* and *Lipaugus vociferans* are similarly gray, but lack the tawny ss cov spots, barred undertail and underwing covs, and rects with tawny tips.

Skull: Completes (n = 4, BDFFP; n = 21, LSUMZ), probably during the FCF.

Brood patch: Probably only ♀♀ attend nest and develop BPs (Walther 2004). The only BP was seen in Nov.

Sex: ♂ = ♀.

Molt: Group 4/Complex Basic Strategy; FPF complete, DPBs complete. FPF complete based on an examination of four specimens with unossified skulls and 17 specimens (LSUMZ) with ossified skulls, all without molt limits and showing the same pseudolimit among the gr cov. Twice seen in molt (Jul and Sep), but seasonal timing unknown.

FCJ   Extremely gaudy, covered in loosely textured cinnamon-orange feathers with black terminal spots thought to mimic a caterpillar for protection (D'Horta et al. 2012, Londoño et al. 2015). Skull is unossified.

(Courtesy of David-Rivera Santiago.)

FPF       Replacing gaudy juvenile body and flight feathers with adult-like feathers. Skull is unossified.

FAJ       Beware of pseudolimits among the gr and med covs; note the consistency in wear and intensity of the gray among these feathers. Orange or yellow on the breast and under the wing appear to be variable in extent, color, and pattern. Skull is ossified.

 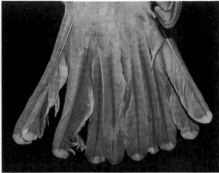

UPB       Replacing adult-like plumage with adult-like plumage. Skull is ossified.

## *Pachyramphus marginatus nanus*

Black-capped Becard • Caneleiro-bordado

Band Size: E

\# Individuals Captured: 3

Wing     62.0–72.0 mm (67.3 ± 5.0 mm; n = 3)

Tail      47.0–52.5 mm (49.8 ± 3.9 mm; n = 2)

Mass    15.3–18.5 g (16.6 ± 1.7 g; n = 3)

Similar species: ♂ from *P. surinamus* and *P. rufus* by gray underparts, bold wing bars, and streaked back. ♀ from *P. surinamus* and *P. rufus* by tawny underparts and bolder wing bars.

Skull: Completes (n = 1), probably during the FCF.

Brood patch: Unknown if both sexes incubate. Timing unknown.

Sex: ♂♂ and ♀♀ distinguishable starting in the FPA. ♂♂ obviously with some (in the first cycle) to completely (in definitive cycles) black and gray overall with bold white edging to the ss covs and terts. ♀♀ with olive, yellowish, and greenish tones throughout with buffy to creamy edging to the ss covs and terts.

Molt: Group 7(?)/Complex Alternate Strategy; FPF complete(?), DPBs complete, PAs limited(?)-partial(?). FPF complete based on one specimen of *P. m. nanus* with entirely fresh plumage and 80% skull ossification (LSUMZ 84993). PAs, based on one *P. m. marginatus* (LSUMZ 69114), appears to include at least the crown, some back, and some less and/or med covs. More study and confirmation is needed. Timing unknown.

FCJ       ♀ = ♂. Like adult ♀, but with loosely textured plumage, slightly narrow rects with less extensive black in the outer r6.

(LSUMZ 102281.)                   (LSUMZ 102281.)

FPF       ♀ = ♂. Replacing juvenile body and relatively fresh flight feathers with adult-like body and flight feathers. Look for differences in rect shape between slightly more pointed retained and more rounded replaced rects. Skull is unossified.

FAJ   ♀ = ♂. FCF like DCB ♀♀, but perhaps with less rufous and more olive in the crown, thus it may be possible to code such birds as FCF. More study is needed. No apparent molt limits in the body or ss covs. Skull unossified or ossified.

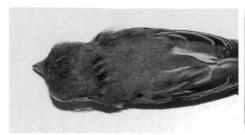

(LSUMZ 133385.)                    (LSUMZ 133385.)

FPA/FCA   ♂ only. With mixed ♀- and ♂-like plumage with a black cap and ♂-like body and/or ss covs. Skull is ossified.

UCA   ♀ only. Look for molt limits among the less and/or med covs. Skull is ossified.

SPB   ♂ only. Black adult-like ♂ flight feathers replacing relatively worn ♀-like flight feathers. Skull is ossified.

UPB   ♀ only. Replacing relatively worn adult-like flight with adult-like flight feathers. Skull is ossified.

DCB   ♂ only. All ♂-like without molt limits in the upperparts or ss covs. Skull is ossified.

DPA/DCA   ♂ only. Mix of darker and lighter black feathers in the upperparts or ss covs. Skull is ossified.

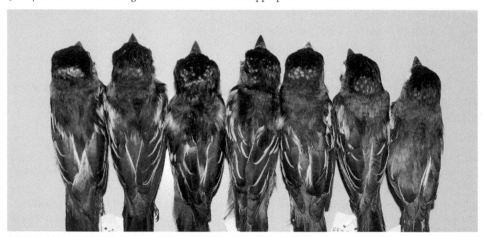

(LSUMZ 110288, 110289, 64256, 70922, 156703, 120095, 133387.)

DPB   ♂ only. Black flight feathers replacing black flight feathers. Skull is ossified.


## Pachyramphus minor (monotypic)

Band Size: Unknown

Pink-throated Becard • Caneleiro-pequeno

# Individuals Captured: 3

Wing   86.0–90.0 mm (88.0 ± 2.0 mm; n = 3)

Tail   61.0–64.0 mm (62.7 ± 1.5 mm; n = 3)

Mass   40.0–46.0 g (43.2 ± 3.0 g; n = 3)

Similar species: Compare ♂ to Rhytipterna simplex (Tyrannidae), Laniocera hypopyrra, and Lipaugus vociferans (Cotingidae). ♀ larger and with more gray above and on crown than other Pachyramphus.

Skull: Completes during the FCF (Pyle 1997), confirmed through an examination of specimens at LSUMZ (n = 12).

Brood patch: Only ♀♀ incubate and develop BPs (as in P. aglaiae; Pyle 1997). Timing unknown.

Sex: ♂ mostly dark gray, but with pink patch on throat, often indistinct. ♀ rusty below and gray above. Not able to distinguish until FPA.

Molt: Group 8/Complex Alternate Strategy; FPF limited-partial(?), DPBs complete, FPAs partial-incomplete, DPAs absent-limited based on an examination of specimens (n = 5, LSUMZ from Boliva, Peru, and Ecuador) and Pyle (1997). FPF includes body feathers and up to a few less and med covs. FPA includes a variable number of back and inner less, med, and gr covs, between 1–7 inner ss, and probably all rects (n = 4, LSUMZ). DPA, as suggested by Pyle (1997), appears to be either absent or more limited to a variable number of less covs, back, and crown feathers. Timing unknown.

FCJ     ♀ = ♂. ♀-like, although with less gray in crown and back. Skull is unossified.

FPF     ♀ = ♂. Replacing body feathers and perhaps less and med covs, but not flight feathers, with ♀-like feathers. Skull is unossified.

FCF     ♀ = ♂. Probably difficult to discern from FCJ or ♀ DCB, but look for contrast in quality and brilliance of feathers between back and some ss covs, particularly among the less and med covs, and retained dull juvenile pp covs. Skull is unossified or ossified.

FPA/FCA  ♀ ≠ ♂. Look for molt limits among the ss with the inner 1–7 ss replaced and others retained. Probably also limits among the gr, med, and less covs with the outer ones retained and inner ones replaced. Obvious in ♂♂ with a mix of gray and brown plumage, but more cryptic in ♀♀ and perhaps not distinguishable from DCA (use UCA) except by greater replacement extent of FPA and by the greater degree of wear in retained outer pp. Skull is ossified.

(LSUMZ 17100, 34620, 83379.)

SPB     ♀ ≠ ♂. Difficult to distinguish from DPB in ♀♀ (use UPB), but look for highly worn outer pp being replaced by fresh inner pp. In ♂♂, gray feathers replacing brown (as in FPA), but also includes pp and outer ss. Skull is ossified.

DCB     ♀ ≠ ♂. Without molt limits among the ss covs and with adult-like pp covs. Skull is ossified.

DPA/DCA  ♀ ≠ ♂. Difficult to distinguish from FCA in ♀♀ (use UCA), but look for relatively fresh outer pp and a less extensive DPA replacement. In ♂♂, look for two generations of adult-like ♂ body feathers and ss covs (and possibly also ss). Skull is ossified.

DPB     ♀ ≠ ♂. Difficult to distinguish from SPB in ♀♀ (use UPB), but in ♂♂ replacing blackish flight feathers with blackish flight feathers. Skull is ossified.

## LITERATURE CITED

D'Horta, F. M., G. M. Kirwan, and D. Buzzetti. 2012. Gaudy juvenile plumages of Cinereous Mourner (*Laniocera hypopyrra*) and Brazilian Laniisoma (*Laniisoma elegans*). Wilson Journal of Ornithology 124:429–435.

Donegan, T. M., A. Quevedo, M. McMullan, and P. Salaman. 2011. Revision of the status of bird species occurring or reported in Colombia 2011. Conservación Colombiana 15:4–21.

Fitzpatrick, J. W., and J. A. Mobley. 2004. Masked Tityra (*Tityra cayana*). Pp. 451 in J. del Hoyo, A. Elliott, and J. Sargatal (editors), Handbook of the birds of the world. Lynx Edicions, Barcelona, Spain.

Howell, S. N. G. 2010. Molt in North American Birds. Houghton Mifflin Harcourt, Boston, MA.

Junge, G. C. A., and G. F. Mees. 1961. The avifauna of Trinidad and Tobago. E. J. Brill, Neider, Netherlands.

Kirwan, G. M., and G. Green. 2011. Cotingas and Manakins. Christopher Helm, London, UK.

Londoño, G. A., D. A. García, and M. A. Sánchez Martínez. 2015. Morphological and behavioral evidence of Batesian mimicry in nestlings of a lowland Amazonian bird. American Naturalist 185:135–141.

Nyári, A. S. 2007. Phylogeographic patterns, molecular and vocal differentiation, and species limits in *Schiffornis turdina* (Aves). Molecular Phylogenetics and Evolution 44:154–164.

Pyle, P. 1997. Identification guide to North American Birds, Part I. Slate Creek Press, Bolinas, CA.

Remsen, J. V., C. D. Cadena, A. Jaramillo, M. Nores, J. F. Pacheco, M. B. Robbins, T. S. Schulenberg, F. G. Stiles, D. F. Stotz, and K. J. Zimmer. [online]. 2016. A classification of the bird species of South America. American Ornithologists' Union. Version 30 November 2016. <http://www.museum.lsu.edu/~Remsen/SACCBaseline.htm>.

Walther, B. 2004. Cinereous Mourner (*Laniocera hypopyrra*). In J. del Hoyo, A. Elliot, J. Sargatal, D. A. Christie, and E. de Juana, editors. Handbook of the Birds of the World Alive. Lynx Edicions, Barcelona, Spain.

Wetmore, A. 1972. The birds of the Republic of Panama. Smithsonian Miscellaneous Collection 150. Smithsonian Institute, Washington, DC.

Wolfe, J. D., T. B. Ryder, and P. Pyle. 2010. Using molt cycles to categorize the age of tropical birds: An integrative new system. Journal of Field Ornithology 81:186–194.

# PART FOUR

# Passerines, the Oscines

# CHAPTER THIRTY-ONE

# Vireonidae (Vireos)

# Species in South America: 27

# Species recorded at BDFFP: 7

# Species captured at BDFFP: 4

The vireos include only four genera confined to the New World. Some are quite wood-warbler-like in plumage, but have an antbird-like structure with a heavy, often hooked bill for eating large insects and sometimes fruit. Several species are surprisingly similar in appearance to syntopic non-vireos, like *Hylophilus ochraceiceps* is to *Myrmotherula menetriesii* (Thamnophilidae), *Vireo olivaceus* is to *Oreothlypis peregrina* (Parulidae), *Vireo huttonii* is to *Regulus calendula* (Regulidae), and *Vireo flavifrons* is to *Setophaga pinus* (Parulidae). Most vireos are generally inconspicuous forest- or shrub-dwelling birds and are often heard before they are seen. Most species are sexually monochromatic. Vireos have 10 pp (the 10th reduced, sometimes greatly so in some species), 9 ss, and 12 rects.

Molt in vireos is similar to that of other passerines, often following a Complex Basic Strategy in nonmigrants or short-distance migrants, and a Complex Alternate Strategy in short-distance and long-distance migrants. Prealternate molts, when they occur, do not generally result in a change of appearance, but may serve exclusively to deal with feather wear and degradation as a consequence of increased year-round exposure to ultraviolet radiation. Preformative molts in most if not all migrants are partial to incomplete (sometimes being eccentric), but can apparently be complete in at least some resident Neotropical species. Red-eyed Vireo creates an interesting question, as highlighted by Howell (2010), because of its many populations across North and northern South America, some of which are migratory and others are not: How important is migration in determining the frequency and extent of molts, particularly the preformative and prealternate molts? We do not yet know if resident tropical Red-eyed Vireos have a prealternate molt, but their preformative molt may be more extensive than in migratory North American populations. This may simply be a consequence of extra time made available by not migrating, thus ensuring higher quality feathers across more feather tracts, and slowing down cell molts may produce higher quality feathers.

Skull ossification is generally complete, but small windows may be retained well into the FCF. Sexes generally share in incubation, but this is dominated by females such that males develop less extensive brood patches, at least in most temperate species. Vireos have 10 pp (the 10th reduced, sometimes greatly so in some species), 9 ss, and 12 rects.

## SPECIES ACCOUNTS

### *Cyclarhis gujanensis gujanensis*
Rufous-browed Peppershrike • Pitiguari

Band Size: G
# Individuals Captured: 21

| | |
|---|---|
| Wing | 65.0–72.5 mm (69.1 ± 2.1 mm; n = 17) |
| Tail | 51.0–58.5 mm (53.5 ± 2.6 mm; n = 16) |
| Mass | 24.0–31.0 g (26.8 ± 2.3 g; n = 18) |
| Bill | 10.5–10.7 mm (10.6 ± 0.1 mm; n = 3) |
| Tarsus | 22.0–24.5 mm (23.3 ± 1.3 mm; n = 3) |

Similar species: None.

Skull: Completes (n = 4), probably late in the FPF or just after its completion.

Brood patch: Unknown if both sexes incubate. Timing unknown, but nesting has been observed in the dry season.

Sex: ♂ = ♀.

Molt: Group 4/Complex Basic Strategy; FPF complete, DPBs complete. Extent of FPF based on examination of specimens at LSUMZ of *C. g. gujanensis* (n = 12), *C. g. dorsalis* (n = 18), *C. g. virenticeps* (n = 16); all with complete skulls and lacking apparent molt limits, except for one *C. g. dorsalis* with 50% skull ossification and also lacking molt limits, one molting *C. g. virenticeps* with r1 and s9 molting and s8 and p1 missing and 10% skull ossification, one molting *C. g. virenticeps* with only s5 remaining and 90% skull ossification. Timing unknown.

FCJ — Head markings are less distinct and the iris is brown (Brewer 2010) or tannish (LSUMZ), but otherwise quite similar to subsequent plumages. Rects relatively narrow and pointed. Skull is unossified.

FPF — Replacing duller juvenile feathers with brighter adult-like feathers. Skull incompletely ossified, but may sometimes ossify near the completion of this molt.

(Courtesy of Philip C. Stouffer.)

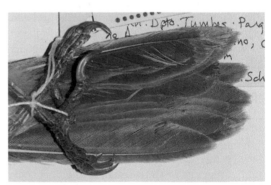

(LSUMZ 183345.)

p10 molting, p1–p9 new, s1–s4 and s6–s9 new, s5 retained juvenile, with skull 90% ossified. (LSUMZ 183347.)

FCF    Like FAJ without molt limits and bright orange iris, but with an incompletely ossified skull.

FAJ    Without molt limits and with bright markings in the head. Iris is variably bright orange to red (Brewer 2010). Skull is ossified. Rects relatively broad and rounded.

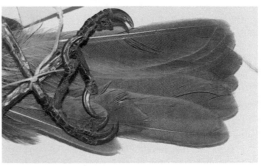

(LSUMZ 183347.)

UPB    Replacing adult-like feathers with adult-like feathers. Skull is ossified. May be very difficult to distinguish from FPF late in the molt (use UPU).

## *Vireo olivaceus vividior/solimoensis* and *V. o. olivaceus*

Red-eyed Vireo • Juruviara

Band Size: E, D

\# Individuals Captured: 8

Wing    BDFFP: 60.0–68.0 mm (62.9 ± 2.9 mm; n = 7)
    *V. o. vividior:* 71–75 mm (n = 4, Trinidad and Tobago; Junge and Mees 1961)
    *V. o. olivaceus:* 72.0–85.0 mm (n = 200; Pyle 1997)

Tail    BDFFP: 39.0–48.0 mm (42.4 ± 3.0 mm; n = 7)
    *V. o. vividior:* 48–52 mm (n = 4, Trinidad and Tobago; Junge and Mees 1961)
    *V. o. olivaceus:* 47.0–60.0 mm (n = 134; Pyle 1997)

Mass    BDFFP: 12.5–15.0 g (13.9 ± 1.0 g; n = 7)
    *V. o. vividior:* 13.5–15.5 g (n = 4, Trinidad and Tobago; Junge and Mees 1961)

*V. o. vividior/solimoensis*

*V. o. olivaceus*

Similar species: From *V. altiloquus* by lack of black malar stripes and smaller bill. During the boreal winter (3 Oct to 14 Apr), *V. o. olivaceus* can inhabit mature interior forest (Stotz et al. 1992), although its presence in second growth may go unnoticed given the abundance of residents, which are largely absent in mature forest. Austral migrant *V. o. chivi* may also be possible from Apr–Nov (B. Whitney, pers. comm.). In the hand, the *V. o. chivi* group (broadly encompassing many South American taxa) has p9 < p5 by 1.0–1.5 mm, wing <75.0 mm, and undertail coverts yellowish (not whitish), although no measurements were taken from specifically *V. o. vividior/solimoensis* (Phillips 1991). Eye color in *V. o. vividior/solimoensis* may average browner than *V. o. olivaceus*, but beware of especially first cycle or occasionally after first cycle northern birds with brown eyes (Parkes 1985).

Skull: Completes in both subspecies (n = 3, BDFFP; Pyle 1997), probably during FCF.

Brood patch: Only ♀♀ incubate in *V. o. olivaceus* (Pyle 1997) and thus may also be true in South American populations. Timing unknown. Should only be observed in *V. o. vividor/solimoensis* because *V. o. olivaceus* breeds in North America.

Sex: ♂ = ♀.

Molt: Group 3/Complex Basic Strategy; FPF is partial (*V. o. olivaceus*)-incomplete and eccentric (*V. o. vividior/solimoensis*), DPBs complete. In *V. o. olivaceus*, FPF includes body feathers, proximal feather of the alula, all gr covs, and 1–3 terts (Mulvihill and Rimmer 1997, Pyle 1997). The FPF initiates at about 15 days old and lasts <42 days in *V. o. olivaceus* (Lawrence 1953); presumably it initiates similarly early in life in resident subspecies. In *V. o. vividior/solimoensis*, the FPF can be eccentric based on one capture, one specimen from Guyana, and one specimen from Peru (LSUMZ); replaces all gr covs, all rects, 1–6 inner ss and/or 1–4 outer ss, and 0–6 (or more?) pp, but not pp covs. In *V. o. olivaceus*, DPBs and possibly FPF suspend during migration (Sep–Dec) and complete on the wintering grounds from Jan–Apr. See Pyle (1997) and Howell (2010) for hypotheses about prolonged winter molts in *V. o. olivaceus*.

Notes: One outlier is not included in the measurement data because it is possible that this bird, weighing 18.5 g, may have been a wintering *V. o. olivaceus*. It was showing wing and tail molt, and the timing (23 Feb) was consistent with the known timing of this subspecies. Unfortunately no wing, tail, or other measurements were taken to corroborate this. Recent genetic evidence of polyphyly suggests that North American and South American *V. olivaceus* may constitute distinct species groups (Slager et al. 2014).

FCJ      Tawny-tinged upperparts and perhaps with less contrast in the face. Skull is unossified. Only possible to see in *V. o. vividior/solimoensis* as *V. o. olivaceus* completes FPF before arriving on wintering grounds, and austral migrant *V. o. chivi* may likely also complete the FPF before arriving on wintering grounds.

FPF      Replacing adult-like juvenile body or outer pp with adult-like formative feathers while retaining inner adult-like juvenile pp. Skull is unossified.

FCF      Look for replaced inner gr covs and/or inner ss contrasting with retained outer gr covs and/or outer ss. FPF can be eccentric in *V. o. vividior/solimoensis* with blocks of outer pp (but not pp covs) and inner ss replaced. Iris is brown. The mouth lining may be paler than in DCB, but this feature may only be useful until about 6 months old. Skull is unossified or ossified.

SPB      Determine by difference in color and wear between juvenile pp covs and formative outer pp. Skull is ossified.

p6–p10 replaced (formative), p4–p5 retained (juvenile), p1–p3 new (second basic, in suspension), pp covs 4–10 retained (juvenile), s1–s3 retained (juvenile), s4–s9 replaced (formative, or s8 second basic?).

DCB      Uniform bright olive upperparts without molt limits among the gr covs, ss, or pp. Skull is ossified.

DPB      Distinguish from SPB by lack of difference in color and wear between retained pp covs and retained pp. Skull is ossified.


## *Hylophilus muscicapinus muscicapinus*

Buff-cheeked Greenlet • Vite-vite-camurça

Band Size: E

\# Individuals Captured: 22

Wing      55.0–63.0 mm (58.6 ± 2.2 mm; n = 18)

Tail        37.5–47.0 mm (41.5 ± 2.8 mm; n = 17)

Mass     8.2–15.0 g (10.5 ± 1.8 g; n = 18)

Bill        7.8–9.6 mm (8.4 ± 0.5 mm; n = 12)

Tarsus    17.4–20.7 mm (18.7 ± 0.9 mm; n = 12)

Similar species: The buff-colored face contrasting with olive upperparts and grayish crown should be unique, unlike the unicolored buffy upperparts of *Phaeothlypis rivularis*.

Skull: Completes (n = 9), but timing unknown.

Brood patch: Unknown if both sexes incubate. Have been seen in Sep and Nov and is probably a dry-season breeder.

Sex: ♂ = ♀.

Molt: Group 4(?)/Complex Basic Strategy; FPF complete(?) based on 15 captures we examined without molt limits, DPBs complete. Timing unknown.

FCJ — Undescribed (Brewer 2010). Skull is unossified.

FPF — Replacing juvenile-like plumage with adult-like plumage. Look for differences in rect shape between old and new feathers, relatively dull retained pp covs, and relatively fresh retained flight feathers. Skull is unossified, but may be ossified before FPF completes.

FAJ — Unknown how to discern from FCJ by plumage, but look for relatively glossy pp covs and relatively rounded rects. Skull is probably ossified or nearly ossified.

UPB — Look for similar shape between retained and replaced rects, relatively glossy pp covs, and relatively worn retained flight feathers. Skull is ossified.

## Tunchiornis ochraceiceps luteifrons

Band Size: D

Tawny-crowned Greenlet • Vite-vite-uirapuru

\# Individuals Captured: 460

Wing — 50.0–61.0 mm (56.0 ± 2.1 mm; n = 283)

Tail — 34.0–45.0 mm (39.5 ± 2.1 mm; n = 272)

Mass — 8.0–12.5 g (10.0 ± 0.9 g; n = 399)

Bill — 7.4–10.0 mm (8.5 ± 0.7 mm; n = 23)

Tarsus — 16.5–18.7 mm (17.8 ± 0.7 mm; n = 22)

Similar species: Most similar to ♀ *Myrmotherula menetriesii*, but note grayish lores with reduced feathering, more conical bill shape, longer tail, and bright tawny above eye.

Skull: Completes (n = 117) probably during FCF.

Brood patch: Probably only ♀♀ incubate and develop BPs (Brewer 2013). Peaks from Jul–Nov, but perhaps occasionally breeds in the early wet season.

Sex: ♂ = ♀. In *H. o. ochraceiceps*, sexes were distinguished correctly 83% of the time using a combination of wing, tail, and bill (anterior edge of nares to tip), but it did a poor job correctly classifying individuals outside of the region within the same subspecies, suggesting regional variation and large overlap between ♂♂ and ♀♀ (Winker et al. 1994).

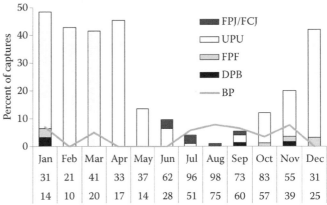

| | Jan | Feb | Mar | Apr | May | Jun | Jul | Aug | Sep | Oct | Nov | Dec |
|---|---|---|---|---|---|---|---|---|---|---|---|---|
| | 31 | 21 | 41 | 33 | 37 | 62 | 96 | 98 | 73 | 83 | 55 | 31 |
| | 14 | 10 | 20 | 17 | 14 | 28 | 51 | 75 | 60 | 57 | 39 | 25 |

Legend: FPJ/FCJ, UPU, FPF, DPB, BP

Molt: Group 4/Complex Basic Strategy; FPF complete, DPBs complete. Most frequent from Dec–Apr.

FCJ     Duller overall with more brown tones, especially in cov edgings. Lacks tawny above eye. Rects are distinctly pointed. Skull is ossified.

FPF     Replacing adult-like juvenile body or flight feathers with adult-like formative body or flight feathers. Skull is unossified, but may ossify before FPF completes, or just after.

Starting FPF with body and ss covs in molt, but not flight feathers.

FAJ     Tawny above eye contrasts with top of head. Olive-yellow overall, especially in cov edgings. Rectrices slightly pointed to rounded. Note that this species nearly always has a juvenile-like gape as an adult and should not be used as an aging clue. Skull is ossified.

DPB     Replacing adult-like body and flight feathers with adult-like feathers. Skull is ossified.

---

## LITERATURE CITED

Brewer, D. 2010. Family Vireonidae (Vireos). Pp. 378–439 in J. del Hoyo, A. Elliott, and D. A. Christie (editors), Handbook of the birds of the world: Weavers to New World Warblers. Lynx Edicions, Barcelona, Spain.

Brewer, D. [online]. 2013. Tawny-crowned Greenlet (*Hylophilus ochraceiceps*). In J. del Hoyo, A. Elliott, J. Sargatal, D., A. Christie, and E. de Juana (editors), Handbook of the birds of the world alive. Lynx Edicions, Barcelona, Spain. <http://www.hbw.com>.

Howell, S. N. G. 2010. Molt in North American birds. Houghton Mifflin Harcourt, Boston, MA.

Junge, G. C. A., and G. F. Mees. 1961. The avifauna of Trinidad and Tobago. E. J. Brill, Neider, Netherlands.

Lawrence, L. K. 1953. Nesting life and behaviour of the Red-eyed Vireo. Canadian Field Naturalist 67:47–77.

Mulvihill, R. S., and C. C. Rimmer. 1997. Timing and extent of the molts of adult Red-eyed Vireos on their breeding and wintering grounds. Condor 99:73–82.

Parkes, K. C. 1985. A brown-eyed adult Red-eyed Vireo specimen. Journal of Field Ornithology 59:60–62.

Phillips, A. R. 1991. The known birds of North and Middle America. Pt. 2—Bombycillidae; Sylviidae to Sturnidae; Vireonidae. A. R. Phillips, Denver, CO.

Pyle, P. 1997. Identification guide to North American birds, Part I. Slate Creek Press, Bolinas, CA.

Slager, D. L., C. J. Battey, R. W. Bryson, G. Voelker, and J. Klicka. 2014. A multilocus phylogeny of a major New World avian radiation: the Vireonidae. Molecular Phylogenetics and Evolution 80:95–104.

Stotz, D. F., M. Cohn-Haft, P. Petermann, J. Smith, A. Whittaker, and S. V. Wilson. 1992. The status of North American migrants in central Amazonian Brazil. Condor 94:608–621.

Winker, K., G. A. Voelker, and J. T. Klicka. 1994. A morphometric examination of sexual dimorphism in the *Hylophilus*, *Xenops*, and an *Automolus* from southern Veracruz, Mexico. Journal of Field Ornithology 65:307–323.

# CHAPTER THIRTY-TWO

# Hirundinidae (Swallows)

# Species in South America: 27

# Species recorded at BDFFP: 7

# Species captured at BDFFP: 1

Swallows and martins are a familiar group of cosmopolitan "songbirds," not for any vocal magnificence (although their twitters and whistles are nothing less than endearing), but for several species that closely associate with man. Purple Martins (*Progne subis*) that breed in North America live almost entirely in human-constructed bird houses and condominiums, and the cosmopolitan Barn Swallow (*Hirundo rustica*) is a familiar bird of porches, fishing piers, and their barn namesakes. Swallows have short but surprisingly wide bills specialized for aerial foraging. Rictal bristles and lore feathers that face forward help guide insect prey toward open mouths and away from delicate eyes. Designed for maneuverability, swallows often have long pointed wings and sometimes forked tails. Their similarity in appearance to swifts is purely evolutionary convergence; swifts are non-passerines and more closely related to hummingbirds. Swallows have 9 visible pp (a 10th is greatly reduced), 9 ss, and 12 rects.

Swallows all seem to have a complete preformative molt, usually resulting in an adult-like plumage (or sometimes duller or with shallower notches in forked tails), but at least in *Progne subis* the adult plumage in males is delayed to the second prebasic molt (Rohwer and Niles 1979). Distinguishing the FCF from DCB is often not simple or possible in many species, but Pyle (1997) recommends examining the distal marginal covs, which are basal to the pp covs and hidden by the alulas; these may be browner or duller in FCF than DCB birds within each sex class.

Skulls appear to ossify, but may retain small windows well into the FCF. Females do the bulk of incubation in most (all?) swallows and develop BPs, whereas males do not.

## SPECIES ACCOUNT

*Atticora tibialis griseiventris*                                                  Band Size: D (CEMAVE 2013)

White-thighed Swallow • Calcinha-branca                          # Individuals Captured: 5

Wing    86.0–95.0 mm (89.2 ± 3.4 mm; n = 5)

Tail    44.0–53.0 mm (49.2 ± 4.9 mm; n = 5)

Mass    9.0–13.0 g (10.5 ± 1.5 g; n = 5)
        9.8 g (Hilty 2003)

Similar species: Only swallow captured at the BDFFP. *Stelgidopteryx ruficollis* is larger, less dark above and below, and tawnier in the throat.

Skull: Completes (n = 2), but timing unknown.

Brood patch: If like other swallows, then only ♀♀ develop BPs. Timing unknown.

Sex: ♂ = ♀.

Molt: Group 4/Complex Basic Strategy; FPF complete(?), DPBs complete. Timing unknown, but one bird was finishing molt in Sep.

FCJ      Like adults, but have pale feather edges on the underparts (LaBarbera 2010). Skull is unossified.

FPF      Replacing juvenile body and flight feathers with adult-like feathers. Skull is unossified, but may ossify before completion.

FAJ      Lacks pale feather edges on the underparts (LaBarbera 2010). Skull is ossified.

UPB      Replacing adult-like body and flight feathers with adult-like feathers. Skull is ossified.

---

## LITERATURE CITED

CEMAVE. 2013. Lista das espécies de aves brasileiras com tamanhos de anilha recomendados. Centro Nacional de Pesquisa e Conservação de Aves Silvestres, Cabedelo, Brasil (in Portuguese).

Hilty, S. L. 2003. Birds of Venezuela (2nd ed.). Princeton University Press, Princeton, NJ.

LaBarbera, K. 2010. White-thighed Swallow (*Atticora tibialis*). In T. S. Schulenberg (editor), Neotropical birds online. Cornell Lab of Ornithology, Ithaca, NY. <http://neotropical.birds.cornell.edu>.

Pyle, P. 1997. Identification guide to North American birds, Part I. Slate Creek Press, Bolinas, CA.

Rohwer, S., and D. M. Niles. 1979. The subadult plumage of male Purple Martins: variability, female mimicry and recent evolution. Zeitschrift für Tierpsychologie. 51:282–300.

# CHAPTER THIRTY-THREE

# Troglodytidae (Wrens)

# Species in South America: 49

# Species recorded at BDFFP: 5

# Species captured at BDFFP: 4

Wrens are familiar birds of gardens, shrub-lands, and forests, and are often quickly recognizable by barred plumage patterns and the way they hold their tail cocked and upright. Except for the Winter Wren, this family is almost entirely found in the New World and is well represented throughout North and South America. Also known for their loud songs, males and females often engage in sophisticated duets to solidify and maintain pair bonds (Harris 2009). Wrens have 10 pp, 9 ss, and 12 rects.

Although a couple of migratory species (*Cistothorus* spp.) follow a Complex Alternate Strategy, the Complex Basic Strategy is probably the most widespread molt strategy among wrens, particularly in the tropics (Pyle 1997, Howell 2010). Preformative molts range from partial to incomplete, (and sometimes eccentric), after involving rectrices, to possibly complete. Juvenile plumages are probably typically quickly replaced. The width and intensity of barring on juvenile feathers often differs from subsequent plumages, which is useful for identifying molt limits in formative-plumaged birds.

Incubation and brood patch development is often exclusive to the female, although males assist with rearing young (Harris 2009). Skull ossification typically completes before the start of the second prebasic molt and is useful for aging.

## SPECIES ACCOUNTS

*Microcerculus bambla bambla*

Wing-banded Wren • Uirapuru-de-asa-branca

| | |
|---|---|
| Wing | 49.0–60.0 mm (55.2 ± 2.2 mm; n = 135) |
| Tail | 17.0–27.0 mm (21.9 ± 1.9 mm; n = 122) |
| Mass | 14.0–19.5 g (16.6 ± 1.1 g; n = 195) |
| Bill | 11.4–12.7 mm (12.2 ± 0.4 mm; n = 12) |
| Tarsus | 23.0–25.7 mm (24.4 ± 0.7 mm; n = 13) |

Similar species: None.

Skull: Completes probably during FCF.

Brood patch: Unknown if both sexes develop BPs. Timing uncertain, but from Aug–Feb, and perhaps to about Jun.

Sex: ♂ = ♀.

Band Size: E

# Individuals Captured: 235

Molt: Group 3/Complex Basic Strategy; FPF partial, DPBs complete. FPF includes body feathers, less and med covs, and about 6–9 gr covs, but not ss, pp, or rects. Molts mainly from Feb–Sep.

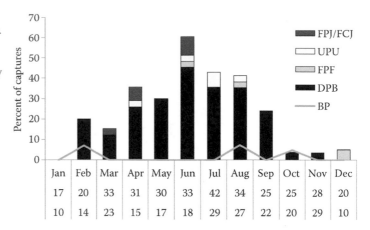

| | Jan | Feb | Mar | Apr | May | Jun | Jul | Aug | Sep | Oct | Nov | Dec |
|---|---|---|---|---|---|---|---|---|---|---|---|---|
| | 17 | 20 | 33 | 31 | 30 | 33 | 42 | 34 | 25 | 25 | 28 | 20 |
| | 10 | 14 | 23 | 15 | 17 | 18 | 29 | 27 | 22 | 20 | 29 | 10 |

FCJ Gray chest without or with limited dark dusky barring, flanks tawny-brown with widely spaced faint barring. Gr covs dark brown with dusky or tawny (not white) tips. Pp covs brownish. Rects relatively pointed. Skull is unossified.

FPF Replacing chestnut-brown juvenile body feathers and ss covs with more olive-toned body feathers and blackish ss covs. Rects relatively pointed. Skull is unossified.

Replacing less and med covs, but gr covs retained juvenile (left). Note difference between texture of juvenile plumage, especially in the head and back (upper right) and more intense and dense barring in flanks (lower right).

FCF With molt limits among the outer gr covs. Replaced inner gr covs black with bold white tips contrasting against retained outer gr covs that are brownish with faint dusky or tawny tips; beware of pseudolimits. Pp covs brownish. Rects relatively pointed. Skull is unossified or ossified.

SPB      Replacing retained brownish juvenile ss, pp, and pp covs with black adult-like ss, pp, and pp covs. Skull is ossified.

DCB      Without molt limits among the outer gr covs; will have a pseudolimit with outer 1–3 gr covs with reduced white tips that often have dusky tips, but the bases of all gr covs will be black or slightly brownish with wear. Pp covs blackish. Rects relatively rounded. Skull is ossified.

DPB      Replacing adult-like brownish-black ss, pp, and pp covs with black adult-like ss, pp, and pp covs. Skull is ossified.

## *Troglodytes aedon clarus*

House Wren • Corruíra

Band Size: C, D (CEMAVE 2013)

\# Individuals Captured: 33

| | |
|---|---|
| Wing | 46.0–53.0 mm (49.5 ± 1.7 mm; n = 21) |
| Tail | 32.0–40.0 mm (36.7 ± 2.3 mm; n = 19) |
| Mass | 10.5–13.5 g (12.0 ± 0.8 g; n = 19) |
| Bill | 9.7–10.4 mm (10.1 ± 0.5 mm; n = 2) |
| Tarsus | 19.5–19.9 mm (19.7 ± 0.3 mm; n = 2) |

Similar species: None.

Skull: No data. Probably completes during FCF.

Brood patch: Only ♀♀ incubate and develop BPs. Perhaps mainly during the dry season but possibly year-round (Kroodsma and Brewer 2013).

Sex: ♂ = ♀.

Molt: Group 3/Complex Basic Strategy; FPF partial(?), DPBs complete. As in North American *T. aedon*, FPF may include body feathers, less covs, med covs, and a variable number of inner gr covs and terts. May molt Dec–Apr.

FCJ      Underparts more heavily scalloped (Kroodsma and Brewer 2013). Skull is unossified.

FPF      Replacing juvenile-like body feathers with adult-like body feathers. Skull is unossified.

FCF      Look for molt limits among the gr covs and perhaps sometimes among the terts. Replaced inner gr covs may be slightly longer than retained outer gr covs. Skull is unossified or ossified.

SPB      Replacing juvenile-like flight feathers and rects with adult-like flight feathers and rects. Skull is ossified.

DCB      Without molt limits among the gr covs or terts. Skull is ossified.

DPB      Replacing adult-like flight feathers and rects with adult-like flight feathers and rects. Skull is ossified.

*Pheugopedius coraya coraya*

Coraya Wren • Garrinchão-coraia

Wing    51.0–61.0 mm (56.4 ± 2.3 mm; n = 65)

Tail    47.0–58.0 mm (52.2 ± 2.8 mm; n = 55)

Mass    13.5–20.0 g (16.7 ± 1.6 g; n = 69)

Bill    9.9–11.6 mm (10.8 ± 0.7 mm; n = 5)

Tarsus    22.4–26.2 mm (24.4 ± 1.4 mm; n = 5)

Similar species: None.

Skull: Completes (n = 21) probably during FCF.

Brood patch: Unknown if both sexes develop BPs, although incubation likely only done by ♀♀ as in other wrens. Probably breeds during the dry season with BPs seen in Jun, Jul, Nov, and Dec.

Sex: ♂ = ♀.

Molt: Group 3/Complex Basic Strategy; FPF partial, DPBs complete. FPF includes the body feathers, less and med covs, and inner 2–5 (or sometimes more or less?) gr covs, but not pp covs, flight feathers (occasionally 1–3 terts?), and no (or occasionally some?) rects. Probably mainly molts during the wet season through Jun.

Band Size: E

# Individuals Captured: 92

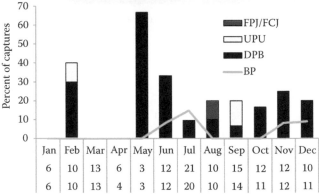

| | Jan | Feb | Mar | Apr | May | Jun | Jul | Aug | Sep | Oct | Nov | Dec |
|---|---|---|---|---|---|---|---|---|---|---|---|---|
| | 6 | 10 | 13 | 6 | 3 | 12 | 21 | 10 | 15 | 12 | 12 | 10 |
| | 6 | 10 | 13 | 4 | 3 | 12 | 20 | 10 | 14 | 11 | 12 | 11 |

FCJ    Less bright brown overall compared to subsequent plumages with lacking or very faint darker barring in the pp and ss. Rects may have slightly broader dark bars relative to the buffy-white bars, but their shape is not notably different from DCB rects. Skull is unossified.

FPF    Replacing brown juvenile body feathers and ss covs (and occasionally one or more terts) with slightly brighter brown feathers. Skull is unossified.

FCF    Molt limits among the gr covs with the inner gr covs brighter rufous-brown than the retained juvenile outer gr covs, which are dusky-brown. Rect and pp/ss pattern as in FCJ. Flight feathers unbarred. Skull is either unossified or ossified.

Inner 4 gr covs replaced, outer gr covs and all flight feathers retained.

SPB    Replacing retained unbarred pp and ss with slightly barred wing feathers. Replaced rects may have relatively narrow dark bars compared to retained juv rects. Skull is ossified.

DCB    Without molt limits among the gr covs, all of which are bright rufous-brown. Rects with dark and whitish barring even in width. Pp and ss are faintly barred. Skull is ossified.

DPB    Replacing retained slightly barred pp and ss with slightly barred wing feathers. Replaced rects may have similarly narrow dark bars compared to retained rects. Skull is ossified.

## *Cyphorhinus arada arada*

Musician Wren • Uirapuru-verdadeiro

Band Size: E

\# Individuals Captured: 359

| | |
|---|---|
| Wing | 55.0–65.0 mm (59.4 ± 2.2 mm; n = 160) |
| Tail | 28.0–41.0 mm (33.4 ± 2.4 mm; n = 155) |
| Mass | 16.0–25.0 g (20.2 ± 1.6 g; n = 280) |
| Bill | 9.0–12.4 mm (11.3 ± 1.3 mm; n = 5) |
| Tarsus | 24.0–25.6 mm (25.0 ± 0.7 mm; n = 4) |

Similar species: None.

Skull: Completes (n = 85) during FCF.

Brood patch: Unknown if both sexes develop BPs, but probably incubation by ♀ as in other wrens. Probably can breed year-round, but perhaps most often from Jan–Feb and Jun–Aug (perhaps also Mar–May?).

Sex: ♂ = ♀.

Molt: Group 3/Complex Basic Strategy; FPF partial, DPBs complete. The FPF includes all body feathers, less and med covs, and 2–4 (sometimes less or more?) inner gr covs, but not pp covs, ss, or pp. Molting peaks in Jun and Jul, but may regularly extend from May–Nov.

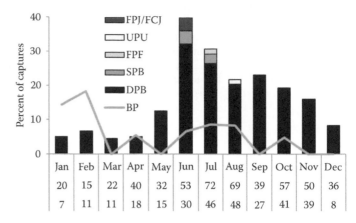

| | Jan | Feb | Mar | Apr | May | Jun | Jul | Aug | Sep | Oct | Nov | Dec |
|---|---|---|---|---|---|---|---|---|---|---|---|---|
| | 20 | 15 | 22 | 40 | 32 | 53 | 72 | 69 | 39 | 57 | 50 | 36 |
| | 7 | 11 | 11 | 18 | 15 | 30 | 46 | 48 | 27 | 41 | 39 | 8 |

FCJ    Like subsequent plumages, but with reduced black and white pattern on nape, and much reduced orange colors in face and throat. Barring on rects relatively diffuse. Skull is unossified.

(Courtesy of Cameron Rutt.)

FPF    Replacing dull chestnut-brown juvenile back feathers and ss covs with more bright olive-brown feathers. Face and throat acquiring bright rusty-orange coloration. Skull is unossified.

FCF     Like DCB, but with molt limits among the gr covs. The replaced inner gr covs are brighter olive contrasting against the chestnut-edged retained outer gr covs and flight feathers. Skull is unossified or ossified.

Inner 5 gr covs replaced, outer gr covs and all flight feathers retained juvenile.

SPB     Replacing faintly and relatively widely spaced barred feathers with those that are more strongly barred with relatively narrow spacing. Replaced rects may have relatively bold dark bars compared to retained juv rects. Skull is ossified.

p1 new, p2 missing, p3–p10 and all ss retained juvenile.

DCB     Without molt limits among the gr covs. Rects relatively boldly barred. Skull is ossified.

DPB     Replaced and retained flight and tail feathers have similarly spaced bold barring. Skull is ossified.

## LITERATURE CITED

CEMAVE. 2013. Lista das espécies de aves brasileiras com tamanhos de anilha recomendados. Centro Nacional de Pesquisa e Conservação de Aves Silvestres, Cabedelo, Brasil (in Portuguese).

Harris, T. 2009. Wrens. Pp. 304–305 in T. Harris (editor), National Geographic complete birds of the world. National Geographic Society, Washington, DC.

Howell, S. N. G. 2010. Molt in North American birds. Houghton Mifflin Harcourt, Boston, MA.

Kroodsma, D., and D. Brewer. [online]. 2013. Southern House Wren (*Troglodytes musculus*). In J. del Hoyo, A. Elliott, J. Sargatal, D. A. Christie, and E. de Juana (editors), Handbook of the birds of the world alive. Lynx Edicions, Barcelona, Spain. <http://www.hbw.com>.

Pyle, P. 1997. Identification guide to North American Birds, Part I. Slate Creek Press, Bolinas, CA.

# CHAPTER THIRTY-FOUR

# Polioptilidae (Gnatwrens and Gnatcatchers)

# Species in South America: 9

# Species recorded at BDFFP: 3

# Species captured at BDFFP: 2

Polioptilidae are a small, New World, mostly tropical family that is probably an offshoot of the Old World Warblers (Sylviidae), and have previously been lumped into the broader Muscicapidae that included not only Old World warblers but also thrushes (Turdidae) and several other Old World groups like monarch flycatchers. The three polioptilid genera are basically grouped into gnatwrens (*Microbates* and *Ramphocaenus*) and gnatcatchers (*Polioptila*). Gnatwrens are typically understory-dwelling, long-billed, and clad in browns, whereas gnatcatchers are often canopy-dwelling, relatively short-billed, and clad in grays and blacks (Rosenberg 2009). Gnatwrens and gnatcatchers have 10 pp, 9 ss, and 10–12 rects.

North American gnatcatchers appear to all follow a Complex Alternate Strategy, in spite of some species

with limited to no migratory tendencies. Therefore, this strategy should be considered for tropical gnatcatchers as well, even though males do not seem to undergo seasonal changes in appearance, and perhaps instead could be used as a strategy to combat constant year-round exposure to strong ultraviolet radiation. *Microbates* and *Ramphocaenus* also do not undergo obvious seasonal plumage transitions, and a prealternate molt may be unnecessarily in the shaded understory of the forest. We suspect that these species instead follow a Complex Basic Strategy, although it would be difficult to rule out a limited cryptic prealternate molt with the information we have so far. Preformative molts typically range from partial to sometimes incomplete and eccentric in some gnatcatchers (Pyle 1997, Pyle et al. 2004).

Both sexes apparently incubate and presumably develop brood patches (Rosenberg 2009). The skull probably usually ossifies before the SPB in adult polioptilids and is a useful aging tool, as molt limits are sometimes subtle and difficult to detect.

## SPECIES ACCOUNTS

*Microbates collaris collaris*

Collared Gnatwren • Bico-assovelado-de-coleira

Wing    46.0–55.0 mm (50.1 ± 1.6 mm; n = 322)

Tail    24.0–36.0 mm (30.3 ± 2.0 mm; n = 288)

Mass    8.0–13.5 g (10.7 ± 0.8 g; n = 489)

Bill    10.6–13.5 mm (12.2 ± 0.7 mm; n = 42)

Tarsus   23.4–27.0 mm (25.4 ± 0.8 mm; n = 43)

Similar species: Superficially like *Corythopis torquatus*, but note long bill and short tail. Also note only 10 rects.

Skull: Completes (n = 116) probably during FCF.

Band Size: D

# Individuals Captured: 584

Brood patch: Unknown if both sexes share in incubation. Probably can breed year-round, but perhaps most often from Nov–May.

Sex: ♂ = ♀.

Molt: Group 3/Complex Basic Strategy; FPF partial-incomplete, DPBs complete. The FPF includes all body feathers, less and med covs, and 0–3 gr covs, but no pp covs, ss, or pp. Some FPF molts (42% of n = 12) included rects. Primarily molts from Nov–Apr, but occasionally molts in other months.

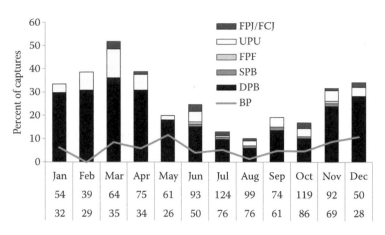

| | Jan | Feb | Mar | Apr | May | Jun | Jul | Aug | Sep | Oct | Nov | Dec |
|---|-----|-----|-----|-----|-----|-----|-----|-----|-----|-----|-----|-----|
| | 54 | 39 | 64 | 75 | 61 | 93 | 124 | 99 | 74 | 119 | 92 | 50 |
| | 32 | 29 | 35 | 34 | 26 | 50 | 76 | 76 | 61 | 86 | 69 | 28 |

FCJ    Patterned like adult, but duller brown above lacking olive-brown tones. Rects moderately pointed. Skull is unossified.

FPF    Replacing dull chestnut-brown juvenile back feathers and ss covs with more bright olive-brown feathers. Skull is unossified.

FCF    Molt limits between the brighter olive-brown replaced less and med covs against the dull brown, chestnut-edged retained gr covs. Rects can be replaced or retained from FCJ. Skull is unossified or ossified.

Second inner gr cov replaced, other gr covs and all flight feathers retained juvenile.

SPB  Look for differences in barb density and quality between retained and replaced pp covs and ss. Retained flight feathers should be quite worn and dull brown, and retained ss should be relatively rusty-edged. Skull is ossified.

DCB  Lacks molt limits among wing covs. Rects rounded. Skull is ossified.

DPB  Differences between retained and replaced pp covs and ss not particularly different, other than in freshness (which can lead to distinct color shifts). Retained flight feathers should be relatively olive-brown and fresh. Skull is ossified.

p8–p10 retained definitive basic, p6–p7 molting, p1–p5 new, s1 new, s2–s6 retained definitive basic, s7–s9 new.

*Ramphocaenus melanurus albiventris*                    Band Size: C, D

Long-billed Gnatwren • Bico-assovelado          # Individuals Captured: 13

| | |
|---|---|
| Wing | 43.0–47.5 mm (45.2 ± 1.4 mm; n = 12) |
| Tail | 43.0–48.0 mm (45.2 ± 1.5 mm; n = 10) |
| Mass | 8.0–11.0 g (9.2 ± 0.8 g; n = 12) |
| Bill | 13.5 mm (n = 1) |
| Tarsus | 22.3 mm (n = 1) |

Similar species: Unique with rufous back and elongated bill and tail.

Skull: Completes (n = 3), probably during the FCF.

Brood patch: Both sexes incubate and develop BPs (Atwood and Lerman 2006). Timing unknown, but one BP with nearby nest found in Jan.

Sex: ♂ = ♀.

Molt: Group 3/Complex Basic Strategy; FPF partial, DPBs complete. Like other Polioptilidae, FPF appears to be partial based on one capture with two retained outer gr covs. Timing unknown.

FCJ  Drabber than adult, and perhaps with faint streaking on face. Skull is unossified.

FPF  Perhaps replaces body feathers and wing covs. Skull is unossified.

| FCF | Look for molt limits among the gr covs, alulas, between the gr covs and med covs, or perhaps sometimes between the gr covs and pp covs. Skull may be unossified or ossified. |

Inner 8 gr covs and a1 replaced, other gr covs, alulas, and all flight feathers retained juvenile. (Courtesy of Cameron Rutt.)

| SPB | Look for differences in shape between retained and replaced rects, relatively dull retained pp covs, and relatively worn flight feathers. Skull is ossified. |
| DCB | Without molt limits among the gr covs or between the med and gr covs. Skull is unossified. |

| DPB | Look for similar shape between retained and replaced rects, relatively glossy retained pp covs, and relatively fresh flight feathers. Skull is unossified. |

---

## LITERATURE CITED

Atwood, J., and S. Lerman. [online]. 2006. Long-billed Gnatwren (*Ramphocaenus melanurus*). In J. del Hoyo, A. Elliott, J. Sargatal, D. A. Christie, and E. de Juana (editors), Handbook of the birds of the world alive. Lynx Edicions, Barcelona, Spain. <http://www.hbw.com>.

Pyle, P. 1997. Identification guide to North American birds, Part I. Slate Creek Press, Bolinas, CA.

Pyle, P., A. McAndrews, P. Veléz, R. L. Wilkerson, R. B. Siegel, and D. F. DeSante. 2004. Molt patterns and age and sex determination of selected southeastern Cuban landbirds. Journal of Field Ornithology 75:136–145.

Rosenberg, G. H. 2009. Gnatcatchers. Pp. 306 in T. Harris (editor), National Geographic complete birds of the world. National Geographic Society, Washington, DC.

# CHAPTER THIRTY-FIVE

# Turdidae (Thrushes)

# Species in South America: 38

# Species recorded at BDFFP: 3

# Species captured at BDFFP: 3

The thrushes are a nearly worldwide group of landbirds, many of which are familiar to humans for their pleasant songs that brighten yards and gardens. Respected and endearing, the Clay-colored Thrush (*Turdus grayi*) and Rufous-bellied Thrush (*Turdus rufiventris*) are the national birds of Costa Rica and Brazil, respectively. The classic thrush or "robin" is round-headed, plump-bodied, and colored with browns, tans, olives, grays, blacks, or sometimes with a brighter rufous or orange. Omnivorous, their bills are moderately slender and somewhat pointed (Clement 2009). Thrushes have 10 pp (the 10th reduced in size), 9 ss, and 12 rects (Pyle 1997).

All North American Turdidae (excluding Muscicapidae; *sensu* AOU 1998) follow a Complex Basic Strategy, even long-distance boreal migrants that spend the nonbreeding season in South America. The juvenile plumage is often with dark spots below and with buffy spots above. The preformative molt is nearly always partial, often involving body feathers, lesser and median coverts, and some greater coverts, but not flight feathers. In some species, terts and/or a few rects may also be replaced (Pyle 1997, Pyle et al. 2004, Howell 2010, Hernández 2012).

In North American and probably all other thrushes, only females incubate and develop brood patches (Pyle 1997, Collar 2013). Skulls generally completely ossify, but in some individuals and species, small windows (<3 mm) can persist into DCB. In some species, the mouth lining may be more yellow or paler in FCF than in DCB.

## SPECIES ACCOUNTS

*Catharus fuscescens* (all subspecies?)

Band Size: E, F

# Individuals Captured: 22

Veery • Sabiá-norte-americano

| | |
|---|---|
| Wing | 90.0–105.0 mm (98.0 ± 4.4 mm; n = 17) 89–106 mm (Pyle 1997) |
| Tail | 63.0–78.0 mm (69.9 ± 4.5 mm; n = 15) 62–79 mm (Pyle 1997) |
| Mass | 24.5–33.5 g (29.4 ± 2.7 g; n = 17) |
| Bill | 10.2 mm (n = 1) |
| Tarsus | 32.7 mm (n = 1) |

Similar species: Uniform bright rufous-brown upperparts and auriculars, tawny wash to throat, and pinking base to mandible is distinct from *C. minimus*.

Skull: Completes during FCF, sometimes as early as 15 Oct. Some DCB will retain small windows (<3 mm) at the rear of the skull (Pyle 1997).

Brood patch: Does not develop BPs at BDFFP (boreal migrant).

Sex: ♂ = ♀. A tail <65 mm and/or a wing <91 mm may indicate ♀, whereas a tail >74 mm and/or a wing >103 mm may indicate ♂ (Pyle 1997).

Molt: Group 3/Complex Basic Strategy; FPF partial, DPBs complete. During the FPF, body feathers, less covs, some to all med covs, and the inner 0–5 gr covs, but no flight feathers or rects are replaced. All molts take place on breeding and/or staging grounds in North America (Pyle 1997).

Notes: Primarily a boreal spring passage migrant; 17 of 22 captures from 2 Mar to 10 Apr, one capture in Jan and Feb each, and three captures from 17 Nov to 28 Nov.

FCJ  Not found at the BDFFP.

FPF  Not found at the BDFFP.

FCF  Molt limits among the gr covs with inner gr covs often slightly longer, and without obvious buff spots on tips of retained outer gr covs. Beware of buffy edging to fresh formative feathers in fall and worn off buffy tips on retained gr covs in spring. Rects relatively pointed. Skull is unossified or ossified (Pyle 1997).

Inner 3 gr covs replaced, all alulas and flight feathers retained juvenile (left); rectrices relatively pointed (right).

SPB  Not found at the BDFFP.

DCB  Without molt limits among the gr covs. Beware of buffy edging to fresh definitive feathers in fall. Rects relatively rounded. Skull is ossified, or sometimes with small windows (<3 mm) at the rear of the skull (Pyle 1997).

DPB  Not found at the BDFFP.

*Catharus minimus* (all subspecies?)

Gray-cheeked Thrush • Sabiá-de-cara-cinza

Band Size: E, F

# Individuals Captured: 18

Wing  95.5–103.0 mm (100.4 ± 2.8 mm;
        n = 10)
        93–109 mm (Pyle 1997)

Tail  65.0–78.5 mm (71.5 ± 4.3 mm;
        n = 10)
        63–79 mm (Pyle 1997)

Mass  26.0–36.0 g (29.5 ± 2.9 g; n = 14)

Bill  9.5–10.6 mm (10.1 ± 0.8 mm; n = 2)

Tarsus  33.2 mm (n = 1)

Similar species: From other *Catharus* by limited to no rufous-brown on upperparts, grayish face, and yellow base to mandible.

Skull: Completes as early as 1 Nov during FCF, although some DCBs retain small windows (<3 mm) above the occipital triangle (Pyle 1997).

Brood patch: Does not develop BPs at BDFFP (boreal migrant).

Sex: ♂ = ♀. A tail <68 mm and/or a wing <96 mm may indicate ♀, whereas a tail >76 mm and/or a wing >106 mm may indicate ♂ (Pyle 1997).

Molt: Group 3/Complex Basic Strategy; FPF partial, DPBs complete. The FPF includes body feathers, less and med covs, and the inner 0–5 gr covs, but not flight feathers or rects. All molts occur on the breeding grounds (Pyle 1997).

Notes: Primarily a boreal fall and spring passage migrant; eight captures from 15 Oct to 23 Nov, one capture in Jan, two captures in Feb, and seven captures 1 Mar to 14 Apr.

FCJ  Not found at the BDFFP.

FPF  Not found at the BDFFP.

FCF  Molt limits among the gr covs with inner gr covs often slightly longer, and without obvious buff spots on tips of retained outer gr covs. Beware of buffy edging to fresh formative feathers in fall and worn off buffy tips on retained gr covs in spring. Rects relatively pointed. Skull is unossified or ossified (Pyle 1997).

Inner 4 gr covs replaced, all alulas and flight feathers retained juvenile (left); rectrices relatively pointed (right).

SPB  Not found at the BDFFP.

DCB  Without molt limits among the gr covs. Beware of buffy edging to fresh definitive feathers in fall. Rects relatively rounded (Pyle 1997).

DPB  Not found at the BDFFP.

*Turdus albicollis phaeopygus*

White-throated Thrush • Sabiá-coleira

# Individuals Captured: 880

| | |
|---|---|
| Wing | 94.0–110.0 mm (102.7 ± 3.2 mm; n = 476) |
| Tail | 70.0–91.0 mm (81.4 ± 4.0 mm; n = 458) |
| Mass | 39.4–61.0 g (49.2 ± 4.0 g; n = 677) |
| Bill | 9.7–12.1 mm (11.0 ± 0.6 mm; n = 60) |
| Tarsus | 31.0–35.8 mm (33.7 ± 1.1 mm; n = 58) |

Similar species: No other *Turdus* is known to occur at the BDFFP, but *T. ignobilis* and *T. leucomelas* occupy farmsteads and urban centers in the region, and should be expected in more open second growth, whereas *T. albicollis* is much more restricted to interior forest.

Skull: Completes (n = 35) during FCF.

Brood patch: Probably only ♀♀ incubate and develop BPs. Primarily breeds in the late dry to early wet season (Dec–Apr), but perhaps occasionally during other months, particularly Sep–Nov.

Sex: ♂ = ♀.

Molt: Group 3/Complex Basic Strategy; FPF partial, DPBs complete. The FPF includes all body feathers, less and med covs, and a variable number of inner gr covs, often 2–5 (sometimes 0–7), but not rects, pp covs, ss, or pp. Primarily molts during the wet season, from Mar–May, but also occasionally from Jan–Aug.

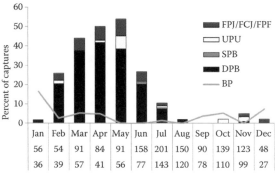

| | Jan | Feb | Mar | Apr | May | Jun | Jul | Aug | Sep | Oct | Nov | Dec |
|---|---|---|---|---|---|---|---|---|---|---|---|---|
| | 56 | 54 | 91 | 84 | 91 | 158 | 201 | 150 | 90 | 139 | 123 | 48 |
| | 36 | 39 | 57 | 41 | 56 | 77 | 143 | 120 | 78 | 110 | 99 | 27 |

Legend: FPJ/FCJ/FPF, UPU, SPB, DPB, BP

FCJ   The chest is variably spotted, and the upperparts and wing covs (except alulas and pp covs) tipped with tawny tear-drop-shaped spots. The iris is dull brown. Rects are pointed. Skull is unossified.

FPF   Replacing spotted dull brown juvenile body feathers and wing covs with solid olive-brown feathers. Skull is unossified.

FCF    Molt limits among the gr covs, with the replaced inner gr covs longer, brighter brown, and lacking a buffy spot at the tips contrasting against retained outer gr covs that often have a distinct buffy spot at their tip (but beware of heavily worn retained gr covs that lack buffy tips). Iris is dull to bright brown. Rects are pointed. Skull is unossified or ossified.

Inner 3 gr covs replaced, all alulas and flight feathers retained juvenile.

SPB    May be difficult to distinguish from DPB (use UPB), but look for differences in shape between retained and replaced rects, and for juvenile-like retained pp covs and flight feathers. Skull is ossified.

DCB    Without molt limits among the gr covs. Iris is bright brown. Rects are rounded. Skull is ossified.

DPB    May be difficult to distinguish from SPB (use UPB), but look for similar shape among retained and replaced rects, and for adult-like retained pp covs and flight feathers. Skull is ossified.

---

## LITERATURE CITED

AOU. 1998. Checklist of North American birds (7th ed.). The Committee on Classification and Nomenclature of the American Ornithologists' Union, Washington, DC.

Clement, P. 2009. Thrushes. Pp. 315–318 in T. Harris (editor), National Geographic complete birds of the world. National Geographic Society, Washington, DC.

Collar, N. [online]. 2013. Family Turdidae (Thrushes). In J. del Hoyo, A. Elliott, J. Sargatal, D. A. Christie, and E. de Juana (editors), Handbook of the birds of the world alive. Lynx Edicions, Barcelona, Spain. <http://www.hbw.com>.

Hernández, A. 2012. Molt patterns and sex and age criteria for selected landbirds of southwest Colombia. Ornitología Neotropical 23:215–223.

Howell, S. N. G. 2010. Molt in North American birds. Houghton Mifflin Harcourt, Boston, MA.

Pyle, P. 1997. Identification guide to North American birds, Part I. Slate Creek Press, Bolinas, CA.

Pyle, P., A. McAndrews, P. Veléz, R. L. Wilkerson, R. B. Siegel, and D. F. DeSante. 2004. Molt patterns and age and sex determination of selected southeastern Cuban landbirds. Journal of Field Ornithology 75:136–145.

# CHAPTER THIRTY-SIX

# Thraupidae (Tanagers)

# Species in South America: 328

# Species recorded at BDFFP: 30

# Species captured at BDFFP: 16

The tanagers and other nine-primaried oscines (mainly Emberizidae, Cardinalidae, Parulidae, Icteridae, and Fringillidae) form a diverse and taxonomically complex (and often taxonomically ambiguous) group of songbirds. Historically, bill shape was often used as a character to assume genetic relationships, but it is now clear that this is a highly plastic and derived characteristic that, in many cases, misleads taxonomic assignments. For example, the tanagers no longer include *Piranga*, which has been lost to the Cardinalidae, and *Euphonia*, which has been lost to Fringillidae, but the Thraupids have acquired *Sporophila* and *Volatinia* from the Emberizidae. Tanagers (or more precisely Thraupidae) include a diverse group of omnivores, some of which are more or less specialists on insects or fruit. The classic tanager is brightly colored, although a few are drab and, given these new taxonomic realities, are almost exclusive to Central and South America. In *terra firme* forests of the BDFFP, they are either canopy species or found in second growth—none exclusively occupy dense forest understories. Tanagers have 9 visible pp, 9 ss, and 12 rects.

Just as this is a family that is diverse in its use of habitat, food resources, and dispersal and migratory capacity, molt strategies can vary considerably and include Complex Basic and Complex Alternate Strategies. The preformative molt can be partial, incomplete, or complete (sometimes even within the same species), and in sexually dichromatic species, formative-plumaged males may have a female-like plumage (at least in some *Thraupis*), an adult-like male plumage (at least some *Tachyphonus*), or a distinct formative plumage that is neither adult female-like or adult male-like (e.g., *Cyanerpes*). We warn that generalizations at the genus level should be avoided; for example, various *Ramphocelus* spp. have been noted to have either incomplete (e.g., *Ramphocelus passerini*) to complete (Mallet-Rodrigues et al. 1995, Wolfe et al. 2009) preformative molts and *Tachyphonus* spp. reportedly have partial to complete preformative molts. Prealternate molts, when they occur, may be limited or partial, and have been documented in at least *Cyanerpes*, *Thraupis*, and *Habia* (Dickey and Van Rossem 1938, Ryder and Wolfe 2009), and here in *Lanio*.

As far as is known (among a relatively small proportion of the family) only females incubate and develop BPs, but this may change as taxonomies are clarified and more information is learned from a wide variety of tropical species (Hilty 2011). Among the species we have been able to confirm, we have only seen females with BPs. Skulls typically ossify.

## SPECIES ACCOUNTS

*Tachyphonus cristatus cristatellus*

Flame-crested Tanager • Tiê-galo

Band Size: E

\# Individuals Captured: 29

| | | |
|---|---|---|
| Wing | ♂ | 73.0–81.0 mm (77.5 ± 3.4 mm; n = 4) |
| | ♀ | 69.5–81.0 mm (74.1 ± 3.4 mm; n = 15) |
| Tail | ♂ | 68.0–72.0 mm (71.0 ± 2.0 mm; n = 4) |
| | ♀ | 61.5–72.0 mm (66.3 ± 3.4 mm; n = 13) |
| Mass | ♂ | 19.5–21.0 g (20.3 ± 0.7 g; n = 5) |
| | ♀ | 15.5–22.0 g (18.6 ± 2.0 g; n = 16) |
| Bill | ♂ | Unknown at the BDFFP |
| | ♀ | 9.6–11.9 mm (10.7 ± 0.9 mm; n = 5) |
| Tarsus | ♂ | Unknown at the BDFFP |
| | ♀ | 19.5–21.2 mm (20.2 ± 0.6 mm; n = 5) |

Similar species: ♂ from *T. surinamus* by presence of a yellow throat patch and more red in the crown. ♀ from *Lanio fulvus* by rufous cap and lacking pronounced "tooth" on maxilla.

Skull: Completes (n = 3), either late in the FPF (n = 1) or during the FCF (n = 1) based on specimens from Bolivia, as molt progresses relatively rapidly in *Tachyphonus*.

Brood patch: The only breeding bird, a ♀, was seen in Jul, and it is likely that only ♀♀ incubate as in *T. surinamus*. Timing is unknown, but may be earlier in the dry season than in *T. surinamus* based on one brood patch, one juvenile, and the timing of molt.

Sex: ♂♂ are almost all black with a red-orange crown, narrow white throat patch, white semiconcealed flanks, and a tawny thigh patch. ♀♀ are tawny-brown with a grayish face and forehead with a rufous cap.

Molt: Group 4/Complex Basic Strategy; FPF complete, DPBs complete. Probably occurs mainly during the wet season, but has been observed as early as Oct and as late as May.

FCJ    ♀ = ♂. ♀-like, but without rufous crown and with grayish-tawny (not gray) face. Skull is unossified.

FPF    ♀ ≠ ♂. Obvious in ♂♂ with a mix of brownish and black feathers. Much more subtle in ♀♀, replacing adult-like juvenile body and flight feathers with adult-like formative body and flight feathers; look for dull brownish retained pp covs and differences in the shape between retained and replaced rects. Skull is unossified, but may ossify before completion of FPF.

FAJ    ♀ ≠ ♂. ♂ unmistakable. ♀ similar to, but brighter and with more rufous tones than FCJ. Rects relatively rounded. Skull is ossified.

UPB    ♀ ≠ ♂. Replacing adult-like body or flight feathers with adult-like body or flight feathers. Skull is ossified.

## *Tachyphonus surinamus surinamus*

Band Size: E

Fulvous-crested Tanager • Tem-tem-de-topete-ferrugíneo

\# Individuals Captured: 454

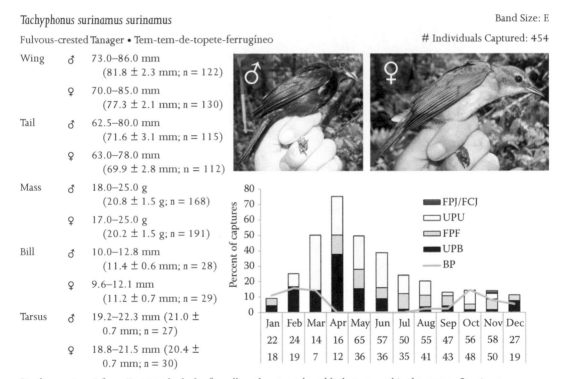

| | | |
|---|---|---|
| Wing | ♂ | 73.0–86.0 mm (81.8 ± 2.3 mm; n = 122) |
| | ♀ | 70.0–85.0 mm (77.3 ± 2.1 mm; n = 130) |
| Tail | ♂ | 62.5–80.0 mm (71.6 ± 3.1 mm; n = 115) |
| | ♀ | 63.0–78.0 mm (69.9 ± 2.8 mm; n = 112) |
| Mass | ♂ | 18.0–25.0 g (20.8 ± 1.5 g; n = 168) |
| | ♀ | 17.0–25.0 g (20.2 ± 1.5 g; n = 191) |
| Bill | ♂ | 10.0–12.8 mm (11.4 ± 0.6 mm; n = 28) |
| | ♀ | 9.6–12.1 mm (11.2 ± 0.7 mm; n = 29) |
| Tarsus | ♂ | 19.2–22.3 mm (21.0 ± 0.7 mm; n = 27) |
| | ♀ | 18.8–21.5 mm (20.4 ± 0.7 mm; n = 30) |

Similar species: ♂ from *T. cristatus* by lack of a yellow throat patch and little to no red in the crown. ♀ unique.

Skull: Completes (n = 96) during FCF, but perhaps occasionally late in the FPF.

Brood patch: BPs have only been seen in ♀♀ (n = 14), between Oct and Mar.

Sex: ♂♂ mostly black with yellow and rust in rump and white shoulders. ♀♀ are green above, buffy below with tawny-yellow in undertail covs and sometimes belly, and have a distinct facial pattern with yellow around eyes contrasting against gray auriculars, crown, and nape. Distinguishable during the FPF.

Molt: Group 4/Complex Basic Strategy; FPF complete, DPBs complete. Probably can molt in any month, but mainly Feb–Aug, peaking Mar–Jun.

FCJ   ♀ = ♂. Adult ♀-like, but with duller olive pp covs and flight feathers, faintly buffy under tail covs, and weaker face markings. Rects relatively pointed. Skull is unossified.

FPF   ♀ ≠ ♂. Obvious in ♂♂ with a mix of green and black feathers. Much more subtle in ♀♀, replacing adult-like juvenile body and flight feathers with adult-like formative body and flight feathers; look for dull brownish-green retained pp covs and differences in the shape between retained and replaced rects. Skull is unossified, but usually ossifies before completion of FPF.

Just beginning the FPF with one black med cov replaced (upper left); p7–p9 retained juvenile, p6 molting, p1–p5 replaced, s1 new, s2 molting, s3–s6 retained juvenile, s7–s8 missing, s9 new (upper right); r5– r6 retained juvenile, r4 molting, r1–r3 new (lower left); same bird as upper right and lower left (lower right).

FAJ ♀ ≠ ♂. ♂ distinct. ♀ similar to, but brighter than FCJ with more intense green in ss covs and flight feathers and brighter undertail covs. Rects relatively rounded. Skull is ossified.

Variation in the extent of yellow-buffy underparts in FAJ ♀♀. It is not understood whether this is due to age or other factors (lower right).

UPB ♀ ≠ ♂. Replacing adult-like body or flight feathers with adult-like body or flight feathers. Skull is ossified.

## Lanio fulvus fulvus

Fulvous Shrike-Tanager • Pipira-parda

Band Size: E, F

# Individuals Captured: 38

| Wing | ♂ | 89.0–94.0 mm (91.3 ± 1.7 mm; n = 12) |
|------|---|--------------------------------------|
|      | ♀ | 82.0–88.0 mm (84.4 ± 1.9 mm; n = 7)  |
| Tail | ♂ | 74.0–82.0 mm (78.0 ± 2.7 mm; n = 11) |
|      | ♀ | 70.5–78.0 mm (72.7 ± 2.7 mm; n = 6)  |
| Mass | ♂ | 24.0–30.0 g (26.4 ± 1.5 g; n = 17)   |
|      | ♀ | 23.0–29.0 g (25.6 ± 1.9 g; n = 12)   |
| Bill | ♂ | 12.5 mm (n = 1)                      |
|      | ♀ | Unknown                              |
| Tarsus | ♂ | 19.7 mm (n = 1)                    |
|      | ♀ | Unknown                              |

Similar species: All ages and sexes from *Tachyphonus* by large "tooth" on side of maxilla.

Skull: Completes (n = 7) probably during FPF or perhaps FCF.

Brood patch: A BP has only been seen in one bird, a ♀, in Oct.

Sex: ♂♂ distinct with black wings, head and tail, and orange back and underparts. ♀♀ unmarked tawny-brown overall. Distinguishable during the FPF.

Molt: Group 7/Complex Alternate Strategy; FPF complete, DPBs complete, PAs partial. PAs appear to include less and med covs, and sometimes up to 3 inner gr covs; the difference in the extent of PAs between FPAs and DPAs is not known, but might be expected to be more extensive in FPAs. Timing not well understood, but perhaps similar to *T. surinamus* with birds in molt observed in Oct, Nov, Jan, and Mar.

FCJ     ♀ = ♂. Adult ♀-like. Rects relatively pointed. Skull is unossified.

FPF     ♀ ≠ ♂. Replacing adult-like juvenile body or flight feathers with adult-like formative body or flight feathers. Obvious in ♂♂ with a mix of brown and black feathers; much more subtle in ♀♀. Skull is unossified.

Just beginning molt with some less covs (left) and body feathers replaced (right).

FAJ     ♀ ≠ ♂. ♂ distinct. ♀ like FCJ, but browner with less tawny tones throughout. Rects relatively rounded. Skull is ossified.

UCA     ♀ ≠ ♂. Both sexes like FAJ, but with molt limits among the inner gr covs, or between the med and gr covs. Skull is ossified (or perhaps nearly ossified in some FCA?).

All med covs and second inner gr cov replaced alternate, other gr covs and all flight feathers retained.

UPB  ♀ ≠ ♂. Replacing adult-like body or flight feathers with adult-like body or flight feathers. Skull is ossified.

p6–p9 retained, p5 missing, p4 molting, p1–p3 new, s1 molting, s2–s7 retained, s8 missing, s9 new.

## *Ramphocelus carbo carbo*

Silver-beaked Tanager • Pipira-vermelha

Band Size: F, G

\# Individuals Captured: 210

| | | |
|---|---|---|
| Wing | ♂ | 71.0–81.0 mm (75.7 ± 2.0 mm; n = 67) |
| | ♀ | 68.0–78.0 mm (72.5 ± 1.9 mm; n = 72) |
| Tail | ♂ | 67.0–79.0 mm (71.5 ± 2.5 mm; n = 63) |
| | ♀ | 65.0–76.0 mm (70.5 ± 2.7 mm; n = 76) |
| Mass | ♂ | 21.2–28.5 g (25 ± 1.5 g; n = 70) |
| | ♀ | 20.0–28.0 g (24.3 ± 2 g; n = 85) |
| Bill | ♂ | 11.2–12.9 mm (11.8 ± 0.5 mm; n = 16) |
| | ♀ | 10.3–12.5 mm (11.3 ± 0.5 mm; n = 16) |
| Tarsus | ♂ | 20.9–24.1 mm (22.7 ± 0.9 mm; n = 16) |
| | ♀ | 20.7–23.8 mm (22.6 ± 0.8 mm; n = 15) |

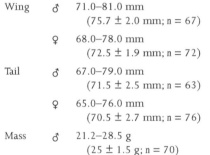

| | Jan | Feb | Mar | Apr | May | Jun | Jul | Aug | Sep | Oct | Nov | Dec |
|---|---|---|---|---|---|---|---|---|---|---|---|---|
| | 11 | 12 | 32 | 15 | 17 | 23 | 24 | 13 | 27 | 14 | 42 | 21 |
| | 11 | 11 | 27 | 14 | 13 | 22 | 24 | 13 | 27 | 12 | 42 | 16 |

Similar species: None.

Skull: Completes (n = 23) perhaps during FPF, but maybe also sometimes during the FCF.

Brood patch: BPs only seen in ♀♀ (n = 8) from Aug to Dec.

Sex: ♂♂ has silvery beak, bright deep red tones in head and underparts, and blackish on back and tail. ♀♀ unmarked rusty-brown overall, slightly brighter below. Distinguishable during the FPF.

Molt: Group 4/Complex Basic Strategy; FPF complete, DPBs complete. May molt in any month, but primarily from Feb–Jun, peaking in Apr.

THRAUPIDAE (TANAGERS)

FCJ    ♀ = ♂. Generally ♀-like, but with stronger rusty edging to ss covs. Some ♂♂ may acquire white in the lower mandible before FPF begins. Iris is brown. Rects are relatively pointed. Skull is unossified.

FPF    ♀ ≠ ♂. Replacing adult-like juvenile body or flight feathers with adult-like formative body or flight feathers. Obvious in ♂♂ with a mix of dull rusty-brown and brighter dark red or blackish feathers and acquisition of silvery lower mandible early in the molt (before p3, and perhaps regularly before wing molt commences). Much more subtle in ♀♀, but retained juvenile ss covs and flight feathers should be edged with chestnut. Skull is unossified, but may ossify before completion.

(Courtesy of Lindsay Wieland.)

FCF    ♀ ≠ ♂. Like FAJ, but with unossified skull. Especially early in FCF, the iris may be less reddish and, in ♂♂, the whitish-silver mandible may be duller (Ryder and Wolfe 2009).

FAJ    ♀ ≠ ♂. ♂ deep red with bright whitish-silver mandible (intensifies later in FCF?). ♀ without silver beak, brown overall, with tawny-brown underparts and minimal to no rusty edging on ss covs. Iris reddish-brown. Rects relatively rounded. Skull is ossified.

UPB    ♀ ≠ ♂. Replacing adult-like body or flight feathers with adult-like definitive body or flight feathers. Skull is ossified.

*Thraupis episcopus episcopus*

Band Size: E (CEMAVE 2013)

Blue-gray Tanager • Sanhaçu-da-amazônia

# Individuals Captured: 5

| | |
|---|---|
| Wing | 84.0–93.0 mm (87.8 ± 3.3 mm; n = 5) |
| Tail | 57.0–63.5 mm (60.3 ± 2.7 mm; n = 5) |
| Mass | 34.0–36.0 g (35.1 ± 1.0 g; n = 4) |

Similar species: None. Although *T. palmarum* has not yet been captured at the BDFFP, it is common in second growth and should be expected.

Skull: No data available.

Brood patch: Only ♀♀ incubate and develop BPs (Hilty 2014). Appears to breed during the dry season, but exact timing has not been determined.

Sex: ♂ = ♀.

(Courtesy of Lindsay Wieland.)

Molt: Group 7(?)/Complex Alternate Strategy(?); FPF complete, DPBs complete, PAs partial(?). Dickey and Van Rossem (1938) followed by Ryder and Wolfe (2009) suggest a Complex Alternate Strategy with a partial FPF. Among 11 specimens (*T. e. episcopus*) at LSUMZ, one was a mostly juvenile-plumaged bird symmetrically molting r1, and the remaining 10 were adult-like birds without any indication of molt limits. In addition, two specimens (one *T. e. caerulea* and one *T. e. mediana*, LSUMZ) from Peru were symmetrically molting flight feathers in sequence, replacing juvenile body feathers and remiges with adult-like feathers. Probably occurs mainly during the late dry and early wet season.

| | |
|---|---|
| FCJ | With reduced bluish-white patch on less covs, more grayish (rather than cerulean blue) on crown and back, and relatively grayish on ss covs and terts. Rects are relatively pointed. Skull is unossified. |
| FPF | Replacing dusky juvenile body feathers, dull brownish-blue ss covs and flight feathers, and dull bluish-gray relatively pointed rects with adult-like bluish body feathers, flight feathers, and more rounded rects. Skull is unossified, but may occasionally ossify before completion of molt. |
| FAJ | Without molt limits and relatively bright throughout. Rects are relatively rounded. Skull is ossified. |
| UPA/UCA | Not certain if prealternate molts occur in this species but, if they do, they may involve body and ss covs. It would be likely difficult or impossible to distinguish FCA from DCA. |
| UPB | Replacing adult-like bluish body and flight feathers will adult-like feathers. Look for relatively bright blue retained pp covs and relatively rounded rects. Skull is ossified. |

*Tangara varia* (monotypic)

Band Size: D

Dotted Tanager • Saíra-carijó

# Individuals Captured: 1

| | | |
|---|---|---|
| Wing | ♂ | 66.0 mm (n = 1) |
| | ♀ | Unknown at the BDFFP |
| Tail | ♂ | 39.0 mm (n = 1) |
| | ♀ | Unknown at the BDFFP |
| Mass | ♂ | 10.7 g (n = 1) |
| | ♀ | Unknown at the BDFFP<br>10 g (Hilty 2011) |

Similar species: ♂ distinct being green with blue wings. ♀ like *Dacnis cayana*, but with a yellow belly patch, legs are not pink, and the head lacks blue.

Skull: Probably completes (n = 1) during FCF, or perhaps late in the FPF.

Brood patch: Unknown if both sexes incubate. Timing unknown.

Sex: ♂♂ are green with blue wings and rects, and dark lores. ♀♀ similar but with green wings and rects, and green lores.

Molt: Group 3(?)/Complex Basic Strategy(?); FPF partial(?), DPBs complete. *Tangara* reportedly have partial FPF molts (Ryder and Wolfe 2009). Cryptic PA molts may be possible but have not been documented in *Tangara* to our knowledge. Timing unknown.

FCJ     Undescribed (Hilty 2011). As in other *Tangara*, rects relatively pointed. Skull is unossified.

FPF     Undescribed. Skull is unossified.

FCF     Look for molt limits among the gr covs, between the gr covs and pp covs, and/or among the terts. Pp covs may be less distinctly blue and black than DCB. Skull is unossified or ossified.

SPB     Replacing juvenile-like pp covs and flight feathers with adult-like feathers. Skull is ossified.

DCB     Without molt limits among the ss covs. Pp covs and flight feathers with relatively broad edging. Skull is ossified.

DPB     Replacing juvenile-like pp covs and flight feathers with adult-like feathers. Skull is ossified.

 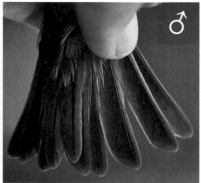

p9 and s6 in molt (more appropriately aged UPB), s7–s9 and inner gr covs missing, but probably adventitiously lost (left); all rects rounded, fresh, and brightly colored (right).

*Tangara punctata punctata*                          Band Size: D

Spotted Tanager • Saíra-negaça                       # Individuals Captured: 12

Wing    54.0–60.5 mm (57.6 ± 1.8 mm; n = 10)

Tail    36.0–44.0 mm (39.5 ± 2.3 mm; n = 10)

Mass    11.4–16.5 g (13.6 ± 1.3 g; n = 11)

Bill    6.5–6.6 mm (6.6 ± 0.1 mm; n = 2)

Tarsus  18.4–19.5 mm (18.9 ± 0.8 mm; n = 2)

Similar species: *T. varia* lack spots or is faintly spotted, never with bold distinct spots through upperparts and underparts.

Skull: Probably completes (n = 1) during FCF.

Brood patch: Unknown if both sexes incubate. Timing unknown.

Sex: ♂ = ♀.

Molt: Group 3(?)/Complex Basic Strategy(?), FPF partial, DPBs complete (Ryder and Wolfe 2009). FPF includes body feathers, less, med, and gr covs, 1–3 terts, and 1 or 2 alula covs, but not rects. Cryptic PA molts may be possible but have not been documented in *Tangara* to our knowledge (see discussion in FCF). Timing unknown.

**FCJ**     Generally similar to adult, but much duller and less obviously spotted underneath. Rects are relatively pointed. Skull is unossified.

**FPF**     Replacing dull juvenile body feathers with brightly patterned adult-like body feathers. Rects as in FCJ. Skull is unossified.

**FCF**     Adult-like, but with molt limits between the pp covs and gr covs, alula tracts, and/or among the terts or between the terts and ss. Edging along the retained ss and pp is narrower than in DCB. Skull is unossified or ossified.

Ss covs and 3 terts replaced, other flight feathers retained. It seems possible that the gr covs are older than the med covs, in which case the med covs, terts, (and possibly inner 2 gr covs?) may have been replaced in a FPA molt.

**SPB**     Look for differences in edging width between retained and replaced ss and pp. Skull is ossified.

**DCB**     Without molt limits between the gr covs and pp covs, among the alula tracts, or among the terts and secondaries. Skull is ossified.

All ss covs and flight feathers appear to be of the same generation.

**DPB**     Similar width in edging between retained and replaced ss and pp. Skull is ossified.

---

### *Tangara chilensis paradisea*

Paradise Tanager • Sete-cores-da-amazônia

Band Size: Unknown

\# Individuals Captured: 4

Wing     68.0 mm (n = 1)

Tail     56.0 mm (n = 1)

Mass     13.0–16.5 g (14.5 ± 1.8 g; n = 3)

Similar species: None.

Skull: Probably completes (n = 1) during FCF.

Brood patch: Unknown if both sexes incubate. Timing unknown.

Sex: ♂ = ♀.

Molt: Group 3 and Group 4(?)/Complex Basic Strategy; FPF partial-incomplete-complete(?), DPBs complete. Based on an examination of specimens of *T. c. chilensis* and *T. c. chlorocorys* from Bolivia and Peru (LSUMZ), the FPF may be highly variable. Of eight birds in FCF (based on unossified skull and/or presence of molt limits), one bird had a complete FPF (or was the skull data incorrect?), one bird had a partial FPF, and the remaining six had incomplete FPFs. Incomplete FPFs involve a combination of 1 pair of inner to all rects, 0–7 inner pp (and associated pp covs), 2–3 terts, and 0–5 outer ss. Cryptic PA molts may be possible but have not been documented in *Tangara* to our knowledge. Timing unknown.

FCJ     Somewhat similar to adult, but distinctly mottled greenish throat, with black speckles in face, and lower back dull orange or yellowish (Hilty 2011). Rects relatively pointed. Skull is unossified.

FPF     Involves body feathers and all ss covs, with a variable amount of flight feathers and rects. Incoming body and flight feathers are clearly blacker than juvenile feathers, although this seems to be less obvious in the tail. Skull is unossified.

FCF     With molt limits among the terts, rects, ss, pp, and/or pp covs and/or an unossified skull. Rects relatively pointed if retained (and brownish) or relatively rounded if replaced (and blackish). Skull is unossified or ossified.

FAJ     Without molt limits and an ossified skull.

SPB     Possible to identify in some individuals with retained juvenile feathers, which should be quite worn by this stage. Skull is ossified.

UPB     Replacing adult feathers with adult feathers. Skull is ossified. Beware of confusion with nearly completed FPF, but those birds should have an unossified skull.

## *Tangara gyrola gyrola*

Bay-headed Tanager • Saíra-de-cabeça-castanha

Band Size: Unknown

# Individuals Captured: 2

Wing     66.0–70.0 mm (68.0 ± 2.8 mm; n = 2)

Tail     45.0–51.0 mm (48.0 ± 4.2 mm; n = 2)

Mass     14.8–16.2 g (15.5 ± 1.0 g; n = 2)

Similar species: None.

Skull: No data.

Brood patch: Unknown if both sexes incubate. Timing unknown.

Sex: ♂♂ may have brighter bay-brown head and blue underparts than ♀♀ (Hilty 2011).

Molt: Group 3(?)/Complex Basic Strategy; FPF partial(?)-incomplete, DPBs complete. Based on two specimens of *T. g. nupera* from Ecuador (LSUMZ), the FPF included 10 or all rects, 4–7 inner pp, all terts, and 0–4 outer ss. As in other *Tangara*, some may have less or more extensive

PF molts. Cryptic PA molts may be possible but have not been documented in *Tangara* to our knowledge. Timing unknown.

FCJ     Mostly greenish with a hint of blue below and a mottled golden head (Hilty 2011). Rects are relatively pointed. Skull is unossified.

FPF     Probably always involves body feathers and all ss covs, with a variable amount of flight feathers and rects. Differences between incoming formative and retained juvenile feathers are subtle. Skull is unossified.

FCF     With subtle molt limits among the terts, rects, ss, pp, and/or pp covs. May average duller than DCB. Rects relatively pointed if retained or relatively rounded if replaced. Skull is unossified or ossified.

| SPB | It is theoretically possible to identify some molting individuals with retained juvenile feathers, retained formative feathers, and incoming definitive basic feathers. The retained juvenile feathers may include outer pp, outer rects, and/or inner ss, which should be quite worn by this stage relative to retained formative feathers. Skull is ossified. |
|---|---|
| DCB | Without molt limits among the terts, rects, ss, pp, and pp covs. May average brighter than FCF. Rects relatively rounded. Skull is ossified. |
| DPB | Replacing adult feathers with adult feathers. Skull is ossified. Beware of confusion with nearly completed FPF, but those birds should have an unossified skull. Some SPB may not be identifiable as such (use UPB). |

## *Cyanerpes caeruleus microrynchus*

Band Size: D (CEMAVE 2013)

Purple Honeycreeper • Saí-de-perna-amarela

# Individuals Captured: 2

| Wing | 52.0–53.0 mm (52.5 ± 0.7 mm; n = 2) |
|---|---|
| Tail | 25.0–26.0 mm (25.5 ± 0.7 mm; n = 2) |
| Mass | 11.0 g (n = 1) |
| | 7.8–14 g (Hilty 2011) |

Similar species: Both sexes from other *Cyanerpes* by yellow (not red or pink) legs. Juvenile plumages of *Cyanerpes* may be difficult to tell apart, but bill length may be helpful with experience. *C. caeruleus* has a bill intermediate in length, *C. nitidus* has the shortest bill, and *C. cyaneus* has the longest bill.

Skull: No data. Probably completes during FCF.

Brood patch: Unknown if both sexes incubate. Timing unknown.

Sex: ♂♂ glossy blue with a black throat and black wings. ♀♀ green above, streaked below, and with a buffy throat.

Molt: Group 7(?)/Complex Alternate Strategy; FPF complete(?), DPBs complete, PAs limited(?)-partial(?). The best description of relating plumage maturation with molt in *Cyanerpes* comes from Dickey and Van Rossem (1938), who focused on *C. cyaneus carneipes*; this was also briefly reviewed by Ryder and Wolfe (2009). Dickey and Van Rossem (1938) suggested that *Cyanerpes* make four plumage transitions in the first cycle. This seems a bit unrealistic; lumping two of those sequences into a single molt may more closely match reality, although a brief auxiliary preformative molt between the juvenile and formative plumages may be possible. Dickey and Van Rossem (1938) also assigned wing molt and tail molt in this first year of life to different molts, which also seems unlikely, although they may not necessarily be replaced concordantly. If we follow molt and plumage transitions in males, it seems likely that the FPF is complete resulting in male-like wing and tail feathers (rather than the wing and tail feathers occurring in two different molts), but otherwise results in a mostly green body plumage mixed with some blue. A limited to partial FPA would then replace the ♀-like body plumage with a ♂-like plumage, perhaps resulting in a few ♀-like feathers remaining. The SPB would then replace all feathers again, resulting in the final adult-like plumage aspect if the FPA retains some ♀-like plumage characters. Timing unknown.

| FCJ | ♀ = ♂. Like ♀, but paler and less well marked (Hilty 2011). Skull is unossified. |
|---|---|
| FPF | ♀ ≠ ♂. Replacing juvenile-like body and flight feathers with ♀-like body and black flight feathers (in ♂♂) or green flight feathers (in ♀♀). Look for differences in rect shape and the quality of pp covs and flight feathers between retained and replaced feathers in both sexes, which should be useful in ♀♀ to distinguish from DPB. Skull is unossified but may ossify before FPF completes. |
| FCF | ♂ only (or ♀♀ with unossified skulls). In ♂♂, black wings and tail with mostly green body feathers. |
| FAJ | ♀ only. Without molt limits. Skull is ossified. |
| FPA/FCA | ♂ only. Mostly ♂-like with a few ♀-like feathers on body. Skull is ossified. |
| SPB | ♂ only. Replacing retained green feathers with black feathers. Skull is ossified. |
| DCB | ♂ only. Without molt limits or evidence of ♀-like plumage. Skull is ossified. |
| DPA/DCA | ♂ only. With two generations of black body feathers. Skull is ossified. |
| UPB | ♀ only. Replacing retained green body and flight feathers with green feathers. Skull is ossified. |
| UPA/UCA | ♀ only. With two generations of green body feathers. Skull is ossified. |

THRAUPIDAE (TANAGERS)

*Chlorophanes spiza spiza*

Green Honeycreeper • Saí-verde

Band Size: E

# Individuals Captured: 12

| | | |
|---|---|---|
| Wing | ♂ | 60.0–70.0 mm (64.8 ± 3.2 mm; n = 7) |
| | ♀ | 54.0–62.0 mm (59.0 ± 4.4 mm; n = 3) |
| Tail | ♂ | 42.0–49.0 mm (45.2 ± 2.7 mm; n = 7) |
| | ♀ | 40.0–43.0 mm (41.7 ± 1.5 mm; n = 3) |
| Mass | ♂ | 14.6–18.0 g (16.0 ± 1.1 g; n = 8) |
| | ♀ | 14.0–17.0 g (15.5 ± 2.1 g; n = 2) |

Similar species: ♂ distinct. ♀ from ♀ *Dacnis* and *Cyanerpes* by being all green and dark gray (not yellow, pink, or red) legs, and from ♀ T. *varia* by longer thinner bill.

Skull: No data. Probably completes during FCF.

Brood patch: Only ♀♀ incubate (Hilty 2011) and develop BPs. Timing unknown.

Sex: ♂♂ are iridescent aquamarine green with a black head. ♀♀ are uniform green throughout. Distinguishable during the FPF.

Molt: Group 3 and Group 4/Complex Basic Strategy; FPF eccentric or complete (D. G. Olaechea, pers. comm.), DPBs complete. Timing unknown, although a pair was captured while finishing molt in late Apr, so possibly during wet season.

FCJ ♀ = ♂. Like ♀, but duller with grayish underparts and brownish iris (Hilty 2011). Skull is unossified.

FPF ♀ ≠ ♂. Replacing juvenile green body and flight feathers, sometimes in an eccentric pattern, with adult-like outer pp (green in ♀♀ and aquamarine blue in ♂♂). Skull is probably unossified.

FCF ♀ ≠ ♂. Look for molt limits among the pp and ss with outer pp and inner ss being adult-like and replaced, contrasting against retained inner pp and outer ss. Skull is unossified or ossified. Or without molt limits and an unossified skull.

FAJ ♀ ≠ ♂. Without molt limits among the pp and ss. Skull is ossified.

SPB ♀ ≠ ♂. With retained juvenile inner pp and/or ss contrasting against a combination of formative outer pp, being replaced by adult-like pp and ss. Skull is ossified.

UPB ♀ ≠ ♂. Replacing adult-like pp and ss with adult-like pp and ss. Skull is ossified.

*Hemithraupis flavicollis flavicollis*

Band Size: D, E (CEMAVE 2013)

Yellow-backed Tanager • Saíra-galega

# Individuals Captured: 7

| | | |
|---|---|---|
| Wing | ♂ | 66.0–74.0 mm (68.8 ± 3.1 mm; n = 5) |
| | ♀ | 71.0 mm (n = 1) |
| Tail | ♂ | 48.0–50.0 mm (48.8 ± 0.8 mm; n = 5) |
| | ♀ | 50.0 mm (n = 1) |
| Mass | ♂ | 12.0–14.0 g (12.5 ± 0.9 g; n = 5) |
| | ♀ | 8.0 g (n = 1) |

Similar species: ♂ distinct. ♀ like generic warbler, but stubbier and thicker bill; note bright yellow undertail covs.

Skull: Completes (n = 1 BDFFP, n = 11 LSUMZ) probably during FCF.

Brood patch: Unknown if both sexes incubate. Timing unknown.

Sex: ♂♂ are black above with a bright yellow rump, bright yellow throat, and white underparts. ♀♀ are olive-yellow above and yellow below.

Molt: Group 7(?)/Complex Alternate Strategy; FPF complete, DPBs complete, PAs limited(?)-partial. FPF based on an examination of 15 males none of which had molt limits from retained juvenile feathers (LSUMZ). PAs appear to include some body feathers and often 1 or 2 inner gr covs (but not usually the innermost). Timing unknown.

FCJ   ♀ ≠ ♂. Both sexes like adult plumage of same sex. ♂♂ with distinct yellow tips to ss covs and terts, and less brightly colored throughout. ♀♀ perhaps only distinguishable by loosely textured and slightly duller plumage. Rects relatively pointed. Skull is unossified.

(LSUMZ 72885.)

FPF   ♀ ≠ ♂. Molting body and wing feathers, replacing juvenile like plumage with adult-like plumage of the same sex. Skull is unossified.

FAJ   ♀ ≠ ♂. Adult-like in all respects. ♂♂ without yellowish-tipped ss covs and terts. Rects relatively rounded. Skull is ossified.

(LSUMZ 133836.)

THRAUPIDAE (TANAGERS)

UPA/UCA ♀ ≠ ♂. Look for molt limits among the inner gr covs and/or the med covs. Otherwise like FAJ. Skull is ossified.

Second and third inner gr cov replaced (left); second inner gr cov replaced (right). (LSUMZ uncatalogued, 133834.)

UPB ♀ ≠ ♂. Replacing adult-like feathers with adult-like feathers. Skull is ossified.

## *Volatinia jacarina splendens*

Blue-black Grassquit • Tiziu

Band Size: C, D (CEMAVE 2013)

# Individuals Captured: 5

| Wing | ♂ | 49.0–51.0 mm (49.7 ± 1.2 mm; n = 3) |
| | | 48–49 mm (n = 3, Junge and Mees 1961) |
| | ♀ | 48.0–53.0 mm (50.5 ± 3.5 mm; n = 2) |
| | | 50 mm (n = 1; Junge and Mees 1961) |
| Tail | ♂ | 36.0–45.0 mm (41.0 ± 4.6 mm; n = 3) |
| | | 40 mm (n = 2; Junge and Mees 1961) |
| | ♀ | 42.0–46.0 mm (44.0 ± 2.8 mm; n = 2) |
| | | 40 mm (n = 1; Junge and Mees 1961) |
| Mass | ♂ | 8.0–13.0 g (10.0 ± 2.7 g; n = 3) |
| | | 7.5–8.5 g (n = 3; Junge and Mees 1961) |
| | ♀ | 10.0–13.0 g (11.5 ± 2.1 g; n = 2) |
| | | 9.5 g (n = 1; Junge and Mees 1961) |

Similar species: ♂ distinct. ♀ *Oryzoborus angolensis* has heavier bill and white underwing.

Skull: Completes (n = 5), but timing unknown.

Brood patch: Unknown if both sexes incubate, although has been seen in two ♀♀ in Mar and May.

Sex: ♂♂ are all glossy blue-black with a small white shoulder mark. ♀♀ are brown, slightly streaked below.

Molt: Group 7/Complex Alternate Strategy; FPF complete, DPBs complete, PAs partial (Dickey and Van Rossem 1938, Ryder and Wolfe 2009). Timing unknown.

FCJ ♂ = ♀. Like adult ♀, but streakier underparts (Rising 2013). Skull is unossified.

FPF ♂ ≈ ♀. Replacing juvenile body and flight feathers with ♀-like body and flight feathers. Incoming flight feathers and rects of ♂ blackish with olive to brownish edgings (Dickey and Van Rossem 1938). Skull is unossified.

FCF ♂ ≈ ♀. ♂♂ are like ♀♀, but with grayer or blacker flight feathers and rects (Dickey and Van Rossem 1938). Not known how to distinguish FCF ♀♀ from DCB ♀♀, except when the skull is unossified. Skull probably ossifies before or early in the FCF.

FAJ ♀ only. Adult-like in all respects, with little to no streaking in underparts. Skull is ossified.

FPA/FCA ♂ only. Mix of brown and black body feathers and ss covs. Flight feathers and rects as in FCF (Dickey and Van Rossem 1938). Skull is ossified.

SPB ♂ only. Replacing brown body feathers and grayish flight feathers with glossy blue-black feathers. Skull is ossified.

DCB ♂ only. Glossy blue-black plumage without molt limits. Skull is ossified.

DPA/DCA ♂ only. Two generations of glossy blue-black plumage with molt limits occurring in the body and ss covs. Skull is ossified.

| | |
|---|---|
| DPB | ♂ only. Replacing adult-like plumage with adult-like plumage. Skull is ossified. |
| UPA/UCA | ♀ only. Two generations of brown plumage with molt limits occurring in the body and ss covs. Skull is ossified. |
| UPB | ♀ only. Replacing adult-like plumage with adult-like plumage. Skull is ossified. |

## *Sporophila castaneiventris* (monotypic)

Chestnut-bellied Seedeater • Caboclinho-de-peito-castanho

Band Size: D (CEMAVE 2013)

\# Individuals Captured: 3

| | | |
|---|---|---|
| Wing | ♂ | 49.5 mm (n = 1) |
| | ♀ | 55.5–56.0 mm (55.8 ± 0.4 mm; n = 2) |
| Tail | ♂ | 38.0 mm (n = 1) |
| | ♀ | 45.0–50.0 mm (47.8 ± 2.8 mm; n = 2) |
| Mass | ♂ | 9.5 g (n = 1) |
| | ♀ | 10.5–14.0 g (12.3 ± 2.1 g; n = 2) |
| Bill | ♂ | 6.3 mm (n = 1) |
| | ♀ | 5.3 mm (n = 1) |
| Tarsus | ♂ | 15.7 mm (n = 1) |
| | ♀ | 15.8 mm (n = 1) |

Similar species: ♂ distinct. ♀ is smaller, darker billed, and more buffy below than other *Sporophila*.

Skull: Based on an examination of 14 specimens at LSUMZ, the skull ossifies, probably during the FCF or perhaps rarely late in the FPF.

Brood patch: Unknown if both sexes incubate. Timing unknown.

Sex: ♂♂ are gray above with gray flanks and chestnut-red below. ♀♀ are buffy-brown overall.

Molt: Group 7/Complex Alternate Strategy; FPF complete(?), DPBs complete, PAs partial (see also Dickey and Van Rossem 1938, Ryder and Wolfe 2009, Wolfe et al. 2010). PAs appear to include body feathers (especially on upperparts) and inner less, med, and gr covs, as well as 0–2 terts. Are SCB males distinct? One bird captured with molt in Feb.

| | |
|---|---|
| FCJ | ♀ = ♂. ♀-like in both sexes, but slightly more ochraceous overall with obviously loose-textured feathers. Rects relatively pointed. Skull is unossified. |
| FPF | ♀ ≠ ♂. In ♂♂, replacing juvenile-like ochraceous feathers with mixed ♂ and ♀-like plumage. In ♀♀, replacing juvenile-like ochraceous feathers with relatively tawny ♀-like plumage. Skull is unossified. |
| FCF | ♂ only. A mix of adult and ♀-like feathers, but all of the same generation. Skull is unossified or ossified. |
| FAJ | ♀ only. Without molt limits. Skull is ossified. |
| FPA/FCA | ♂ only. Molt limits among the less, med, gr covs, and/or terts with the retained feathers mixed ♂ and ♀-like. Skull is ossified. |
| SPB | ♂ only. Body plumage mottled with brown, gray, and chestnut, as all body and flight feathers are being replaced, resulting in the adult-like ♂ plumage. Skull is ossified. |
| DCB | ♂ only. Adult-like without molt limits among the less, med, gr covs, and/or terts. Skull is ossified. |

(LSUMZ 110883.)

DPA/DCA ♂ only. Molt limits among the less, med, gr covs, and/or terts with the retained feathers adult ♂-like. Skull is ossified.

(LSUMZ 116298.)

DPB ♂ only. Replacing adult-like body and flight feathers with adult-like feathers. Skull is ossified.

UPA/UCA ♀ only. Two generations of brown plumage with molt limits occurring in the body and ss covs. Skull is ossified.

UPB ♀ only. Replacing adult-like plumage with adult-like plumage. Skull is ossified.

## *Oryzoborus angolensis torridus*

Chestnut-bellied Seed-Finch • Curió

Band Size: D, E

\# Individuals Captured: 38

| Wing | ♂ | 55.0–59.5 mm (56.1 ± 1.2 mm; n = 10) |
| | ♀ | 52.0–56.0 mm (54.7 ± 1.3 mm; n = 11) |
| Tail | ♂ | 49.5–52.0 mm (51.2 ± 0.9 mm; n = 9) |
| | ♀ | 46.0–57.0 mm (49.4 ± 3.0 mm; n = 13) |
| Mass | ♂ | 12.0–15.0 g (12.8 ± 0.8 g; n = 14) |
| | ♀ | 11.5–14.5 g (12.7 ± 0.9 g; n = 16) |
| Bill | ♂ | 8.9–10.1 mm (9.3 ± 0.7 mm; n = 3) |
| | ♀ | 8.4–10.0 mm (9.0 ± 0.7 mm; n = 4) |
| Tarsus | ♂ | 17.2–19.2 mm (18 ± 1.1 mm; n = 3) |
| | ♀ | 16.4–19.1 mm (18 ± 1.1 mm; n = 4) |

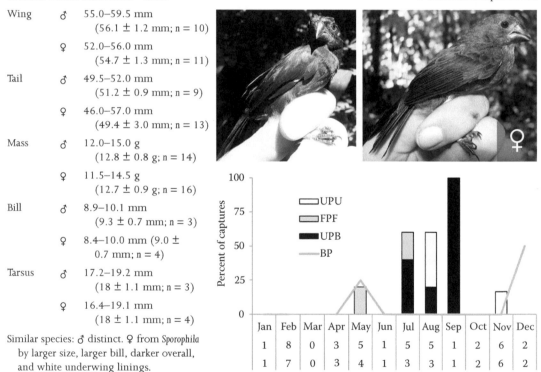

Similar species: ♂ distinct. ♀ from *Sporophila* by larger size, larger bill, darker overall, and white underwing linings.

Skull: Completes (n = 6) probably during FCF.

Brood patch: Unknown if both sexes incubate, but only seen in ♀♀ (n = 2). Possibly a wet-season breeder.

Sex: ♂♂ are black with chestnut belly and undertail. ♀♀ unmarked brown with chestnut tinge to belly and undertail. Distinguishable during the SPB.

Molt: Group 7/Complex Alternate Strategy; FPF complete, DPBs complete, PAs limited(?)-partial. A review of specimens (LSUMZ) showed 16 nonjuvenile ♂♂ in ♀-like plumage, and 13 of these had unossified skulls, suggesting that the FPF is rapid and complete, and that ♂♂ have delayed plumage maturation. PA molts also appear to be partial and ♀-like in the FPA based on two specimens (LSUMZ) and at least one capture at the BDFFP. Has a well-defined season, probably peaking Jul–Sep (May–Nov?).

FCJ       ♀ = ♂. Like ♀, but duller and with distinctly loosely textured plumage. Skull is unossified.

FPF       ♀ = ♂. Look for relatively fresh flight feathers and loosely textured body plumage being replaced by adult-♀-like feathers. Skull is unossified.

FCF       ♀ = ♂. Like FAJ, but with unossified skull.

FAJ       ♀ = ♂. ♀-like without molt limits. ♂♂ and ♀♀ only distinguishable by presence of breeding characters. With more study, it may be possible to distinguish FCF and DCB greater covs. Skull is ossified.

UPA/UCA   ♀ = ♂. With molt limits among the less, med, and/or gr covs. ♂♂ and ♀♀ only distinguishable by presence of breeding characters. Skull is ossified.

With less and inner med covs replaced. Note this individual has an adventitiously replaced p4 indicating this is a ♂ FCA, and also note the gradient between s1 and s6 with inner ss becoming progressively darker, potentially useful for distinguishing other ♂ FCA. (Courtesy of Angelica Hernández Palma.)

SPB       ♂ only. ♀-like body and flight feathers being replaced with ♂-like plumage. Skull is ossified.

DCB ♂ only. Entirely black above with chestnut underparts and white bases to the pp. Skull is ossified.

(Courtesy of Lindsay Wieland.)

DPA/DCA ♂ only. Like DCB, but look for relatively fresh and bright ss covs. Skull is ossified.

DPB ♂ only. Replacing adult-♂-like black and chestnut body feathers with adult-♂-like feathers. Skull is ossified.

UPB ♀ only. Replacing adult-♀-like body feathers with adult-♀-like feathers. Skull is ossified.

## *Coereba flaveola minima*

Band Size: D, C

Bananaquit • Cambacica

# Individuals Captured: 30

Wing 48.0–57.0 mm (52.8 ± 2.7 mm; n = 25)

Tail 26.0–35.0 mm (29.6 ± 2.4 mm; n = 21)

Mass 7.3–11.0 g (9.2 ± 0.9 g; n = 26)

Bill 9.9 mm (n = 1)

Tarsus 17.0 mm (n = 1)

Similar species: None.

Skull: Completes (n = 4), probably during the FCF as molt proceeds rapidly in this species (Prŷs-Jones 1982).

Brood patch: Only ♀♀ incubate and develop BPs (Hilty 2011). One captured in Aug with a BP and probably a dry season breeder.

Sex: ♂ ≈ ♀. ♀♀ may be slightly paler in the face than ♂♂ (Junge and Mees 1961).

Molt: Group 4/Complex Basic Strategy; FPF complete, DPBs complete. Complete FPF based on description of C. f. *bartholemica* on Dominica (Prŷs-Jones 1982). No evidence of a PA plumage in a breeding condition bird, but a limited PA would be difficult to rule out without further study. Timing not well understood, but probably follows breeding, during the late dry and/or wet seasons.

FCJ A duller version of adults with more gray in the face and crown, browner on the back, and duller yellow underneath (Hilty 2011). Rects relatively pointed. Skull is unossified.

FPF Replacing juvenile body and flight feathers with adult-like body and flight feathers. Skull is unossified.

FAJ    Adult-like without molt limits. Rects relatively rounded. Skull is ossified.

DPB    Replacing adult-like body and flight feathers with adult-like body and flight feathers. Skull is ossified.

## *Saltator grossus grossus*

Band Size: G

Slate-colored Grosbeak • Bico-encarnado

# Individuals Captured: 26

Wing    ♂    91.0–98.0 mm (95.2 ± 2.5 mm; n = 7)

        ♀    91.0–96.0 mm (93.2 ± 1.9 mm; n = 7)

Tail    ♂    78.0–87.0 mm (83.7 ± 3.3 mm; n = 10)

        ♀    79.0–87.0 mm (82.6 ± 2.5 mm; n = 7)

Mass    ♂    41.0–50.0 g (44.9 ± 2.6 g; n = 15)

        ♀    40.0–47.0 g (44.4 ± 2.7 g; n = 8)

Bill    ♂    13.1 mm (n = 1)

        ♀    Unknown at the BDFFP

Tarsus  ♂    25.1 mm (n = 1)

        ♀    Unknown at the BDFFP

Similar species: None.

Skull: Completes (n = 2), but timing unknown. Beware of some older FCF or DCB with small retained windows.

Brood patch: Unknown if both sexes develop BPs. No data, but probably breeds late dry season into early wet season based on the molt schedule.

Sex: ♂♂ have an extensive black border around the bill and bold white throat patch, whereas ♀♀ have a limited white patch on the throat not bordered by black.

Molt: Group 3/Complex Basic Strategy; FPF partial, DPBs complete. Based on an examination of three *S. g. grossus* specimens at LSUMZ from Peru and Bolivia with unossified skulls, FPF includes all body feathers, 6–7 gr covs, a1, and sometimes cc (33%). Probably occurs mainly during the late wet and early dry seasons, from about Mar–Jul.

Notes: The genus *Saltator* has had a long history of taxonomic uncertainty, having been placed in Cardinalidae and Emberizidae, but recent genetic information suggests a placement within Thraupidae (Klicka et al. 2007, Barker et al. 2013) as originally proposed by Sushkin (1924).

FCJ    White throat less extensive, particularly in males, and more grayish or cinnamon in underparts (Restall et al. 2006, Orenstein and Brewer 2011). Bill orange or yellowish, perhaps sometimes with dusky or blackish maxillary ridge. Rects relatively pointed. Skull is unossified.

FPF    Replacing juvenile-like brownish-gray or cinnamon body feathers with grayer feathers (♂♂) or brighter grayish-brown feathers (♀♀). Does not apparently involve flight feather replacement. Bill perhaps more orange or yellowish. Rects relatively pointed. Skull is unossified.

FCF    With molt limits among the gr covs and between a1 and a2, or sometimes among the terts or between the gr covs and pp covs. Cc may also be replaced, even with some outer gr covs retained. Rects relatively pointed. Skull may be unossified or ossified.

(LSUMZ 102964.)                                    (LSUMZ 171078.)

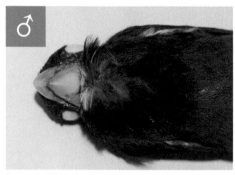

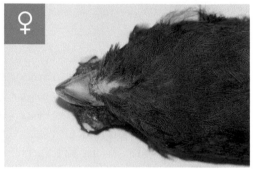

(LSUMZ 102964.)                                    (LSUMZ 171078.)

SPB    Replacing juvenile-like rects and flight feathers with adult-like feathers. Skull is ossified.

DCB    Without molt limits among the gr covs, alulas, or terts. Beware of pseudolimits among the terts. Rects relatively rounded. Skull is ossified.

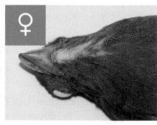

(LSUMZ 156816.)              (LSUMZ 156816.)                   (LSUMZ 117442.)

DPB    Replacing adult-like rects and flight feathers with adult-like feathers. Skull is ossified.

p7–p9 retained definitive basic, p6 missing, p5 molting, p1–p4 new, s1 new, s2 molting, s3 missing, s4 retained definitive basic, s5–s6 missing, s7–s9 new.

*Saltator maximus maximus*

Band Size: G

Buff-throated Saltator • Tempera-viola

# Individuals Captured: 4

Wing     88.5–98.5 mm (92.8 ± 4.8 mm; n = 4)

Tail     80.0–89.5 mm (85.5 ± 4.9 mm; n = 3)

Mass     34.3–43.0 g (38.4 ± 4.4 g; n = 3)

Similar species: None.

Skull: Apparently completes (n = 2) probably during FCF.

Brood patch: Unknown if both sexes incubate. Timing unknown.

Sex: ♂ = ♀.

Molt: Group 3/Complex Basic Strategy; FPF partial, DPBs complete. Dickey and Van Rossem (1938) and subsequently Ryder and Wolfe (2009) concluded the FPF includes most or all body feathers and ss covs, but not pp covs, pp, rects, or ss. An examination of LSUMZ specimens from Bolivia with unossified skulls (n = 9) suggest that two (33%) or three terts, the cc, and a1 (and sometimes a2 [22%]) are also replaced. Timing unknown.

FCJ     Throat mottled, supercilium with a hint of olive, crown dull olive-green, underparts browner than subsequent plumages, and a paler bill (Stiles and Skutch 1989, Orenstein and Brewer 2011). Rects relatively pointed. Skull is unossified.

FPF     Replacing juvenile body feathers and ss covs with adult-like feathers. Rects relatively pointed. Skull is unossified.

FCF     Very much like DCB, but look for molt limits between the gr covs and pp covs, among the inner secondaries, and between a1 and a2. Pp covs greenish, but with slightly less intense olive edging and slightly browner interiors than DCB pp covs. Rects relatively pointed. Skull may be ossified or unossified.

s8–s9 replaced, other flight feathers retained juvenile (left); all rects retained juvenile (right). (LSUMZ 133803.)

SPB     Look for contrastingly pointed retained rects. Once all rects are replaced, it is probably extremely difficult to distinguish from DPB (use UPB). Contrast between retained juvenile flight feathers and incoming feathers probably greater than in DPB, but may require experience to properly assess. Skull is ossified.

DCB    Without molt limits between gr covs and pp covs and among ss. Pp covs relatively bright green when fresh and probably difficult to differentiate from juvenile rects when worn. Rects relatively rounded. Skull is ossified.

DPB    Look for similarity in rect shape between old and incoming rects. Once all rects are replaced, it is probably extremely difficult to distinguish from FPF (use UPB). Contrast between retained adult flight feathers and incoming feathers probably less than in FPF, but may require experience to properly assess. Skull is ossified.

## LITERATURE CITED

Barker, F. K., K. J. Burns, J. Klicka, S. M. Lanyon, and I. J. Lovette. 2013. Going to extremes: contrasting rates of diversification in a recent radiation of New World passerine birds. Systematic Biology 62:298–320.

CEMAVE. 2013. Lista das espécies de aves brasileiras com tamanhos de anilha recomendados. Centro Nacional de Pesquisa e Conservação de Aves Silvestres, Cabedelo, Brasil (in Portuguese).

Dickey, D. R., and A. J. Van Rossem. 1938. The birds of El Salvador. Field Museum of Natural History, Zoological Series, Chicago, IL.

Hilty, S. [online]. 2014. Blue-gray Tanager (*Thraupis episcopus*). In J. del Hoyo, A. Elliott, J. Sargatal, D. A. Christie, and E. de Juana (editors), Handbook of the birds of the world alive. Lynx Edicions, Barcelona, Spain. <http://www.hbw.com>.

Hilty, S. L. 2011. Family Thraupidae (Tanagers). Pp. 46–329 in J. del Hoyo, A. Elliott, and D. A. Christie (editors), Handbook of the birds of the world: Weavers to New World Warblers. Lynx Edicions, Barcelona, Spain.

Junge, G. C. A., and G. F. Mees. 1961. The avifauna of Trinidad and Tobago. E. J. Brill, Neider, Netherlands.

Klicka, J., K. Burns, and G. M. Spellman. 2007. Defining a monophyletic Cardinalini: A molecular perspective. Molecular Phylogenetics and Evolution 45:1014–1032.

Mallet-Rodrigues, F., G. D. A. Castiglioni, and L. P. Gongaza. 1995. Mude e seqüência de plumagens em Ramphocelus bresilius na restinga de Barra de Maricá, Estado do Rio de Janeiro (Passeriformes: Emberizidae). Ararajuba 3:88–93 (in Portuguese).

Orenstein, R. I., and D. Brewer. 2011. Family Cardinalidae (Cardinals). Pp. 330–427 in J. del Hoyo, A. Elliott, and D. A. Christie (editors), Handbook of the birds of the world: Tanagers to New World Blackbirds. Lynx Edicions, Barcelona, Spain.

Prŷs-Jones, R. P. 1982. Molt and weight of some landbirds on Dominica, West Indies. Journal of Field Ornithology 53:552–562.

Restall, R. L., C. Rodner, and R. M. Lentino. 2006. Birds of northern South America: an identification guide. Yale University Press, New Haven, CT.

Rising, J. [online]. 2013. Blue-black Grassquit (*Volatinia jacarina*). In J. del Hoyo, A. Elliott, J. Sargatal, D. A. Christie, and E. de Juana (editors), Handbook of the birds of the world alive. Lynx Edicions, Barcelona, Spain. <http://www.hbw.com>.

Ryder, T. B., and J. D. Wolfe. 2009. The current state of knowledge on molt and plumage sequences in selected Neotropical bird families: a review. Ornitología Neotropical 20:1–18.

Stiles, F. G., and A. F. Skutch. 1989. A guide to the birds of Costa Rica. Comstock Publishing Associates, Ithaca, NY.

Sushkin, P. P. 1924. On the Fringillidae and allied groups. Bulletin of the British Ornithologists' Club 45:36–39.

Wolfe, J. D., P. Pyle, and C. J. Ralph. 2009. Breeding seasons, molt patterns, and gender and age criteria for selected northeastern Costa Rican resident landbirds. Wilson Journal of Ornithology 121:556–567.

Wolfe, J. D., T. B. Ryder, and P. Pyle. 2010. Using molt cycles to categorize the age of tropical birds: an integrative new system. Journal of Field Ornithology 81:186–194.

# CHAPTER THIRTY-SEVEN

# Emberizidae (Sparrows)

# Species in South America: 60

# Species recorded at BDFFP: 2

# Species captured at BDFFP: 2

Emberizid sparrows are found on five continents (except Australia and Antarctica) and a few are often familiar yard birds. Most, however, are birds of fields, grasslands, shrublands, and dense forest understory. Most tropical species are sedentary, but many temperate species are short-distance migrants. Even so, relatively short wings and long tails are a uniting feature. They exhibit complex plumage patterns, sometimes with bold colorful patches (as in *Arremon* and *Pipilo*) or intricate streaking and barring (as in *Melospiza* and *Ammodramus*). Although this is an incredibly diverse and widespread family, their diversity is often lowest in forests, notably in tropical forests (Jaramillo 2009, Orenstein and Bonan 2013), and only one species, *Arremon taciturnis*, occurs in the primary *terra firme* forest of the BDFFP (Cohn-Haft et al. 1997). Sparrows have 9 visible pp, 9 ss, and 12 rects (Pyle 1997).

Sparrows exhibit either a Complex Basic or Complex Alternate Strategy. The preformative molt is often partial, but can be incomplete or complete in some species especially in open grasslands or saltmarshes. Prealternate molts, when they exist, are often limited, sometimes partial, and occasionally include central rects—again the extent is apparently associated with environmental conditions that may include a combination of abiotic (e.g., sun exposure) and biotic (e.g., resource availability) factors. Sexual dichromatism is rare in the family, although males may average slightly brighter than females in several species (e.g., *Ammodramus henslowii* and *Melospiza georgiana*; Pyle 1997, Howell 2010).

Females incubate the young and develop brood patches, and males in most species assist in rearing young, especially in the Neotropics (Pyle 1997, Orenstein and Bonan 2013). Skulls typically completely ossify.

## SPECIES ACCOUNTS

*Ammodramus aurifrons aurifrons*
Yellow-browed Sparrow • Cigarrinha-do-campo

| | |
|---|---|
| Wing | 52.0–60.0 mm (55.8 ± 1.8 mm; n = 65) |
| Tail | 36.0–47.0 mm (42.3 ± 2.2 mm; n = 58) |
| Mass | 14.0–20.0 g (16.5 ± 1.2 g; n = 72) |
| Bill | 7.8–8.9 mm (8.3 ± 0.3 mm; n = 32) |
| Tarsus | 21.3–24.4 mm (22.8 ± 0.9 mm; n = 32) |

Similar species: None.

Band Size: E

# Individuals Captured: 85

Skull: Completes (n = 12) probably during FCF.

Brood patch: Probably only ♀♀ incubate and develop BPs. Timing unknown. Surprisingly, in 93 captures, none have been seen with a BP.

Sex: ♂ = ♀.

Molt: Group 3 and Group 4/Complex Basic Strategy; FPF incomplete-complete, DPBs complete. FPF complete based on an examination of specimens from Peru and Bolivia, but as in some North American *Ammodramus* (see Klicka and Spellman 2007 for discussion of polyphyly), the occasional bird can have

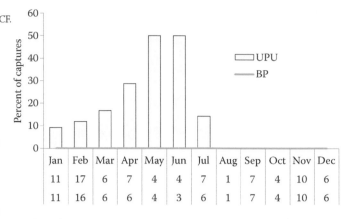

| | Jan | Feb | Mar | Apr | May | Jun | Jul | Aug | Sep | Oct | Nov | Dec |
|---|---|---|---|---|---|---|---|---|---|---|---|---|
| | 11 | 17 | 6 | 7 | 4 | 4 | 7 | 1 | 7 | 4 | 10 | 6 |
| | 11 | 16 | 6 | 6 | 4 | 3 | 6 | 1 | 7 | 4 | 10 | 6 |

retained blocks of rects or perhaps inner ss resulting from an incomplete FPF. PA molts are not known for this species, but may be possible given its preference for open habitat; however, this may be more relevant in migratory populations and species, unlike *A. aurifrons*. Molts Jan–Jul, peaking Apr–Jun.

FCJ    Crown streaking lacking gray tones, lores with only traces of yellow (not bright yellow), wrist without bright yellow, pp covs dull brownish without olive edging, and chest faintly to distinctly streaked (Jaramillo 2011). Skull is unossified.

FPF    Replacing juvenile body and flight feathers with adult-like feathers. Probably becomes very adult-like early in the FPF, relatively fresh and brownish pp and pp covs contrasting against replaced olive-edged pp and pp covs may be useful. Skull is unossified, but may occasionally ossify before the FPF completes.

FCF    With one or more retained rects and/or ss. Note that retained rects may be slightly longer than replaced formative rects, and that retained ss may be slightly shorter than replaced formative ss. Skull is ossified, or unossified and without molt limits.

With symmetrically retained juvenile r2 and r4, but also left r6 and right r3.

FAJ    Crown streaking brown on grayish-brown tones, lores with extensive bright yellow, wrist with bright yellow, pp bright brownish covs with olive edging, and chest lacking streaking. Skull is ossified.

UPB    Replacing adult body and flight feathers with adult-like feathers. Look for relatively worn and olive-brownish pp and pp covs contrasting against replaced olive-edged pp and pp covs. Skull is ossified.

*Arremon taciturnus taciturnus*

Pectoral Sparrow • Tico-tico-de-bico-preto

Wing     65.0–77.0 mm (70.2 ± 3.9 mm; n = 19)

Tail      50.0–63.0 mm (57.2 ± 4.0 mm; n = 19)

Mass    21.0–29.5 g (24.3 ± 3.0 g; n = 17)

Bill      9.0–11.6 mm (9.9 ± 0.7 mm; n = 4)

Tarsus  25.0–27.8 mm (26.9 ± 1.3 mm; n = 4)

Similar species: None.

Skull: Apparently completes (n = 3, BDFFP; n = 4, LSUMZ from Suriname), probably during the FCF.

Brood patch: Only ♀♀ incubate and develop BPs (Valdez-Juarez and Londoño 2011). Timing unknown.

Sex: ♂ with black band below throat, which ♀ lacks.

Molt: Group 3/Complex Basic Strategy; FPF partial, DPBs complete. Based on an examination of *A. a. taciturnus* from Surinam (n = 7), the FPF seems to include all gr covs and a1, but not flight feathers, pp covs, or rects. Timing unknown.

(Courtesy of Aída Rodrigues.)

FCJ      ♂ ≠ ♀. Dark olive crown, more dusky face with less contrast, and lacking yellow on wing (Valdez-Juarez et al. 2012). Rects are relatively pointed. Skull is unossified.

(LSUMZ 178524.)

FPF      Replacing juvenile body feathers and ss covs with adult-like feathers (head and bright yellow wrist patch apparently obvious early in FPF); does not apparently involve flight feathers. Rects are relatively pointed. Skull is unossified.

FCF    Molt limits are subtle, but should be between the gr covs and pp covs, and a1–a2. Pp covs average slightly browner with less olive edging, but differences between DCB are subtle. Relatively pointed rects may be the most useful clue. Skull is unossified or ossified.

SPB    Replacing juvenile-like flight feathers and rects with adult-like flight feathers and rects. Once all rects are replaced, may be extremely difficult to distinguish from DPB (use UPB). Skull is ossified.

DCB    Without molt limits and adult-like in all respects. Rects are relatively rounded. Skull is ossified.

(LSUMZ 178520.)

DPB    Replacing adult-like flight feathers and rects with adult-like flight feathers and rects. Once all rects are replaced, may be extremely difficult to distinguish from SPB (use UPB). Skull is ossified.

## LITERATURE CITED

Cohn-Haft, M., A. Whittaker, and P. C. Stouffer. 1997. A new look at the "species-poor" central Amazon: the avifauna north of Manaus, Brazil. Ornithological Monographs 48:205–235.

Howell, S. N. G. 2010. Molt in North American birds. Houghton Mifflin Harcourt, Boston, MA.

Jaramillo, A. 2009. Buntings and American Sparrows. Pp. 355–357 in T. Harris (editor), National Geographic complete birds of the world. National Geographic Society, Washington, DC.

Jaramillo, A. [online]. 2011. Yellow-browed Sparrow (Ammodramus aurifrons). In J. del Hoyo, A. Elliott, J. Sargatal, D. A. Christie, and E. de Juana (editors), Handbook of the birds of the world alive. Lynx Edicions, Barcelona, Spain. <http://www.hbw.com>.

Klicka, J., and G. M. Spellman. 2007. A molecular evaluation of the North American "grassland" sparrow clade. Auk 124:537–551.

Orenstein, R., and A. Bonan. [online]. 2013. Family Emberizidae (Buntings and New World Sparrows). In J. del Hoyo, A. Elliott, J. Sargatal, D. A. Christie, and E. de Juana (editors), Handbook of the birds of the world alive. Lynx Edicions, Barcelona, Spain. <http://www.hbw.com>.

Pyle, P. 1997. Identification guide to North American birds, Part I. Slate Creek Press, Bolinas, CA.

Valdez-Juarez, S. O., and G. A. Londoño. 2011. Nesting of the Pectoral Sparrow (Arremon taciturnus) in Southeastern Peru. Wilson Journal of Ornithology 123:808–813.

Valdez-Juarez, S. O., T. S. Schulenberg, and A. Jaramillo. [online]. 2012. Pectoral Sparrow (Arremon taciturnus). In T. S. Schulenberg (editor), Neotropical birds online. Cornell Lab of Ornithology, Ithaca, NY. <http://neotropical.birds.cornell.edu>.

# CHAPTER THIRTY-EIGHT

# Cardinalidae (Cardinal Grosbeaks, *Piranga* Tanagers, and Allies)

# Species in South America: 30

# Species recorded at BDFFP: 3

# Species captured at BDFFP: 2

The cardinal grosbeaks form a group of New World songbirds that have large bills and are often brightly colored. Many exhibit sexual dichromatism, although not in *Caryothraustes*, for example. They have colonized many different kinds of habitats with their large bills capable of eating fruit, large seeds, and insects (Dittmann and Cardiff 2009). Most forests in the Western Hemisphere have at least one cardinal grosbeak, and *terra firme* forest at the BDFFP is no exception with two regularly occurring year-round residents (*Piranga rubra* is an additional rare nonbreeding visitor during the boreal winter; Cohn-Haft et al. 1997, Rising and Bonan 2013). Cardinal grosbeaks have 9 visible pp, 9 ss, and 12 rects.

Cardinal grosbeaks exhibit either Complex Basic or Complex Alternate Strategies. The extent of the preformative molt is variable across and within species, ranging from partial to incomplete or eccentric to complete. Some species are thought to have an extra molt between the prejuvenile and preformative, known as the auxiliary preformative. This has been documented in at least *Cardinalis*, *Passerina*, and *Cyanocompsa*, although we have yet to confirm this in *Cyanocompsa cyanoides* at the BDFFP (Thompson and Leu 1994, Pyle 1997, Howell et al. 2003). Some species exhibit delayed plumage maturation, achieving an adult-like plumage upon the second prebasic molt, thus appearing female-like in their first breeding season (Pyle 1997, Howell 2010). The prealternate molt, when present, is also quite variable, although usually limited to partial. Exceptions include *Dolichonyx oryzivorus*, which is one of two North American species that has a complete prealternate molt (*Leucophaeus pipixcan* is the other; Pyle 1997, 2008), and *Passerina cyanea* has been documented with an eccentric prealternate molt (Wolfe and Pyle 2011).

Generally, only the female incubates and develops a brood patch, but both parents attend to the young (Rising and Bonan 2013). Skulls typically completely ossify (Pyle 1997).

## SPECIES ACCOUNTS

*Caryothraustes canadensis canadensis*
Yellow-green Grosbeak • Furriel

Band Size: G, H
# Individuals Captured: 2

Wing      85.0–86.0 mm (85.5 ± 0.7 mm; n = 2)
Tail       63.0–66.0 mm (64.5 ± 2.1 mm; n = 2)
Mass     33.0–33.5 g (33.3 ± 0.4 g; n = 2)

Similar species: None.

Skull: Based on examination of LSUMZ specimens, may typically ossify during FPF.

Brood patch: Probably only ♀♀ incubate and develop BPs. A BP was seen in one bird in Oct, and breeding may occur during late dry to the middle of the wet season, Oct–Apr (Orenstein and Brewer 2011).

Sex: ♂ = ♀.

Molt: Group 3 and Group 4/Complex Basic Strategy; FPF incomplete-complete, DPBs complete. An examination of seven specimens from LSUMZ found three in wing molt, and two (including one with all other ss replaced) with s8 retained, suggesting that some FPF may be incomplete. Timing unknown, but may occur following breeding from about Apr–Aug based on LSUMZ specimens from Guyana and supposed timing of breeding.

FCJ    Apparently similar to adult, but with sooty mask and all dark bill (Restall et al. 2006), but considered undescribed by Orenstein and Brewer (2011). Other plumage features and rect shape may be ambiguously adult-like. Skull is unossified.

FPF    Perhaps very difficult to distinguish from UPB as intensity of retained juvenile wing edging and shape of rectrices may be quite adult-like (use UPU). Perhaps contrast in wear between retained and replaced feathers would be less than in UPB. Skull is unossified, but may ossify just before FPF completes.

FCF    Unknown if readily distinguishable from FAJ, but look for retained s8 or other middle ss. Skull is likely ossified, but perhaps small windows can be retained.

(LSUMZ 25269.)

FAJ    Without molt limits among the flight feathers. Skull is ossified.

(LSUMZ 175538.)

UPB    Perhaps very difficult to distinguish from FPF (use UPU). Contrast in wear between retained and replaced feathers may be greater than in FPF. Skull is ossified.

## *Cyanocompsa cyanoides rothschildii*

Blue-black Grosbeak • Azulão-da-amazônia

Band Size: E

# Individuals Captured: 143

| Wing | ♂ | 72.0–85.0 mm (80.1 ± 2.4 mm; n = 39) |
|---|---|---|
| | ♀ | 71.0–79.0 mm (75.3 ± 2.2 mm; n = 48) |
| Tail | ♂ | 62.0–71.0 mm (66.7 ± 2.9 mm; n = 34) |
| | ♀ | 59.0–68.0 mm (63.1 ± 2.2 mm; n = 46) |

| Mass | ♂ | 21.5–31.0 g (26.4 ± 2.9 g; n = 52) |
| | ♀ | 21.9–30.0 g (25.9 ± 1.7 g; n = 64) |
| Bill | ♂ | 13.6 mm (n = 1) |
| | ♀ | 12.8–13.6 mm (13.2 ± 0.4 mm; n = 3) |
| Tarsus | ♂ | 21.6–22.9 mm (22.3 ± 0.9 mm; n = 2) |
| | ♀ | 21.7–23.0 mm (22.3 ± 0.5 mm; n = 5) |

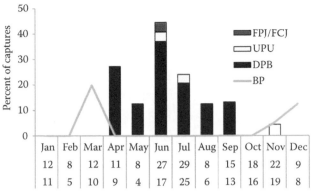

Percent of captures

| | Jan | Feb | Mar | Apr | May | Jun | Jul | Aug | Sep | Oct | Nov | Dec |
|---|---|---|---|---|---|---|---|---|---|---|---|---|
| | 12 | 8 | 12 | 11 | 8 | 27 | 29 | 8 | 15 | 18 | 22 | 9 |
| | 11 | 5 | 10 | 9 | 4 | 17 | 25 | 6 | 13 | 16 | 19 | 8 |

Legend: FPJ/FCJ, UPU, DPB, BP

Similar species: None.

Skull: Completes (n = 12) probably during FCF.

Brood patch: BPs have only been seen in ♀♀ (n = 4); ♂♂ likely do not incubate. Probably breeds mainly Nov–Mar.

Sex: ♂ all blue, ♀ all brown. Not consistently distinguishable until the SPB. Some FCF ♂♂ will have scattered blue feathers, but some older DCB ♀♀ may have an occasional blue feather.

Molt: Group 3/Complex Basic Strategy; FPF partial, DPBs complete. FPF includes body, less and med covs, some (or occasionally all?) gr covs, and 0–3(?) terts, but not rects, pp covs, pp, or ss. Molts from Apr–Sep, peaking in Jun.

| FCJ | ♂ = ♀. Brown and adult ♀-like. Gr covs probably darker brown without rich brown edging or tones. Rects relatively pointed. Skull is unossified. |
| FPF | ♂ ≈ ♀. Replacing dull brown body feathers and ss covs with bright brown body feathers and ss covs. Some ♂♂ may have a few blue incoming feathers, especially in the head. Skull is unossified. |

FCF    ♂ ≈ ♀. Both sexes all brown and adult ♀-like with molt limits in the gr covs (or occasionally between the replaced gr covs and retained pp covs?). Retained pp covs blackish without rich brown edging. Some (all?) ♂♂ can have a few blue feathers on the body, but the absence of blue may not reliably indicate a ♀. Rects are relatively pointed. Skull is unossified or ossified.

All less and med covs, and inner 6 gr covs replaced, all other gr covs and flight feathers retained juvenile (left); retained juvenile rects (right). Some males have a few scattered blue feathers (upper right).

SPB    ♂ ≠ ♀. ♂ replacing brown body and flight feathers with blue. ♀ replacing brown body and flight feathers with brown; look for differences in the shape between retained and replaced rects, and dull blackish-brown retained pp covs. Skull is ossified.

DPB     ♂ ≠ ♀. ♂ replacing blue body and flight feathers with blue. ♀ replacing brown body and flight feathers with brown; look for similar shape between retained and replaced rects, and relatively brownish retained pp covs. Skull is ossified.

p5–p9 retained definitive basic, p4 molting, p1–p3 new, s1 molting, s2 missing, s3–s5 retained definitive basic, s6–s8 new, s9 missing (left, Courtesy of Philip C. Stouffer); p9 retained definitive basic, p8 missing, p6–p7 molting, p1–p5 new, s1–s2 new, s3 molting, s4 missing, s5–s6 retained definitive basic, s7–s8 new, s9 molting (right).

DCB     ♂ ≠ ♀. Without molt limits among the gr covs or between the gr covs and pp covs. Pp covs blackish-brown with rich brown edging. Rects are relatively rounded. Skull is ossified.

---

## LITERATURE CITED

Cohn-Haft, M., A. Whittaker, and P. C. Stouffer. 1997. A new look at the "species-poor" central Amazon: the avifauna north of Manaus, Brazil. Ornithological Monographs 48:205–235.

Dittmann, D. L., and S. W. Cardiff. 2009. Grosbeaks and Allies. Pp. 362–363 in T. Harris (editor), National Geographic complete birds of the world. National Geographic Society, Washington, DC.

Howell, S. N. G. 2010. Molt in North American Birds. Houghton Mifflin Harcourt, Boston, MA.

Howell, S. N. G., C. Corben, P. Pyle, and D. I. Rogers. 2003. The first basic problem: a review of molt and plumage homologies. Condor 105:635–653.

Orenstein, R. I., and D. Brewer. 2011. Family Cardinalidae (Cardinals). Pp. 330–427 in J. del Hoyo, A. Elliott, and D. A. Christie (editors), Handbook of the birds of the world: Tanagers to New World Blackbirds. Lynx Edicions, Barcelona, Spain.

Pyle, P. 1997. Identification guide to North American birds, Part I. Slate Creek Press, Bolinas, CA.

Pyle, P. 2008. Identification guide to North American birds, Part II. Slate Creek Press, Bolinas, CA.

Restall, R. L., C. Rodner, and R. M. Lentino. 2006. Birds of northern South America: an identification guide. Yale University Press, New Haven, CT.

Rising, J., and A. Bonan. [online] 2013. Family Cardinalidae (Cardinals). *In* J. del Hoyo, A. Elliott, J. Sargatal, D. A. Christie, and E. de Juana (editors), Handbook of the birds of the world alive. Lynx Edicions, Barcelona, Spain. <http://www.hbw.com>.

Thompson, C. W., and M. Leu. 1994. Determining homology of molts and plumages to address evolutionary questions: a rejoinder regarding emberizid finches. Condor 96:769–782.

Wolfe, J., and P. Pyle. 2011. First evidence for eccentric prealternate molt in the Indigo Bunting: possible implications for adaptive molt strategies. Western Birds 42:257–262.

# CHAPTER THIRTY-NINE

# Parulidae (Wood-Warblers)

# Species in South America: 67

# Species recorded at BDFFP: 5

# Species captured at BDFFP: 1

Most wood-warblers are small insectivorous birds that feed by gleaning surfaces of leaves, branches, and/or bark, although some also sally like small flycatchers. Many species are forest dwelling and their diversity peaks in the northern temperate zone, where all are also migratory (Chartier 2009). Most, however, do not migrate as far south as the central Amazon, and with only one resident species at the BDFFP, this family is relatively poorly represented here (Cohn-Haft et al. 1997, Curson and Bonan 2013). Wood-warblers have 9 visible pp, 9 ss, and 12 rects (Pyle 1997).

Molt strategies in wood-warblers generally align with migratory distance, with many migratory species following the Complex Alternate Strategy and other short-distance migrants and most non-migratory species following the Complex Basic Strategy. Males of some migratory species add color to their plumage during the prealternate molt, creating a variety of bright patterns often involving yellows, oranges, greens, blues, and blacks. Even so, sexual dichromatism is widespread in other temperate species with limited or absent prealternate molts. Resident understory tropical species are often drabber, with more olive, brown, and buff tones, and they often lack sexual dichromatism and a prealternate molt. The juvenile plumage of wood-warblers is soft and fluffy, and is replaced quickly after leaving the nest, often with a partial preformative molt. In some species, preformative molts can include terts or rects, and can even be eccentric or complete (Pyle 1997, Howell 2010).

Typically, females build the nests and incubate (thus they develop brood patches), but both sexes care for young; however, breeding strategies in many tropical species are poorly known (Chartier 2009). Skulls typically completely ossify.

# SPECIES ACCOUNT

*Phaeothlypis rivularis mesoleucus*

Riverbank Warbler • Pula–pula–ribeirinho

Band Size: D, E

\# Individuals Captured: 58

Wing     58.5–66.5 mm (61.5 ± 1.9 mm; n = 31)

Tail      50.0–59.0 mm (55.0 ± 2.5 mm; n = 27)

Mass     11.4–15.0 g (13.0 ± 0.9 g; n = 38)

Bill      8.5–9.3 mm (8.9 ± 0.4 mm; n = 3)

Tarsus    25.5–26.4 mm (26.0 ± 0.5 mm; n = 3)

Similar species: None.

Skull: Completes (n = 18) during FCF.

Brood patch: Probably only ♀♀ incubate and develop BPs. Timing unknown as BPs have never been observed, but juveniles have been caught from Jun–Nov and may breed into the early wet season based on the molt schedule.

Sex: ♂ = ♀.

Molt: Group 3/Complex Basic Strategy; FPF partial, DPBs complete. During the FPF, body feathers, less and med covs, and the inner 2–8 gr covs are replaced, but not the flight feathers or rects. Has been observed molting Mar–Aug.

FCJ     Darker in the face, throat, and chest compared to subsequent plumages. Facial markings less pronounced. Rects are relatively tapered. Skull is unossified.

FPF     Replacing brownish-olive edged ss covs with bright olive-edged ss covs. Skull is unossified.

FCF     Molt limits among the gr covs with inner gr covs often slightly longer, and more olive–green than browner retained outer gr covs. Rects are relatively pointed. Skull is unossified or ossified.

Less and med covs, and inner 3 gr covs replaced, other gr covs and all flight feathers retained juvenile.

SPB     Look for brownish pp covs being replaced by brighter olive bb covs and differences between retained tapered rects and replaced rounded rects. Skull is ossified.

DCB     Without molt limits among the gr covs. Rects are relatively rounded. Skull is ossified.

DPB     Replacing dull olive pp covs with brigher olive pp covs. Retained and replaced rects are both relatively rounded. Skull is ossified.

# LITERATURE CITED

Chartier, A. 2009. Wood-Warblers. Pp. 349–351 in T. Harris (editor), National Geographic complete birds of the world. National Geographic Society, Washington, DC.

Cohn-Haft, M., A. Whittaker, and P. C. Stouffer. 1997. A new look at the "species-poor" central Amazon: the avifauna north of Manaus, Brazil. Ornithological Monographs 48:205–235.

Curson, J., and A. Bonan. [online]. 2013. Family Parulidae (New World Warblers). In J. del Hoyo, A. Elliott, J. Sargatal, D. A. Christie, and E. de Juana (editors), Handbook of the Birds of the World Alive. Lynx Edicions, Barcelona, Spain. <http://www.hbw.com>.

Howell, S. N. G. 2010. Molt in North American birds. Houghton Mifflin Harcourt, Boston, MA.

Pyle, P. 1997. Identification guide to North American birds, Part I. Slate Creek Press, Bolinas, CA.

# INDEX

# STUDIES IN AVIAN BIOLOGY
Series Editor: Kathryn P. Huyvaert
http://americanornithology.org

30. *Fire and Avian Ecology in North America.* Saab, V. A., and H. D. W. Powell, editors. 2005.

31. *The Northern Goshawk: A Technical Assessment of its Status, Ecology, and Management.* Morrison, M. L., editor. 2006.

32. *Terrestrial Vertebrates of Tidal Marshes: Evolution, Ecology, and Conservation.* Greenberg, R., J. E. Maldonado, S. Droege, and M. V. McDonald, editors. 2006.

33. *At-Sea Distribution and Abundance of Seabirds off Southern California: A 20-Year Comparison.* Mason, J. W., G. J. McChesney, W. R. McIver, H. R. Carter, J. Y. Takekawa, R. T. Golightly, J. T. Ackerman, D. L. Orthmeyer, W. M. Perry, J. L. Yee, M. O. Pierson, and M. D. McCrary. 2007.

34. *Beyond Mayfield: Measurements of Nest-Survival Data.* Jones, S. L., and G. R. Geupel, editors. 2007.

35. *Foraging Dynamics of Seabirds in the Eastern Tropical Pacific Ocean.* Spear, L. B., D. G. Ainley, and W. A. Walker. 2007.

36. *Status of the Red Knot (Calidris canutus rufa) in the Western Hemisphere.* Niles, L. J., H. P. Sitters, A. D. Dey, P. W. Atkinson, A. J. Baker, K. A. Bennett, R. Carmona, K. E. Clark, N. A. Clark, C. Espoz, P. M. González, B. A. Harrington, D. E. Hernández, K. S. Kalasz, R. G. Lathrop, R. N. Matus, C. D. T. Minton, R. I. G. Morrison, M. K. Peck, W. Pitts, R. A. Robinson, and I. L. Serrano. 2008.

37. *Birds of the US–Mexico Borderland: Distribution, Ecology, and Conservation.* Ruth, J. M., T. Brush, and D. J. Krueper, editors. 2008.

38. *Greater Sage-Grouse: Ecology and Conservation of a Landscape Species and Its Habitats.* Knick, S. T., and J. W. Connelly, editors. 2011.

39. *Ecology, Conservation, and Management of Grouse.* Sandercock, B. K., K. Martin, and G. Segelbacher, editors. 2011.

40. *Population Demography of Northern Spotted Owls.* Forsman, E. D. et al. 2011.

41. *Boreal Birds of North America: A Hemispheric View of Their Conservation Links and Significance.* Wells, J. V., editor. 2011.

42. *Emerging Avian Disease.* Paul, E., editor. 2012.

43. *Video Surveillance of Nesting Birds.* Ribic, C. A., F. R. Thompson, III, and P. J. Pietz, editors. 2012.

44. *Arctic Shorebirds in North America: A Decade of Monitoring.* Bart, J. R., and V. H. Johnston, editors. 2012.

45. *Urban Bird Ecology and Conservation.* Lepczyk, C. A., and P. S. Warren, editors. 2012.

46. *Ecology and Conservation of North American Sea Ducks.* Savard, J.-P. L., D. V. Derksen, D. Esler and J. M. Eadie, editors. 2014.

47. *Phenological Synchrony and Bird Migration: Changing Climate and Seasonal resources in North America.* Wood E. M. and J. L. Kellermann, editors. 2015.

48. *Ecology and Conservation of Lesser Prairie-Chickens.* Haukos, D.A. and C. Boal, editors. 2016.

49. *Golden-winged Warbler Ecology, Conservation, and Habitat Management.* Streby, H. M., D. E. Andersen, and D. Buehler, editors. 2016.

50. *The Extended Specimen: Emerging Frontiers in Collections-based Ornithological Research.* Webster, M. S., editor. 2017.

T - #0484 - 071024 - C412 - 254/178/18 - PB - 9780367657635 - Gloss Lamination